Frank Waldo

Modern Meteorology 1893

Frank Waldo

Modern Meteorology 1893

ISBN/EAN: 9783742858788

Manufactured in Europe, USA, Canada, Australia, Japa

Cover: Foto ©berggeist007 / pixelio.de

Manufactured and distributed by brebook publishing software
(www.brebook.com)

Frank Waldo

Modern Meteorology 1893

MODERN
METEOROLOGY:

*AN OUTLINE OF THE GROWTH AND PRESENT
CONDITION OF SOME OF ITS PHASES.*

FRANK WALDO, Ph D

MEMBER OF THE GERMAN AND AUSTRIAN METEOROLOGICAL SOCIETIES, ETC.
Late Junior Professor, Signal Service, U.S.A.

WITH 112 ILLUSTRATIONS.

LONDON:
WALTER SCOTT, LTD.,
24, WARWICK LANE, PATERNOSTER ROW.

1893.

PREFACE.

THE material which I have here brought together will be of interest mainly to those persons who take a special delight in meteorology but who have little access to original sources of information ; among this class are the great majority of the small army of Meteorological Observers and many teachers of physical geography and general physics. The general reader may find it partly supplementary to such works as Scott's *Meteorology,* Abercromby's *Weather,* or the articles in the *Encyclopædia Britannica.*

Living at a distance of several hundred miles from any considerable meteorological library, it was impossible for me to consult many original authorities and to give a complete historical treatment of the various subjects ; and it must be remarked that, as I have been mainly a student of what may be termed the German school of Meteorology, I probably have not brought into sufficient notice the names and work of French, English, and Italian meteorologists ; recognising this, I have tried not to bring prominently forward names in those connections where justice to others would probably require mention of their names also. It may be added that the various lines of meteorology have been so rapidly developed within the past decade or two, that a com-

prehensive treatment of any one of the half a dozen
divisions of the subject which are brought into notice
here would have required a volume longer than the
present : so that variety of topics is obtained only at
the expense of the completeness of detail which is
necessary to render the treatment of a subject generally interesting.

The main object of this little book is to bring the
reader into closer contact with the work which has
been and is actually engaging the attention of working meteorologists rather than to present finished
results.

. Chapter I. is mainly devoted to the mention of
some of the principal sources of information concerning the recent progress in meteorological science.

Chapter II. contains a history and description of
some important meteorological instruments and the
methods of using them.   In the preparation of this
chapter the freest use has been made of Professor
Abbe's treatise on Meteorological Apparatus and
Methods (printed by the U.S. Signal Service). In the
section on thermometers I would have made a
more extended use of Guillaume's *Thermométrie*
if it had been available at the time of writing.
As there is no work published in English which
gives an adequate description of meteorological
instruments, I have given more prominence to this
topic than would otherwise be warrantable ; the
subjects of thermometers, barometers, and wind-
measuring instruments receiving the most attention.
In fact, the relatively large amount of space devoted
to these has compelled me to make but a brief
mention or omit altogether other less common
instruments.   The illustrations of observatories will

serve to show that while meteorology may not have
reached the dignified position occupied by astronomy,
yet these evidences of its stability indicate the firm
foothold it has obtained among state - subsidised
sciences. Special attention is called to the details
of equipment and routine of the Pawlowsk (Russia)
Observatory, which are available through the courtesy
of Director Wild and Dr. Leyst. I know of no
similar published account of the work of an observatory.

Chapter III. is mainly made up of an abstract of
the important memoirs on Thermodynamics of the
Atmosphere, recently communicated to the Berlin
Academy of Sciences by Professor von Bezold. I
consider his ideas of the greatest importance in the
. development of atmospheric mechanics and the study
of atmospheric conditions. It is improbable that
von Bezold's ideas will creep into elementary
English text-books for a number of years to come,
and so the necessity for their presentation here
seemed the more pressing, although it may require
more time than the average reader will give in order
to understand portions of the reasoning. I have
endeavoured to follow closely von Bezold's own
presentation of the subject.

Chapter IV. contains a partial outline of the history
of the development of theories of the general atmospheric circulation, with a brief statement of
Professor Ferrel's completed theory and an account
of some recent contributions to the subject by
Möller, Oberbeck, von Siemens, and von Helmholtz.
In these also I have followed closely the text of the
authors named, so far as I have presented their ideas.

Chapter V. is devoted to a historical sketch and

partial explanation of the secondary atmospheric circulation to which the local character of winds can usually be referred—at least in the middle latitudes. Special attention is called to the attempt of von Bezold to unite the older and newer theories, as explained in his communication to the Berlin Academy of Science; and the results of his study occupy a considerable portion of the present chapter.

Chapter VI. contains some of the principal results obtained by Dr. Brückner in his recent discussion of the results of meteorological observations from the time of their commencement to the present time. The points mainly dwelt on are the long-period oscillations of climatic factors. The second part of the chapter, which treats of the application of meteor-ology to agriculture, I have been permitted by the Assistant Secretary of Agriculture to abstract from an unpublished essay prepared by me for the Department of Agriculture, U.S.A.

I wish to express my thanks to the various authors and publishers who have permitted the use of their illustrations; due credit is given in the proper place. I am indebted to Professor Wild for the photograph of the Pawlowsk Observatory, and to Professor Whipple for the photographs of the Kew Observatory and apparatus.

In closing I must not omit to make mention of the substantial assistance rendered by my wife in the preparation of this book.

FRANK WALDO.

PRINCETON, NEW JERSEY, U.S.A.

CONTENTS.

—◦◦◦—

CHAPTER III.

THERMODYNAMICS OF THE ATMOSPHERE.

CHAPTER IV.

GENERAL MOTIONS OF THE ATMOSPHERE.

there is no axial rotation of the earth—Modifying
effects of the earth's rotation on atmospheric motions—
East and west velocities of air motions—Easterly
motion of the upper poleward air current—Limit of
easterly velocities of air currents—Vertical motions of
the air—Relation of easterly motion of air in high latitudes to that of westerly motion in low latitudes—
Effects of friction on east-westerly air motions—Resultant air motion in higher latitudes—Region of slight
air motion—Trade winds—Air motions near the equator
—Equatorial west wind—Flow of air from northern to
the southern hemisphere—Effect of the general motions
of the atmosphere on the atmospheric pressure and
isobaric surfaces—Dependence of surface gradients on
east-westerly motions—Barometric gradient—Maximum
air pressure at latitude 30°—Minimum air pressure at
the equator—Surface of maximum air pressure—Maximum air pressure in high altitudes at the equator—Air
pressure at the earth's surface in general—Interchange
of air between the two hemispheres—Sprung's table of
meridianal distribution of the air pressure and temperature—Graphical representation of the air circulation as
outlined by Ferrel—Computation of east-westerly velocities of the air from the pressure gradient—Ferrel's
conception of the atmospheric circulation viewed as a
whole.

CHAPTER VI.

APPLIED METEOROLOGY.

LIST OF ILLUSTRATIONS.

———◦◦◦———

MODERN METEOROLOGY.

CHAPTER I.

Sources of Modern Meteorology.

§ 1.—*Development.*

IN giving an idea of the present status of the science of meteorology it would be best to trace its general growth from the condition in which we find it about the year 1860. At this period many of the new departures which have aided in the wonderful progress of the past two or three decades originated, although most of the important work has been accomplished within the last twenty years. Ferrel's pioneer papers (1856–58–59) on the theory of atmospheric motions had been written, and the important law of Buys Ballot [1] concerning the wind's direction discovered, just before 1860; and Schoch had worked out the application of spherical functions to the representation of the mean temperature over the surface of the earth (1856). Le Verrier had begun to issue the reports of simultaneous meteor-

---

[1] Erman had arrived at a conclusion similar to this some years earlier (*Poggendorf's Annalen*, lxxxviii.).

ological observations in the *Bulletin International de l'Observatoire* (1856). Welsh had mounted a normal barometer at the Kew Observatory (1856), which was the first instrument, of other than ordinary construction, set up to serve as a standard of reference for other barometers intended for use in observatories. Regnault's normal barometer, constructed ten years earlier, had been mainly for laboratory use.

These and many more innovations had been made before the publication of Schmid's great work on meteorology, but their importance was not yet felt, and so they remained unmentioned by him. This special mention is made of Schmid's *Meteorologie* (Leipzig, 1860), because in this one volume the author brought together the main facts of meteorological science as they were accepted by most of the scientists of that period. The volume occupied seven years in the making, and is a veritable treasury of facts and references. While it is not so original as Kaemtz's *Meteorologie*, published twenty-five years earlier, yet it is far more valuable as a work of reference. In fact, the publication of Schmid's book makes the year 1860 a very definite epoch in meteorological history. Buchan's *Handy Book of Meteorology*, published a few years later (1867), presents very strongly the new tendency of meteorological investigation, and perhaps it may be regarded as the first of the modern treatises which have since appeared at short intervals.

Government meteorological organisations established. —The modern rapid development of meteorology has been due to the usefulness of meteorological data in practical life rather than to a desire on the part of its supporters to encourage a study of it as a science.

Without government support of meteorological work as a separate department, the rapid growth of the past twenty years would have been impossible. Shortly after the period (1860) which has just been mentioned, the different governments began to establish meteorological organisations which had no direct dependence on other institutions, and which could thus be dealt with in a manner not possible so long as they were merely appendages to some other organisations. There were then few specialists in meteorology, for there had been scarcely any positions in which the incumbents were permitted to devote their whole time to this one department of science ; but the aid derived from the State has allowed the enlistment of investigators of the best abilities in the cause of meteorological science.

Introduction of dynamical methods.—Not only was new life infused into meteorological work by the increased support which it received, but new methods were introduced. The older method of investigation has been termed the statistical, and the new departure the dynamical method. The latter has not displaced the former, but it is made accessory to it. By the old method, data were accumulated and discussed singly, and certain facts were deduced ; by the new they are examined in the aggregate, and the relation of these facts worked out both as to cause and progressive development. Notwithstanding the fact that there are certain general characteristics of weather and climate which are brought to light by the discussion of observations extending over long periods of time, yet the great variability and rapid succession of changes of weather conditions show that if we will investigate their cause, it must be looked for in immediate connection with the event. By the older methods this was attempted by considering the time-relation merely, but the new meteorology shows that we must also take into account the space-distribution, both geographically and, to a relatively small amount, also in altitude.

Sub-departments of meteorological organisation.— While the meteorological services of different countries do not present the same characteristics in the details of organisation, yet there is a normal plan of organisation which, to a certain extent, fits them all. The service controlled by the general government consists of a central office and outlying observing stations managed by a scientific director, who has several assistants having immediate charge of the various departments into which the work is, for convenience, sub-divided. These assistants are, or soon become, specialists in their line of work, and perform their duties or have their headquarters at the central office. These sub-departments may be named nearly as follows: Supervision of Observing Stations; Climatology; Weather Predictions, Storm Warnings, and Summaries of Current Weather; Management of the Central Observatory; Care of the Standard Instruments, and verification of Instruments to be used at the various observing stations; and, in some cases, Maritime Meteorology.

Establishment of stations of different classes.—In the larger organisations these assistants have a force of computers or observers to assist them in their work. The central meteorological observatory is of the first class; that is, observations of the chief elements are made every hour, or self-registering instruments are in operation. A detailed account of the work of this central office is given in another chapter. The out-lying observing stations may be classified as follows :

Additional, but less completely equipped and manned, stations of the first class.

Second-class stations, at which the barometric pres-sure, temperature, rainfall, humidity, wind (direction and force), and cloudiness are observed twice or thrice daily, and miscellaneous observations are also made, such as thunder-storms, frosts, hail, &c. ; these stations are distributed over the country as uniformly as may be convenient, and are in charge of skilled observers, who, if they are not already employed by the govern-ment, usually receive pay for their services.

Third-class stations are those at which temperature, precipitation, and some miscellaneous observations are made. The observers are usually persons who, for different reasons, are sufficiently interested in meteorology to be willing to make and record two or three observations at stated times daily, but who are at liberty to discontinue the work at any time, as it is entirely voluntary.

Special stations for the observation of rainfall and thunder-storms have also become numerous in most countries, and they are likewise conducted by a volunteer corps of observers.

All of these observers send copies of their records to the central office, usually at intervals of a month, where the observations are made use of by the proper departments. In cases, however, where the obser-vations are required for use in making weather predictions, they are telegraphed to the central office with little delay.

§ 2.—*Meteorological Publications.*

Professional papers.—While the smaller meteorological services have usually had to be content
with publishing merely the bi- or tri-daily instrumental readings or the averages of the observations
made under their control, nearly all of the
larger or national services have given us additional publications. The most important of these
are 'the volumes of memoirs, in which the meteorologists of the service, usually of the staff at the
central office, work up the observations which have
been made by the service and which in many cases
can be used only by persons having access to the
archives in which are stored the records of individual
observations. Some of the services are provided with
the means of still further usefulness, which permit
these professional papers to include discussions of
methods and results in general ; these often involve
long-continued experiments, or the gathering together
of material obtained from various parts of the globe.
Such memoirs were of necessity formerly infrequent,
because the investigators were obliged to publish their
work mostly in scattered transactions, where only a
limited amount of space could be assigned for the use
of any one science, and where, as in the case of
journal publications, numerical tables and expensive
illustrations and charts are not inserted unless of
great value.

The Indian *Meteorological Memoirs*, the French
Annales du Bureau Central Météorologique de France,
the Italian *Annali dell' Officio Centrale di Meteor.
Italiana*, the German *Aus dem Archiv der Deutschen
Seewarte*, the *Abhandlungen d. Kön. Preussischen*

Meteor. Instituts, the *Professional Papers* (formerly issued) and *Annual Reports of the United States Weather Bureau*, the Argentine *Annales de la Oficina Meteorologica Argentina*, and the Russian *Repertorium für Meteorologie* (sometimes called *Wild's Repertorium*), are the chief examples of this class of publications. The Russian *Repertorium für Meteorologie* outranks the others, especially as regards the number of papers and the variety of topics treated.

Monthly Weather Reports.—The monthly division of time has been the plan usually followed in the presentation of meteorological data, but it was not until after the establishment of Weather Services that the need was felt for the frequent publication of these data.

The immediate use of observations for weather predicting, the increasing demand of the public for information in regard to current weather conditions for practical applications, the study of the courses of storms which could be followed for only a few days at the utmost, the greatly increased number of observing stations, which made the mass of data too bulky to manipulate properly if left to accumulate until the end of the year, the more general appreciation of the fact that short periods of time had distinctive characteristics in the seasonal changes of the year, and the impatience of the users of meteorological data, who ordinarily had to wait a couple of years or more after they were recorded before they could receive them in tabulated form, all combined to necessitate the publication of this data at frequent intervals.

The month was chosen as a proper subdivision on account of its convenience. It was no simple matter to decide on just the form this new mode of publica-

tion should take, as it was to be merely accessory
to the usual annual *résumé* of observations.' The
Monthly Weather Review of the United States Signal
Service at Washington (established 1872) was the
first [1] of these publications as now issued, covering
a great extent of territory ; and its original form was,
I believe, due to Professor Cleveland Abbe, although
its more recent extension has been carried out under
the guidance of Lieutenant Dunwoody. While this
publication has served in a great measure as a pattern
for similar ones in Europe, yet it gives far more information than any of them, containing, as it does
(for instance, August, 1891), the results obtained from a
discussion of reports for the month from 2,575 regular
and voluntary observers, and covering not only the
United States, but Canada and a portion of the Atlantic Ocean. The monthly averages of temperature
and precipitation are given for about 2,000 observing
stations in the United States ; and averages for all
of the usual meteorological elements for about 160
stations. A special feature of this *Weather Review*
is the report concerning the passage of areas of high
and low atmospheric pressure across the United
States and a portion of the Atlantic Ocean. The
charts showing the course taken by storms, and the
distribution of atmospheric pressure, temperature,
rainfall, and winds for the month, are of great
scientific and practical value.

The monthly report of the *Deutsche Seewarte*, at
Hamburg, is the most important European publication
of this class. It was at first prepared by Dr. Köppen,
but afterwards by Dr. van Bebber, who still has it

[1] Monthly reports had been previously published for single stations
and limited areas.

in charge. At first this report stood alone, on the Continent, but the gradual introduction of the system of a monthly publication of local weather characteristics into many of the other weather services, has caused Director Neumayer of the *Seewarte* to give the Review more of an international character. So that now it is not published for several months after the observations are made, and it contains a record of the storm-tracks and weather for the North Atlantic Ocean, as well as Europe. These two monthly Weather Reviews cover the whole region from the west coast of North America to the eastern boundary of Europe.[1] It is not necessary to enter into an explanation of the purely tabular publications.

General Treatises.—Those students of meteorology who were obliged to acquire their knowledge at first-hand, that is, from the original sources, must be somewhat envious of the facilities now offered for a rapid survey of what has been accomplished in various directions. For a time all the compilations which were published after 1860 were of the school-book type, and this continued until Blanford's *Indian Meteorologist's Vade-Mecum* appeared in 1876–77. This is the first of the treatises which introduce us to some of the ideas of the modern school of dynamical meteorology. Its publication in India, together with its considerable cost, deprived the book of a wide circulation, and it has not, outside of India, accomplished the good which it was capable of doing; it was, for many years, undoubtedly the best treatise on modern meteorology in any language.

[1] Each publication may be obtained at an annual subscription of six shillings.

Blanford's book was but the forerunner of the important compilations which began to appear some years later. Foremost of these came Hann's *Climatologie* in 1883. This most excellent work gives an account of the climatic factors in general, and then gives the special features of the climate of all parts of the globe, and the causes which produce them. Few tables of data are given, as it has been the author's object to trace out the philosophy of the climatic conditions in a way which will attract the general reader. This book was the natural outcome of the frequent contribution of tables of climatic data given in the *Öster. Zeitschrift für Meteorologie* during a period of nearly twenty years. Next there appeared in Germany Sprung's *Lehrbuch der Meteorologie*, in 1885, in which the author has confined himself mainly to the discussion of actual atmospheric occurrences, and has sought to explain these by means of physical-mechanical laws. He has, indeed, added an appendix on the subject of the registration of meteorological phenomena; but this is suggestive, rather than an account of apparatus in common use. Sprung has first of all explained the principal mechanical and physical laws, which are then used as a basis for subsequent development of the theories concerning the statics and dynamics of the atmosphere. His is the only attempt which has been made to fit together the various theories of the atmospheric motions, the history of which subject is traced in another chapter. This is the main feature of the book, although the latest views on other topics, such as the influence of solar radiation, the daily period and the variability of the meteorological elements, are also given. The book is well adapted for university instruction and for study by professional meteorologists ; but most of it cannot be comprehended by the casual reader.

In 1884, van Bebber, of the *Deutsche Seewarte*, published Part I. of his book on *Witterungskunde*. This, together with Part II., published a year or two later, forms a compendium of what has been done from the earliest times in the matter of investigating the cause of weather changes.   The various questions, varying in degree from the proverbs of folk-lore to the modern systems of weather-maps and forecasting of weather, based on the telegraphic transmission of observations made thousands of miles distant, are all discussed in a historic as well as scientific manner. While the work is of the greatest value to students of meteorology, who will for a long time to come go to it as a storehouse of knowledge on this special subject, yet the author has, by the fulness of his text, made it of popular interest as well.

Ferrel's book on *Recent Advances in Meteorology* (1885) next appeared, and it differs in one respect from the other works here mentioned, in the fact that it was published entirely under Government auspices. It has, therefore, not been obtainable by purchase ; and its circulation has not depended on its importance, but upon the will of the chief of the United States Signal Service at Washington.   Naturally, few persons except those on the "exchange list " of this department have been able to obtain a copy.   This work was prepared for professional use, and while portions of it could be understood by the general reader, the difficulty in reading it as a whole is such, that its inaccessibility to the public is not so much to be deplored as would have been the case had it been written more popularly.   Its chief value is that we find given in a single volume of moderate compass, and in a connected manner, the most important results of the author's thirty years' labour in the field of meteorology.

The fact that most of our meteorological publications are weak in the department of apparatus, and also that many, perhaps most, of the descriptions of original ·inventions and improvements have been published in scattered journals, proceedings, and reports, made it desirable that a comprehensive account of them should be given in a single volume. This was in a large measure accomplished by Abbe's *Treatise on Meteorological Apparatus and Methods* (Washington, 1888), which was prepared and published at the expense of the United States Government, and gives us not only a historical sketch of the development of the subject, but also shows its condition at the present time. He has given special attention to the theory of the principal instruments, and we find reproduced about everything of importance that has been done in this line of investigation. The numerous illustrations form an especially valuable feature of the work, which has been rendered as complete as possible. Having access to the great collections of books which have been gathered together at Washington, and having as a source of guidance the manuscript catalogue of meteorological literature which has been accumulated at the Washington Meteorological Office, mainly by the international co-operation of some individual meteorologists with the hope of its speedy publication, the author enjoyed facilities not elsewhere obtainable for rendering his work exhaustively complete. While the diagrams and descriptions of instruments which

are given will interest all who are students of meteorology to only a limited degree, yet the main importance
of the work is to the professional meteorologist, who
will find the need of a mathematical training in order
to read it.

Woeikoff's work on Climatology (*Die Klimate der
Erde*, 1887) must also be mentioned in this list, for
in it he has given us an excellent picture of climatic
conditions and their effects as viewed by a physical
geographer. Portions of the book are, indeed, somewhat similar to Hann's work, mentioned above ; but
its main value is the general climatic relations which
it gives, rather than the climate of any special places,
so that it is largely supplementary to Hann's *Climatologie*.

Recognising the difficulty not only in reading, but
also in obtaining his above-mentioned work, Professor
Ferrel has written out in a popular manner his theories of atmospheric motions and their attendant
phenomena, and has also given a careful comparison
of these theories with the observed facts.   While this
book, which bears the title, *A Popular Treatise on
the Winds* (New York, 1889), is by no means light
reading, yet mathematical analysis has been so far
omitted that only an elementary knowledge of algebra
and trigonometry are requisite for comprehending it,
and it is certainly the most complete statement of the
subject that has yet been made.

Blanford has presented some of the results of his
many years' labour in India in his book on the
Climates and Weather of India (London, 1889) ; and
while it is, as a work on local climatology, at present
of more value to the English people than any other,
yet the majority of meteorologists, who have no

opportunity of seeing the original Indian Meteorological Memoirs, have welcomed this inexpensive summary of the important contributions made by English meteorologists in a region which presents the most diverse field of topographic and climatic characteristics of any limited region on the earth's surface.

I must also mention the German memoirs recently published in Penck's Geographical Series at Vienna : *Luftdruck in Europa*, by Hann ; *Klimaschwankungen*, by Brückner ; and *Schneedecke*, by Woeikoff.

The following partial list of recent popular works on meteorology is given : Marie Davy's *Météorologie Générale* (Paris, 1877); Mohn's *Grundzüge der Meteorologie* (Berlin, 1883) ; Scott's *Elementary Meteorology* (London, 1890); van Bebber's *Lehrbuch der Meteorologie* (Stuttgart, 1890) ; and the works on *Weather*, and *Seas and Skies of Many Lands* (London, 1888), by Abercromby ; and *American Weather*, by Greely (New York, 1888). On the subject of Weather Predicting, van Bebber's little book, *Wettervorhersage* (Stuttgart, 1891), and Scott's older book, *Weather Charts and Storm Warnings* (London, 1887), give quite full information. Mention must also be made of Hann's *Atlas der Meteorologie*, which forms part of the last edition of Berghaus' *Physikal. Atlas*. This work is composed of numerous charts, which are invaluable to the student of meteorology.

There is still one general topic which remains to be written on, and this is the reduction and use of meteorological observations. There are many finished memoirs which may serve as patterns for the manipulation of special observations, such, for

instance, as Wild's great works on the temperature
and rainfall of the Russian Empire, but if an observer
wishes to know just what is the best thing to do with
any particular series of observations, he is many
times at a loss to find reliable guidance. There are
throughout the world many thousand observers who
are willing to take the pains to make regular observations, which they usually reduce to monthly averages
before handing them over to some central observatory
where the averages may be published, but where the
individual observations are consigned to the archives
of the institution, there to lie buried and unused.
There are perhaps a few score of men who have
given in detail the best possible methods of investigating observations, and it is the combined knowledge of these men that should be put into a
volume accessible to all observers, so that each one
can work up his own observations in the most systematic way.   Such a compendium would undoubtedly
be the cause of the rapid completion of a detailed
meteorological survey of all civilized lands, for it
would multiply a hundred-fold the present working
force of climatologists.

   Periodical Literature. — We usually make the
periodical literature of a subject the measure of its
condition and progress, and so this literature relating
to meteorology requires more than a passing mention.
Our modern methods of work and research require
a convenient channel for the timely interchange of
ideas between specialists not only of the same
country, but of different parts of the globe. But
a subject cannot be specialized to the extent of supporting technical journals until it has reached a
certain stage of cutting loose from the apron-string

of some mother-subject under whose wing it has
been sheltered until able to stand alone. Thus
meteorology was fostered by both astronomy and
general physics, and the earlier writings on
meteorology are to be found mainly in the journals nominally devoted to these two topics. There
had been a number of attempts made to establish
technical meteorological journals before the period
of which I am now writing ; but the lack of support
had rendered them of short life, a single exception
being the *Annuaire de la Société Météorologique de
France*, which really dates from 1849. The functions of this journal have been to report the discussions which have taken place at the local meetings of the society, and to promote the interests of,
and report on the meteorology of France. This journal
is, then, chiefly of interest to those who wish to know
what French meteorologists are doing. I think
that this localising tendency has prevented this
monthly publication from exercising as great an
influence as it ought on the general progress of
meteorology.

In 1866 the *Österreichischen Zeitschrift für Meteorologie* was established, and as the society which
publishes it holds only annual meetings, it can be
easily imagined that it bears little resemblance to the
just-mentioned French journal. The Austrian journal
had a separate existence for twenty years, when it
was combined (1886) with the newly (1884) established German *Meteorologische Zeitschrift*, and is now
published under the joint control of the Austrian and
German societies. Dr. Hann and Dr. Hellmann,
of Berlin (succeeding Dr. Köppen, of Hamburg),
Austrian and German meteorologists of world-wide

reputations, are the editors of this, the most important and influential meteorological publication in existence. It has always been under an exceptionally able editorship, which has prevented worthless matter from appearing in its columns, although of course many contributions, especially of the earlier volumes, are seemingly unimportant as we look at them from the present position of meteorology. Dr. Hann was early associated with Dr. Jalinek in the publication of the first volumes, when the character of the journal was being formed, and when the former assumed the entire control, which he exercised during the years of the most rapid elevation of the level of meteorological work, we find, if any change can be noted, an improvement on what was so well done before. Published at first bi-monthly, and later as a monthly, this journal furnishes a ready means for publishing, without the delay usually associated with official government or institutional publications, original papers, some of considerable length, contributed by meteorologists from all parts of the world. While these contributors are mainly those who use the German language, yet national sectionalism is nowhere to be remarked, for apart from editorial work more of the contributed matter comes from Germany than from Austria, the nominal home of the journal. In addition to the valuable original leading articles, there are given reviews and notices of important meteorological publications and undertakings, and also many topics of interest are treated in the so-called *Referat*, which in this case is usually a kind of editorial review, in which the notice of some important paper is made the excuse for bringing together the most valuable ideas which have

been previously advanced on the subject, the
reviewer himself adding something to our knowledge of it. This class of reviews requires an
immense amount of time in preparation, and it is
remarkable that Dr. Hann has been able to do so
much in this line in addition to his other duties.
Still another feature is the mass of climatological
data which has been gradually accumulated, mainly
through the personal labour of Dr. Hann ; for it will
be found that almost every number of the journal
contains tabulated average data for the principal
elements for one or more out-of-the-way places, or
the results obtained by many years observations at
some so-called normal station. These tables have
proved to be of great service in many original
researches undertaken by meteorologists, and a
searcher for information concerning the climate of
any portion of the globe will not be likely to turn
away unrewarded from a file of this journal. We
find in its pages an almost complete account of the
progress of meteorology during the past twenty-five
years, and it is safe to affirm that no one unac-
quainted with its contents can claim to be thoroughly
informed of the progress of meteorology during the
period when it passed from a sub-science to the
dignity of a separate science, demanding the entire
attention of any one who will master its details.
Of all the meteorological publications this is the
most international and widely read, and it is on
this account that its columns are used by writers
of so many nationalities, and from all parts of the
world, who may wish to secure for their ideas the
widest circulation. There is, however, one department, that of instruments and apparatus, in which

there are not given sufficient details to enable one to obtain a good idea of the subject from this publication alone.

Since 1884 Germany has also given us a more popular meteorological monthly magazine, known as *Das Wetter.*

Great Britain furnishes three journals : *The Quarterly Journal of the Royal Meteorological Society, The Quarterly Journal of the Scottish Meteorological Society,* and *Symons' Meteorological Magazine.* The two former contain papers by the members of the societies, and they are usually quite elaborate, being similar to the leading articles in the Austro-German journal. Symons' magazine is more popular in nature, and the articles are somewhat like those in the German *Wetter.*

The American Meteorological Journal, published monthly, was established in 1884, and has contributed largely to the recent advancement of meteorological science in the United States. This journal is a private enterprise, and was instituted by Professor M. W. Harrington, of Ann Arbor, Michigan.

The Italian meteorologists also publish a journal, bearing the title *Bolletino Mensuale Moncalieri;* this was established in 1880.

§ 3.—*International Meteorology.*

The rapid spread of an active interest in meteorology in various lands, and the resulting marked increase in the amount of meteorological data published, had caused frequent utterance of the idea that a greater uniformity in the methods of

procedure adopted in different countries was not
only desirable but indispensable to the proper use of
this material.   No definite step was taken however
until, in the spring of 1872, a circular, signed by
Bruhns of Leipsic, Wild of St. Petersburg, and
Hann of Vienna, was issued to those practically
interested in meteorology, stating that a conference would be held at Leipsic in August, 1872, for
the purpose of studying the needs of international
meteorology.   In order that something definite
might be accomplished at this meeting a list was
proposed of twenty-six questions, which were to be
discussed at this meeting.   Fifty-two persons re-
sponded to this invitation, and the conference met
on August 14, 1872, at Leipsic.   The report of the
proceedings of this meeting is extremely interesting,
and shows the wide divergence of opinion which then
prevailed concerning the collection of meteorological
data, both as regards methods of observation and the
manner in which they ought to be treated.

Only two of the points brought out at this meeting
will be mentioned, viz., that it was desirable for each
country to publish in full the tri-daily observations for
a number of its selected stations and in a uniform
manner, the form then in use in Russia being considered most desirable; and that a more systematic and
more extensive use should be made of telegraphic
weather reports.

The Leipsic meeting, however, was merely preliminary to, and to open the way for, the International Meteorological Congress which it was
decided to hold in Vienna in 1873.   This congress
was convened on September 2, 1873, and all the
principal European Countries (except France), the

United States of America and China were represented by delegates whose official positions and
scientific attainments signified the importance of
the meeting, and indicated that it would be a
veritable turning - point in the development of
meteorology. Among the important results accomplished, at this meeting may be mentioned the
following : Appointment of a Permanent Meteorological Committee with self-perpetuating power; and
the proposal by the delegate from the United States
of America for the establishment of single daily
simultaneous observations as nearly as possible over
the whole northern hemisphere. Many questions of
great importance were left to be decided on by the
Permanent Committee, and the reports of sub-com-
mittees and discussions concerning apparatus and
methods form the most valuable *résumé* we possess of
the worth of instruments and methods in use at that
time.

Meetings of this Permanent Committee were held
in 1874, London 1876, Utrecht 1878 ; and in 1874 a
Conference on Maritime Meteorology was held in
London by the sub-committee appointed by the
Permanent Committee of the Vienna Congress of
1873. At these meetings many questions concerning
the present status and suggested improvements in
meteorological methods were discussed, and enough
matter demanding formal action was accumulated to
warrant the calling of a second International Meteorological Congress at Rome in 1879. At this meeting
the work accomplished by the Permanent Committee
since 1873 was fully considered ; the value of the
changes which had been directly and indirectly the
results of the Vienna meeting of 1873 was carefully

discussed, as well as further questions of importance. A special feature of this meeting was the numerous reports which were presented on various topics by those who were recognised authorities ; these reports having been requested by the Committee. A more permanent organisation of the Permanent Committee was effected by which nine members were appointed to it, with only one from any one country. This Committee from this time has been known as the International Meteorological Committee. It first met at Rome, April 22, 1879 ; again at Berne, August, 1880 ; Copenhagen, August, 1882; Paris, September, 1885; Zurich, September, 1888.

As it was found that another early International Meteorological Congress was not desirable, the Permanent Committee agreed upon a dissolution at the Zurich meeting ; the President, Wild, of St. Petersburg, and the Secretary, Scott, of London, being charged with taking steps for convening, at a suitable time, an international meeting of the representatives of the various Meteorological Services.

Such a meeting was held in Munich from August 26 to September 2, 1891. In response to a circular issued by Messrs. Wild and Scott, there were present at this meeting the heads of the various European meteorological services, and other prominent meteorologists from these and distant lands. The following countries were represented : France, Great Britain, Germany, Russia, Finland, Norway, Sweden, Denmark, Hungary, Roumania, Netherlands, Switzerland, Spain, Bulgaria, Queensland, United States, and Brazil. The subjects discussed were, briefly stated : the best methods of making observations ; the adoption of absolute standards of reference ; methods of calculation or reduction of observations ; the forms for publication of observations, and the methods for discussing them ; the improvements to be made in telegraphing weather conditions and making weather forecasts, and especially in obtaining a knowledge of the weather conditions to the westward of Europe ; methods of making and reducing observations of the earth's magnetism and atmospheric electricity ; co-operation in observing cloud movements ; the means for best securing international co-operation in meteorological work ; the question of the organisation and work of a proposed International Meteorological Bureau ; and finally, the formation of a new International Committee composed of seventeen members. This last was carried out and the members chosen. An informal meeting is expected to be held at Chicago, U.S., in 1893, but it was decided that a formal meeting should take place at Paris within five years.

Reports (in English) of these various meetings can be obtained from the Meteorological Office, London, for one or two shillings each.

In October, 1879, through the action of the Permanent Committee, an International Polar Conference was held at Hamburg, for the purpose of planning the work of the proposed temporary Circumpolar Meteorological-Magnetic Stations.   In April, 1880, mainly by the action of the executives of the Permanent Committee, there was held at Vienna an International Conference for Agriculture and Forest Meteorology.

Only a very limited general statement can be given of the international benefits derived from these meetings ; but they may be stated as follows :

The classification of meteorological observing stations into orders according to the number and variety of observations made, and the establishment of a form for the printing of their results.

The general adoption of either the meter-centigrade or the inch-Fahrenheit scales, and recognised standards of reference.

A European system of weather telegraphy with uniform code of exchange.

The daily simultaneous international observations made at 0h. 43m., Greenwich mean time. (Published by the United States Government, but now discontinued.)

The establishment of the International Circumpolar Stations during 1882–83.

The publication of International Reduction Tables for meteorological observations.

The systematic reduction and publication of Ocean Meteorological data.

CHAPTER II.

APPARATUS AND METHODS.

§ 1.—*Thermometers.*

HISTORICAL.—The history of the early development of the thermometer is not known with certainty, but in the latter part of the sixteenth century Galileo used at Padua a kind of thermoscope which consisted of a glass bulb (*a*) with a narrow bore tube (*b*), a couple of spans in length, attached. A general idea of the construction is given in Fig. 1. After the air in the bulb had been rarified by heating, the open end of the tube was inserted into water in an open vessel (*c*), and the tube was held upright. The water would rise to some extent in the tube, as the air in the bulb cooled ; and the level of the water in the tube would oscillate somewhat with any change of temperature to which the air bulb would be subjected. In this first instrument the fluctuations of the atmospheric pressure would of course introduce some errors, but the principle of the modern thermometer had been discovered, and its further completion did not need the genius of Galileo to develop it. It is not certain who suggested the next step in the gradual improvement of the instrument. Abbe ascribes it to Galileo, and Hellmann thinks that the search now

being made by Javaro concerning Galileo's career
may bring to light some further information.   At
present, however, we must say
(according to Hellmann) that a
French   physician,   Jean   Rey,
about the year 1631, conceived
the idea of inverting the Galileo
thermometer and filling it with a
liquid and observing the expansion of this.   Sometime before
1641, the Grand Duke Ferdinand
II. had adopted the plan of using
alcohol as the fluid, and closing
the open end of the tube so as
to exclude the air.

The members of the Acca-
demia del Cimento, most of
whom were pupils of Galileo,
devised several forms of
thermometer, the most
important of which was
known as the "Little

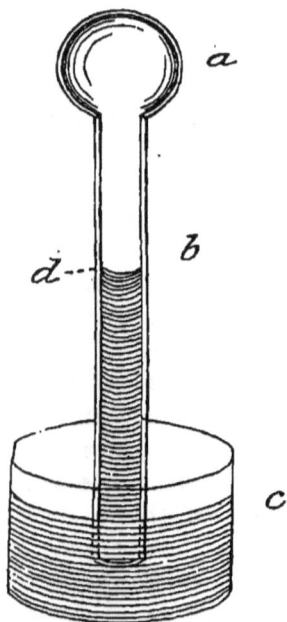

FIG. I.

Florentine Thermometer."   This instrument,
as illustrated in Fig. 2, had a fixed scale with
the zero at about the point shown by the
freezing mixture of salt and water, and the
13.5 division mark at the freezing point of
water.   The graduations which were arbitrary
were small glass particles about the size of a
pin head attached to the glass tube.

It is worthy of note that Galileo's thermometer was introduced into medical practice by
Sanctorius at some time previous to 1624.

FIG. 2.

The history of the development of the modern

mercurial thermometer must be passed over, with the remark that Fahrenheit was the first who fully accepted the necessity of having two fiducial points, the freezing and the boiling of water, and he settled down to the use of mercury in the construction of the thermometer. It has been suggested that the term "degrees" as applied to the thermometer readings was derived from an early form of thermometer in which the tube was bent into a circular form (or arc), and the degrees of the circle were used to show the position of the substance in the tube. It is hardly necessary to enter into an account of the Fahrenheit, Reaumur, and Centigrade (Celsius) thermometer scales, as it is given in every text-book of general physics, and in encyclopædias ; but it may be remarked that the Centigrade scale is in use among meteorologists in nearly all but the English-speaking countries, and in these the Fahrenheit scale is still so firmly established that the international assemblages of meteorologists have been obliged to sanction the publication of temperatures in either scale. Usage has therefore prevented the general adoption of the Centigrade scale, although it is in every way preferable to its rival.

In giving an idea of the present state of thermometric science as applied to meteorology, only a brief mention can be made of the question of the various errors of construction of thermometers and their final reference to normal instruments ; for while, in the laboratory, errors of hundredths or even thousandths of a degree are being taken into account, yet in the practical use of thermometers for determining air temperature there are still discrepancies in various systems of instrument exposure which may amount to one or more degrees.

Standard Thermometers.—It has been shown by physicists that the changes in volume or pressure of an ideal gas will give an absolute scale for measuring temperatures. But as we have no ideal gas, Sir William Thomson proposed for use an absolute scale which should have a temperature of 273°·7 at the freezing point of water, and 373°·7 at the boiling point of water, thus fitting it to the divisions of the Centigrade scale. This scale agrees with the readings of the actual gas thermometer at the two points mentioned, but at intermediate points, and above and below these limits, a correction has to be applied to the observed temperatures in order to reduce them to the absolute scale.

The following table shows the corrections necessary to reduce some gas thermometers to the absolute scale. All but Thomson's first series are under constant volume.

ABSOLUTE SCALE.	CENTI-GRADE.	THOMSON AIR THERMOMETER.		WEINSTEIN AIR THER-MOMETER.	CARBONIC ACID GAS.
		Constant pressure.	Constant volume.		
		°C.	°C.	°C.	°C.
273°·7+ 0°	0°	·000	·000	·000	·000
+20°	20°	—·040	—·030	—·013	—·034
+40°	40°	—·048	—·040	—·018	—·051
+60°	60°	—·047	—·037	—·018	—·050
+80°	80°	—·028	—·022	—·012	—·032
+100°	100°	·000	·000	·000	·000

But for most practical purposes it is necessary to use mercurial thermometers, and we must know how their readings stand with regard to gas thermometers.

Recent comparisons, at the International Bureau

at Sèvres, of standard mercurial thermometers with gas thermometers, show that the following corrections must be applied to the readings of the former instruments in order to reduce them to the latter standards:

Corrections.—To reduce hard-glass thermometers by Tonnelot (Paris) to thermometers of—

HYDROGEN GAS.		NITROGEN GAS.		CARBONIC ACID GAS.	
At Temp. ° C.	Correction.	At Temp. ° C.	Correction.	At Temp. ° C.	Correction.
—25° C.	+·233° C.	—25° C.	+·216° C.		
—20	+·172	—20	+·159	—19° C.	+·094° C.
—15	+·119	—15	+·109	—15	+·069
—10	+·073	—10	+·067	—10	+·041
— 5	+·034	— 5	+·030	— 5	+·019
0	·000	00	·000	0	·000
20	—·085	20	—·075	20	—·042
40	—·107	40	—·097	40	—·048
60	—·090	60	—·085	60	—·037
80	—·050	80	—·052	80	—·019
100	·000	100	·000	100	·000

Wild, in 1888, referred to the scale of the Hydrogen thermometer as the International Temperature Scale.

Comparison of Standard Thermometers.—Although ultimately the absolute scale will be the standard to which all thermometric work will be referred, the actual working standard will probably be the gas thermometer adopted by the International Bureau of Weights and Measures at Sèvres. The gradual distribution of the best standard mercurial thermometers, after a thorough investigation, from Sèvres to various countries, will no doubt introduce practical uniformity in all laboratory thermometric work the world over. A collection of such standard thermometers is represented in Fig. 3. At present a number of standard thermometers are in use. Abbe gives a little summary for a few instruments which have come

to his notice at Washington. The Signal Service
(Washington) air thermometer was found to be
0°·01 C. lower than
Rowland's air thermo-
meter at Baltimore.
Above 0° C. the Signal
Service air thermometer
agreed with the mer-
curial standard, after
the latter has had ap-
plied to it the ordinary
corrections to reduce its
readings to those of an
air thermometer. Also
from + 32° F. to − 38°
F. the Kew and Signal
Service mercurial ther-
mometer agree, but at
− 38° F. the Signal Ser-
vice alcohol thermo-
meters are 0°.6 F. below
the Kew alcohol ther-
mometers. Also the
Signal Service air ther-
mometer agrees with
the Sèvres hydrogen
thermometer (in 1885)
to within ±·01° C.

Wild, at the Central
Physical Observatory,
has three thermometers
(made of Thuringian glass) which he has been using
as a standard. In 1888 these were compared from
0° C. to 40° C. with a Tonnelot mercurial standard

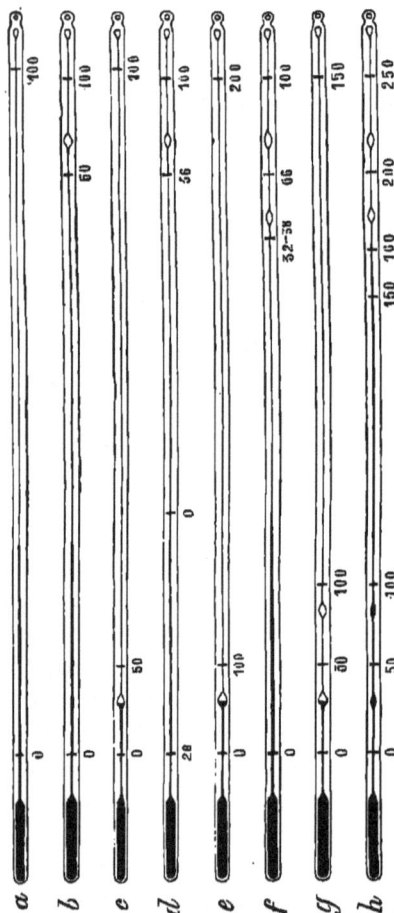

FIG. 3.

(of hard glass), which had been compared with the hydrogen gas thermometer at Sèvres. The following table shows the result of this comparison :

CORRECTIONS OF ST. PETERSBURG STANDARD THERMOMETERS.

| Temp. | GEISSLER No. 2. | | GEISSLER No. 10. | | FUESS No. 1. | |
	Tonnelot.	Hydrogen Sèvres.	Tonnelot.	Hydrogen Sèvres.	Tonnelot.	Hydrogen Sèvres.
0° C.	·00° C.	0° C.	·00° C.	·00° C.	·00° C.	·00° C.
10	— ·0	—·06	—·03	—·08	—·01	—·07
20	—·03	—·11	—·05	—·13	—·02	—·11
30	—·05	—·15	—·05	—·16	—·03	—·13
40	—·07	—·18	—·05	—·16	—·04	—·15

In 1875 the corrections of Geissler No. 10, as referred to the mercurial scale were :

Temperature ...	0° C.	10° C.	20° C.	30° C.	40° C.
Correction ...	·00	—·03	—·04	—·07	—·07.

These show that for at least fifteen years this observatory has had practically standard thermometer readings.

Fig. 4 shows the usual method of comparing the readings of mercurial thermometers, when the instruments are immersed in a water bath, which may be made to vary in temperature by means of a lamp beneath.

Mercurial Thermometers. — The mercurial stem thermometer of the best construction consists of a slender glass tube with capillary bore terminating at one end in a thin glass bulb filled with pure mercury, and having at the other end, and connected with the capillary bore, a small empty reservoir called

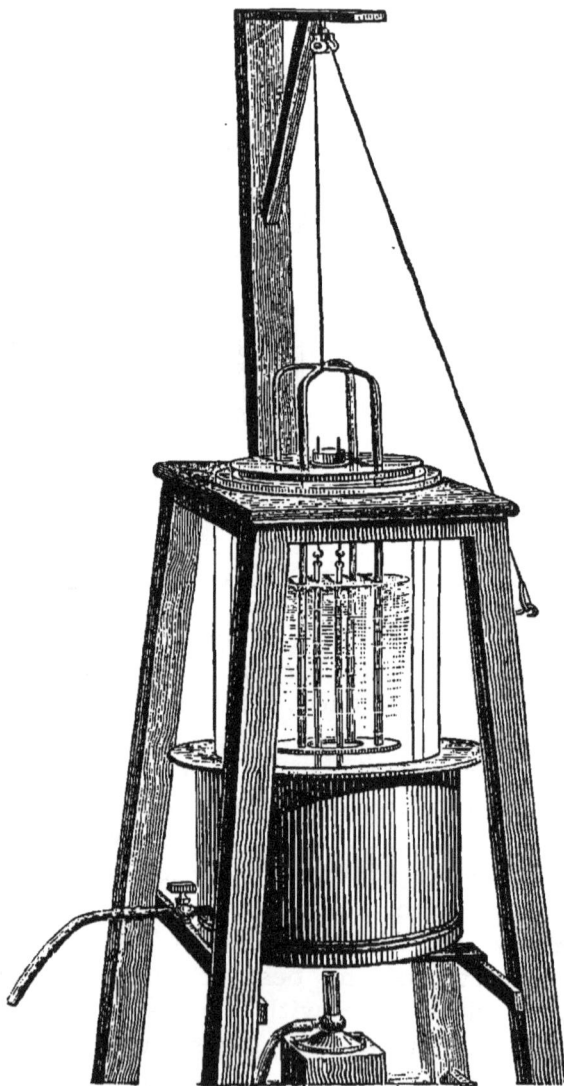

FIG. 4.

the calibrating chamber (Fig. 5). The stem has marked on it, first, the boiling point, and second, the freezing point, after which, in the case of the Centigrade thermometer (which we shall alone consider), the distance between these two parts is divided into one hundred equal spaces called degree divisions. These subdivisions are either ruled directly on the stem, as in the case of the Tonnelot (Paris) thermometers, or they are marked off on a piece of white enamel glass, which is placed behind the thermometer stem with the zero and hundred degree points directly opposite the corresponding points on the stem, as in the case of the Fuess (Berlin) patent barometer. These graduations are extended slightly above, and considerably below, these two fiducial points for ordinary use. In the thermometers shown in Fig. 3 it will be noticed that enlargements of the capillary bore occur. These are mainly for the purpose of allowing a much larger scale of degrees to be used than would be possible otherwise, but it necessitates the

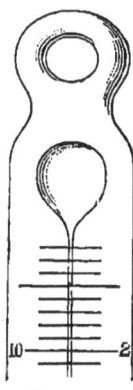

Fig. 5.

use of several thermometers in order to get the complete scale. Temperatures can be measured until the enlarged portion of the bore is reached, but this must be filled and the mercury pass up into the narrow bore above, before additional readings can be made. The enlargement is so made as to require an increase of a certain number of degrees of heat in order to fill it, and if temperatures within this limit are to be read, another thermometer with the enlargement in a different part of the scale must be used. By this means the fiducial points may be determined on each thermometer, no matter what range of temperature the graduations may include. We will consider briefly the various errors which may be met with in such a thermometer of the best construction and the methods of determining the corrections necessary to allow for them. These are as follows:

1. *Correction for Parallax and Refraction.* — In either form of construction of the scale the divisions are not exactly adjacent to the mercury thread in the tube, and consequently the reading will vary somewhat with the direction of the line of sight. In case the thermometer can be carefully adjusted so that the reading can take place at the angle of normal incidence, it should be done, thus avoiding this error, but it will frequently be the case that this is impracticable, in which case the correction must be computed; in which computation the index of refraction of the glass (from the bore to the exterior surface of the stem) must be included in the ordinary formula for parallax. In cases where it can be done, it is better to eliminate the refraction by taking the mean of the readings when the scale is on the side nearest the observer, and again when placed opposite.

2. *Scale and calibration corrections.*—In thermometers of this highest class the scale divisions should be accurate, far within the limits of any readings that are to be made; but in case they are not, this error of graduation is combined with that of calibration about to be described. It is impossible to get a tube or stem that is exactly cylindrical, and in which the size of the bore is absolutely uniform. It is customary therefore, in originally selecting the tube for the thermometer, to put a short thread of mercury in it, and to observe the length of this thread in different

portions of the tube.   If the bore is quite uniform the
thread of mercury will be of about the same length
throughout.   This work is of course done by the
maker.   After the thermometer has been constructed
and the scale attached, a similar but more refined
process is gone through, in order to give quantitative values to the corrections to be made for
the combined errors of graduation and inequality of
bore.   By an artificial heating or cooling of the
thermometer bulb the mercury in the stem can be

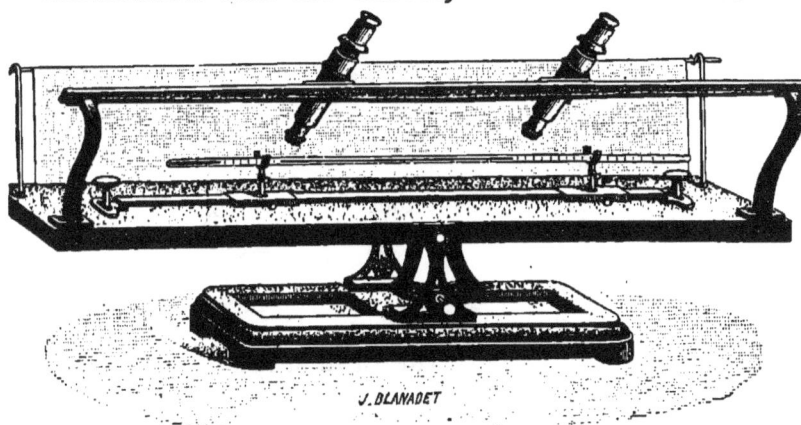

FIG. 6.

brought to any desired length, so that upon a slight
jarring of the instrument a thread of mercury of any
required length will be separated from the main
mass of mercury.   This thread is at first chosen, say
5° in length.   The thermometer being kept at constant temperature, the length of the thread (of
constant volume) is observed in terms of the scale
divisions at various portions of the stem from one
end to the other.   The process is repeated with
threads of 10°, 15°, . . . 95° length, in order to insure

a greater accuracy, and to prevent the accumulation of errors due to the use of a short thread alone. In order to secure ease of manipulation and accuracy of work, the thermometer is placed in a holder, as shown in Fig. 6, and the graduations read by means of microscopes. By the insertion of these observed lengths in a system of symmetrical equations, which must be excluded here, the calibration and scale corrections for various portions of the tube are computed. The following little table shows these corrections for two of Tonnelot's normal thermometers :—

Scale Reading.	Calibration Correction.	
	No. I.	No. II.
°C.	°C.	°C.
0	0·0000	— 0·000
20	— 0·0569	—0·250
40	+ 0·0709	—0·464
60	+ 0·1968	—0·372
80	+ 0·1667	—0·471
100	0·0000	— 0·000

3. *Determination of the fiducial points. a. The boiling point.* The thermometer must be hung inside of a double casing, with the bulb suspended at a little distance above the boiling water. The whole thermometer must be enveloped in the steam, and a hole in the outer casing allows the free escape of this vapour. All air in the inner casing must be replaced by steam. A manometer must show the difference in pressure between the steam inside the apparatus and the air outside. A model apparatus for determining the boiling point is shown in Fig. 7, in which T is the thermometer, L the reading telescope, A the vapour chamber, and C the water reservoir. Whenever a reading is made of the thermometer a corresponding reading must also be made of the manometer, and also of a barometer,

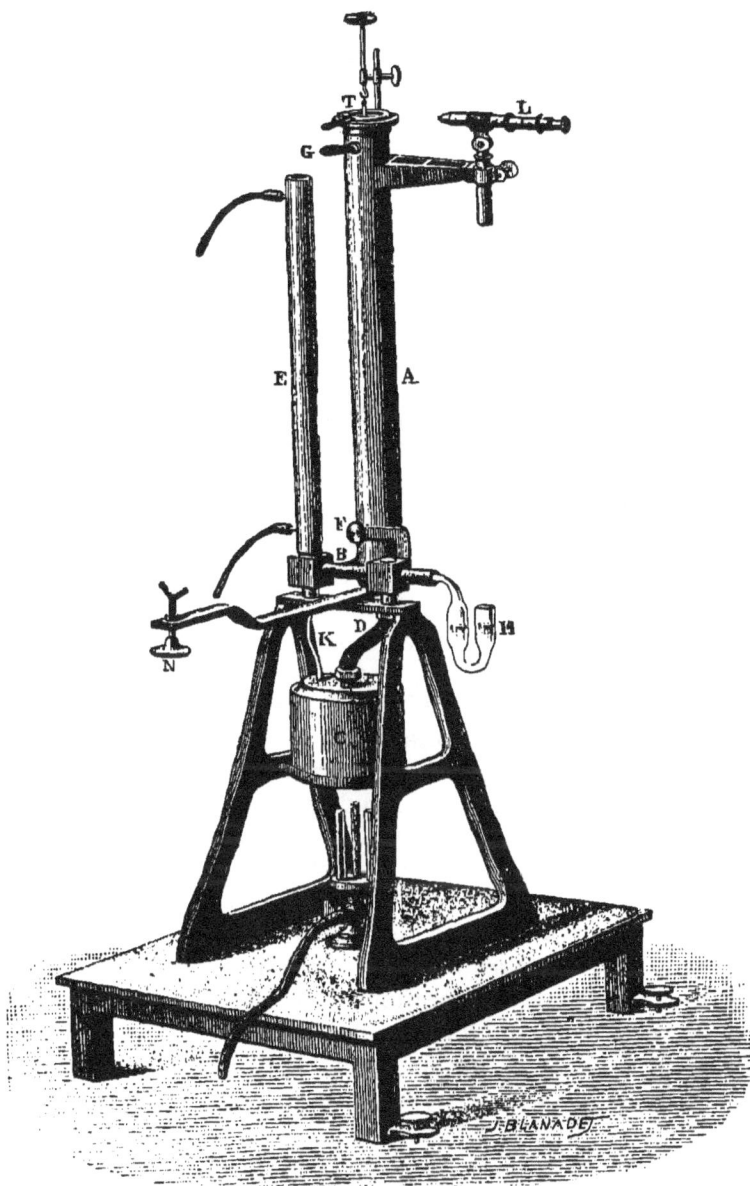

FIG. 7.

If the mercury in the thermometer is carefully observed as it rises up to the boiling point, it will be found to reach a height which it does not permanently retain, and if the boiling is continued for a sufficiently long time, the mercury will be found to have slightly receded from its highest reading. This lower reading is the desired boiling point, and this slight lowering is due to a gradual change in the capacity of the thermometer bulb. It has been found that the true boiling point can best be determined by taking the thermometer from the apparatus, and making a determination of the freezing point, after which a new determination of the boiling point must be again made. If these two successive determinations agree, we may consider that the true boiling point has been obtained. It is necessary that the experiment be performed with the thermometer horizontal as well as vertical. Various standards of atmospheric pressure and geographical position have been adopted by thermometer makers of different countries, but that at present in international use is the boiling point which corresponds to an atmospheric pressure of 760 mm. of mercury at zero degrees C. at the latitude of 45°, and sea-level. In determining this pressure the manometer reading must be applied to the observed barometer reading in order to get the true pressure inside of the apparatus.

b. The freezing point. This is determined by subjecting the thermometer to the temperature of melting ice. The water which has been frozen must be pure, and if ice is used the water must have previously been distilled; but freshly fallen snow furnishes a very good substitute. Not only the bulb but the stem of the thermometer should be immersed

in the ice, the freshly melted water flowing over the
bulb.   An illustration of this process conducted with

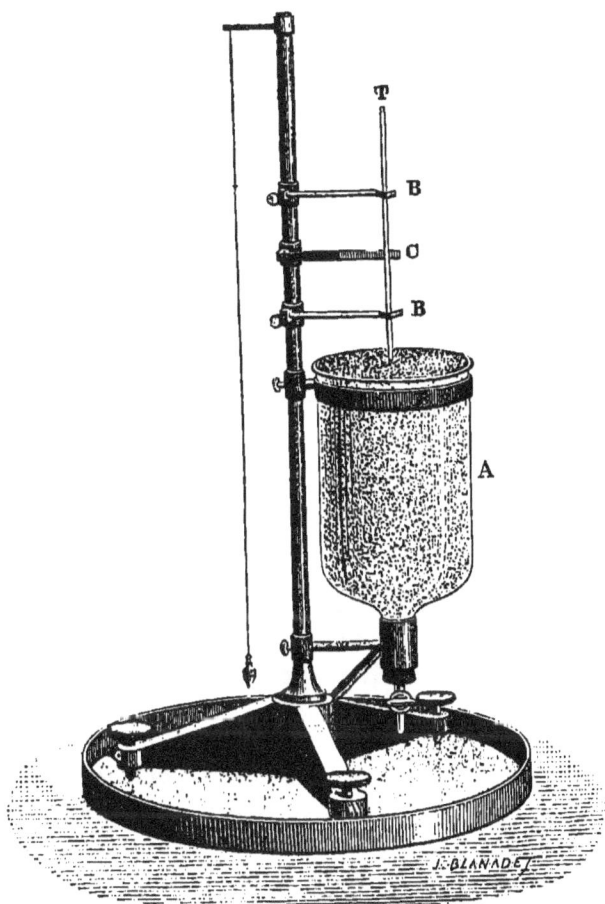

FIG. 8.

proper apparatus is shown in Fig. 8, in which T is the
thermometer.   When the thermometer is first inserted

in the finely chipped ice, the mercury will rise, since the
glass bulb contracts the more quickly, but a little later
the mercury falls rapidly, and soon comes to a stand-still
at the *depressed freezing point*. This point depends
on the condition of the bulb just before it was put
into the melting ice. If the thermometer is kept at
this temperature the mercury slowly rises, and in the
course of twenty-four to ninety-six hours it will reach a
point called the *raised zero*, which can be attained more
rapidly by warming up and cooling the thermometer
slowly ; but if the latter is subjected to the ordinary
air temperature immediately after the depressed zero
point has been noted, and is kept at this temperature
for a year or so, then at the end of this time it will
show a slightly higher reading than it would have if it
was kept at the freezing temperature. If now it is
again put into the melting ice, then what is called the
zero of long repose is at once attained, and will be
found to differ very slightly from the just mentioned
raised zero. These changes are slower and of less
extent, the more nearly the thermometer is kept at
the boiling point of water. The higher the temperature which the thermometer had before it was put
into the melting ice, the lower will be the observed
zero. The observed zero of a thermometer depends
not only upon the highest temperature to which it
has recently been subjected, but also upon the time
which has elapsed since it was at that temperature ;
so that for very accurate work in thermometry the
depressed zero point must be determined immediately
after any measurements of temperature. In order to
facilitate the determination of the freezing point,
where it must be found frequently, a rapidly stirred
mixture of one part water and five parts of slowly

melting ice will be found preferable to the ordinary melting snow or chipped ice.

When making an observation of the depressed zero point it is usual to note the thermometer reading several times near the point ; as the mercury approaches it, reaches it, and recedes from it. Pernet found that this depressed zero corresponding to any temperature varies as the square of the temperature above 0° C. It is the custom at the International Bureau of Weights and Measures to determine the freezing point of a thermometer in a room having a temperature of from 6° to 12° C. ; for an error of 0°·01 C. may be expected if the room is at ordinary temperature. Change of pressure has also a very slight effect on the temperature of melting ice, an increase of pressure lowering the freezing point.

It is also necessary to place the thermometer in a horizontal as well as in a vertical position in determining the freezing point.

The distance between the depressed boiling point and the depressed freezing point, when the latter is determined within, say, an hour after the former, can be determined within 0°·01 C., or 0°·02 C. from a single trial, when the precautions already mentioned have been taken, and Guillaume has shown that the constancy of this distance for *hard* glass thermometers is within the limits of the errors of observation.

Whenever any very accurate measurement of temperature is to be made the fiducial points must be determined at the time, and used as reference points for the observed temperature. Pernet has worked out a plan for this which is very much more accurate than the older methods by which discrepancies amounting to 0°·2 C. might arise. Pernet's method is represented by

such a simple formula that its introduction here will
doubtless be excused.   It is as follows :

$$t = (X - Z_t)\frac{100}{S_{100} - Z_{100}}$$

Here t is the required temperature ; X is the observed thermometer reading for t ; Z_t the depressed
freezing point which is determined immediately *after*
observing X ; S_{100} is the depressed boiling point determined immediately after Z_t ; Z_{100}, the depressed freezing point determined just after finding S_{100}. And,
also, $\frac{100}{S_{100} - Z_{100}}$ is the value of one of the degrees on the
thermometer in terms of true Centigrade degrees.

4. *Corrections for different temperature of scale or
stem.*—In the case of thermometers with brass scales
in which the bulb, scale, and stem have different
temperatures, errors of several hundredths of a degree
C. may be made by not allowing for the expansion
of the brass ; errors of one or two-hundredths of
a degree by not allowing for the expansion of the
glass stem ; and errors amounting up to a quarter or
half a degree by not allowing for the expansion of
the mercury thread in the capillary bore, and in case
alcohol or ether is used in place of mercury this
last error would be increased seven and ten-fold
respectively.

What is known as Poggendorf's correction (so named
from the physicist who first applied it in 1826),
takes account of the variation in the capacity of the
capillary bore due to changes of temperature. The
standard adopted is for the thermometer stem at
100° C., while the bulb is at the temperature to be
determined. The amount of this correction is
0·00° C. at 100° C. and at 0° C. ; the maximum may be
from 0°·05 to 0°·08 C. at 50° C., and has a negative sign ;

at —40° C. the correction may be from 0°·11 to 0°·17 C. with a positive sign ; the limits given are for different coefficients of expansion of glass.

There is still an error due to the irregularities in the relative expansions of the bulb and the mercury in it, but usually the correction for this is included with that for the Poggendorf correction. The combined correction is 0°·00 C. at 0° C. and 100° C., and has a maximum of —0°·207 C. at 50° C. But no set of theoretical corrections can be assigned to convert the scale for mercury into that of the gas thermometer. That the observed differences are mainly due to different qualities of the glass used is shown by similar corrections being obtained for the same kinds of glass. Wiebe and Guillaume maintain that mercurial thermometers can be constructed which will agree with the gas thermometer, and in which the readings are proportional to the temperatures.

5. *Correction for pressure on the bulb.*—There is an external and internal pressure on the thin glass of the bulb. As a standard for reduction, a pressure of one atmosphere inside and outside may be adopted, according to Guillaume.

If the external pressure in the case of a cylindrical bulb Tonnelot thermometer is increased 1 atmosphere it increases the reading 0°·09 C. ; for 0·6 of an atmosphere the increase is 0°·055 C. It would have been more for a spherical bulb on account of thinner glass. At high elevations above the sea-level this quantity must be taken into account in ordinary work, unless the fiducial points were determined at a pressure like that at the altitude of observation.

The part of the internal pressure of the bulb which consists of the hydrostatic pressure of the column (or

thread) of mercury, will vary with the temperature and
with the inclination of the stem to the vertical, and
will have a maximum in the vertical position.

Different kinds of glass possess different elastic
properties, and the effect of pressures on the bulbs
consequently is not the same for all thermometers ; but it is practically constant for the same kinds
of glass, and knowing the correction for one kind of
glass, that for another kind of known composition
can be computed. Guillaume has given a formula
expressing the relation between hard glass and crystal
glass.

There is a strong capillary action in thermometer
tubes. It will cause a downward pressure of about
one-sixth of an atmosphere for a bore of 0·10 mm. ;
and this is to be added to the weight of the column
of mercury in stating the inside pressure. In an
extreme case the sum of the two amounted to 0°·15 C.
In order to be free from this capillary error *all* of the
readings must be made with a rising of the mercury.
Where, however, the differential capillary error does
enter, in the case of a stationary temperature, it may
be found by taking half the difference between the
temperature for a rising and falling column of mercury. When the thermometer changes from a rising
to a falling temperature, it fails to indicate the change
of double the amount of this correction, and this
" dead space " may amount to 0°·07 C. for a bore of
0·03 mm., according to Pernet. Guillaume found that
a difference of 0°·04 C. existed in the determination
of the boiling point when it was made with a rising
and rounded meniscus, and again with a falling and
flat meniscus.

6. *Sluggishness.*—It requires an exposure of some

duration for a thermometer to attain its true reading, and cylindrical bulbs are more sensitive than spherical bulbs ; but the time also depends on the kind of exposure. When a circulation exists in the surrounding medium, as, for instance, a current of air, the true reading is obtained with the minimum rapidity. For a quiet medium the correction for sluggishness can be determined by experiment for each thermometer at rising and falling temperatures.

7. *Thermal reaction and change of zero.*—Bellini, in 1832, and later, Faraday, Regnault, and others, investigated the slow permanent changes in the position of the zero of thermometers. The amount of this change varies for different kinds of glass, and may even reach $0°\cdot8$ C. for apparently excellent thermometers when made of some ordinary kinds of glass. The greater part of this change takes place, however, within two or three years after the thermometer has been made.

The fact that the readings of thermometers depend somewhat on the temperatures to which they have recently been exposed makes the spring readings of an out-of-doors thermometer a little too high after the winter's cold ; and the fall temperatures a little too low after the summer's heat. This seasonal variation may reach even $0°\cdot10$ C.

In order to overcome this elastic or thermal reaction, many experiments have been made to obtain a glass in which the change is a minimum. The best composition yet found is that obtained at the Jena Laboratory, by Abbé. The composition of this glass is as follows : Quartz $67\cdot5$, soda $14°\cdot0$, oxide of zinc $7\cdot0$, lime 7, clay $2\cdot5$, borax $2\cdot0$ per cent. With this glass the depression of the freezing point amounts to only

0°·05 C.; and the change of the zero point will not exceed 0°·01 C. in a long time.

Thermographs.—The self-registering thermometers in use are by no means as satisfactory as is desirable. There are many different kinds, such as the photographic, the metal spiral, the electric, the automatic position, and the air thermographs. Of all these the photographic, as used principally at Kew and other English observatories, is undoubtedly the most accurate, but it has the disadvantage, under which all photographic registrations must come, that its manipulation is difficult, and requires the expenditure of much time and skill. In this form the mercurial thermometer is made with such a long stem that the bulb can be exposed in the thermometer shelter or screen, while the upper part of the mercury column is inside of the photographic box or the dark room. A small air bubble is introduced into the capillary tube of the thermometer, and separates the thread of mercury by an air space. A light is thrown through this space, by means of reflectors and condensors, on to a revolving drum bearing the photographic paper. The amount of rise and fall of the successive records, thus made, is carefully measured, and when compared at stated intervals with the direct readings of an adjacent thermometer, they can be readily reduced to the ordinary thermometric scale.

The metal spiral thermometer, as used in the Hipp, Hottinger, Wild-Hassler, and other thermographs consists of a compensated spiral, the expansion or contraction of which with higher or lower temperatures is communicated to a pointer whose varying position is registered on a recording sheet.

In the electric contact thermometer, as used in the

Secchi, Theorell, and Hough thermographs, the upper end of the thermometer tube is open, and a fine wire is made to descend into it until contact with the mercury, when a circuit is closed since it has the other pole connected with the mercury in the thermometer's bulb. The distance the wire is obliged to descend from a fixed position to the mercury is automatically measured and recorded.

In the air thermograph a vessel of metal is filled with dry air, which is allowed to raise or lower the

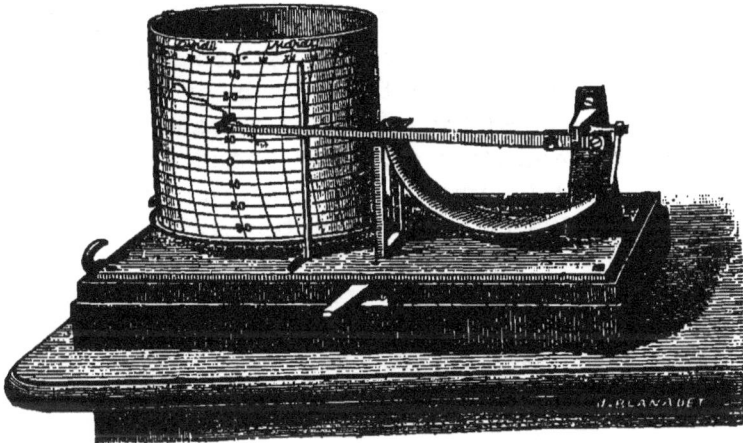

FIG. 9.

pressure in some form of manometer, by being connected with it by means of a metal tube of very small bore. The variations of pressure, as shown by the manometer, are recorded by various devices, such as are to be found in the instruments of Schreiber, Sprung, &c.

In the Richard thermograph, as shown in Fig. 9, a curved tube on the Bourdon plan is hermetically sealed after being filled with alcohol. One end of the tube is fixed (at the left end, as shown in the figure) and the other end curves up or straightens out with the contraction or expansion of the alcohol. This motion of the free end is communicated to the long arm carrying the pencil, which records the motion on the cylindrical register.

The automatic position thermograph, or "upset" thermometer, is so constructed that when the thermometer is inverted at any designated time by the automatic action of clockwork, a portion of the mercurial thread is broken off at a narrow point of the tube, and the mercury runs to the top (inverted) of the thermometer, where the length of the broken thread is measured by means of a scale. If twelve of these thermometers are so arranged that successive ones shall be inverted at successive hours, and if the apparatus is set in operation twice daily, an hourly record of the temperature can be obtained.

As in the case of other records made by means of delicate apparatus, the accuracy attainable by means of these various forms of thermograph depends very much on the care in manipulation and reduction of the observations.

For instruments like the Sprung or Wild-Hassler thermographs the readings are usually accurate to within 0·2° or 0°·3 Fahrenheit, and for the photographic or Kew form the error is only about half of this amount.

In all of these forms where a thermogram gives the variations in temperature, a bi- or tri-daily reading of an adjacent mercurial thermometer must be made in order to reduce the record to true degrees.

Exposure of Thermometers.—In the construction,

investigation, and reading of thermometers the utmost refinement may be practised, and hundredths of a degree considered ; but this care is all lost unless proper precautions are taken to so expose the thermometers that their readings give the true air tem-

peratures. In order to find a proper exposure for the thermometer, the various sources of error must first be considered. The thermometer receives heat by conduction from the surrounding air (and in this convection plays a most prominent part), and by radiation from surrounding objects; it, in turn, gives off heat by radiation and conduction. The question to be considered is the means by which the proper balance between these can be so

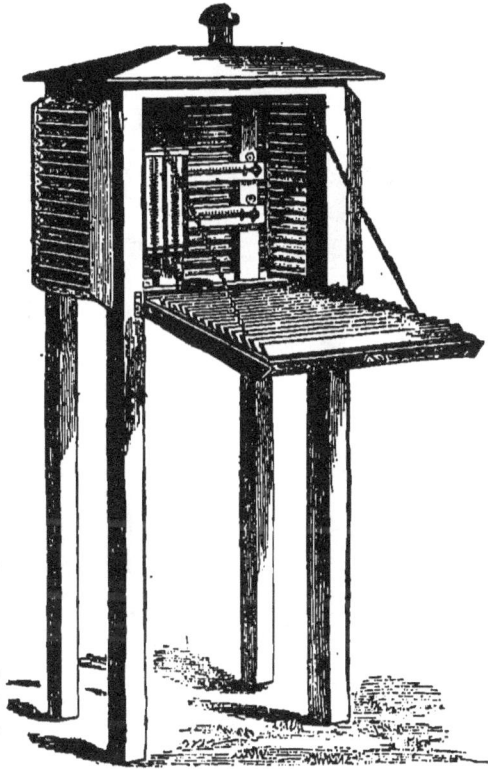

FIG. 10.

regulated that the thermometer shall read neither too high nor too low. It must be also taken into account, that in meteorology we do not wish to obtain the temperature of a small quantity of air, which may

5

itself be subject to local influences, but we desire the temperature of the main body of air.

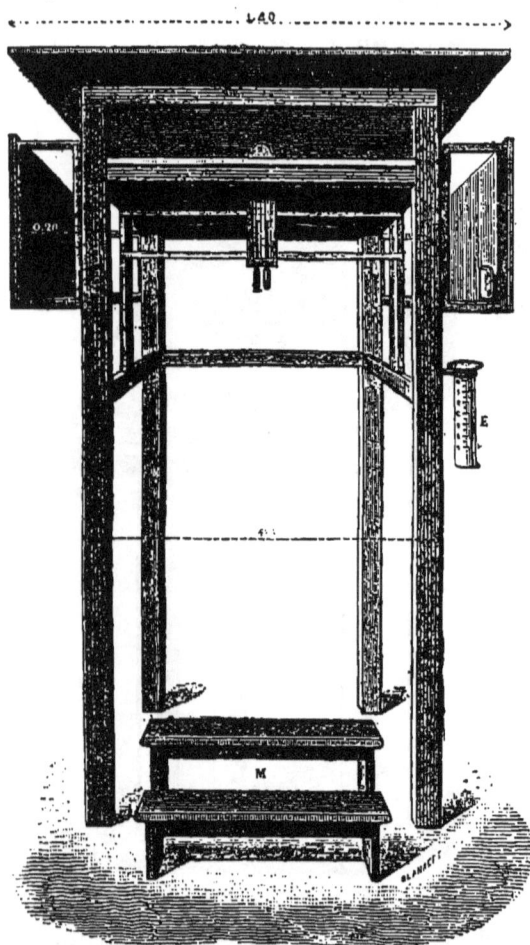

FIG. 11.

Many forms of fixed thermometer shelters or screens have been devised which, as was considered by the originators, had solved the problem of the true thermometer exposure. The most widely known and used of these are the Glaisher, Stevenson, Stow, Renou, Wild, and Signal Service shelters. The main plan on which most of these have been constructed is a box-like structure of wood, with the sloping top of solid wood, while the sides and floor are of various designs according to the ideas of the proposers. There is no doubt that different climates require some differ-

ences of details of construction for these screens, and
so no one form can be recommended for all regions.
In England the Stevenson screen (see Fig. 10) is

FIG. 12.

much used, and it has double louvred sides and floor,
thus permitting the free entrance of the air, but excluding rain and preventing the vitiating action of

radiation. The thermometer is placed about four feet above the sod.

In France the Renou screen, as shown in Fig. 11, is used. It consists essentially of a double roof of wood or sheet zinc, inclined about 30° to the horizon, with an air space between the outer and the inner cover. The thermometers are placed near the inner cover, and are about seven feet from the ground, which must be covered with a sod. On each side is a board or similar shield to prevent the sun's rays from striking the thermometers in the morning or afternoon hours. One shield is sufficient if it is shifted from the west side to the east side at evening, and back again at noon.

In Russia the Wild shelter is used; this is shown in Fig. 12. It is of wood, and has a double board roof and south side, with open space between the outer and inner covering. The sides are double louvred to admit the air freely, and the floor and north side are open. Within this wooden shelter is placed a zinc broken cylindrical case, which contains the thermometers, this case being so arranged as to freely admit the air, as is seen in Fig. 13, where the cylinder is opened to permit a reading of the instruments. At

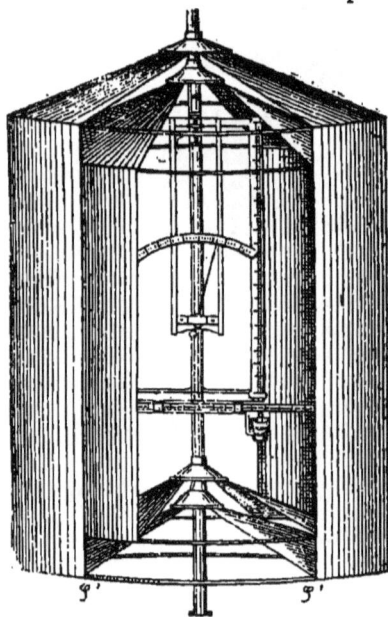

FIG. 13.

the lower end of this inner metal shelter, and replacing the conical bottom shown in the diagram, is a fan-like arrangement, by means of which the air is forced through it (and over the thermometers) for two minutes before each observation. For the ordinary dry bulb thermometer Wild finds that the ventilated shelter and the unventilated shelter give the same results to within about $0°·1$ C. The thermometer is placed about ten feet above sodded ground.

The Hazen thermometer shelter, as used by the United States Weather Bureau, is a cube three and a half feet long, three feet high, and three feet deep. The top and the bottom are of solid boards ; the roof being sloping and double, with an air space of six inches between the outer and inner boards. The sides are to be single louvred with wide slats. In order to thoroughly mix the air inside of the shelter the wet and dry bulb thermometers are whirled rapidly in a circle eight inches in diameter, just before reading. When mounted on a roof the shelter must be nine feet above the platform, and when at the ground, sixteen feet above the sod, and supported in all cases by four posts.

In Fig. 14 the method of exposing the bulbs of the self-registering dry and wet bulb thermometers is shown as used at Kew, where the shelter is attached to a wall. The stems pass through the wall to the photographic apartment within the building.

Numerous comparisons of various forms of thermometer shelters have been made, not only to see how they differ among themselves, but also to determine their absolute accuracy. The Royal Society's experiments made by Griffith at Strathfield Sturges, Hampshire, which extend over the period from November, 1868, to April, 1870, and in which about eighteen different methods of thermometer exposure were compared, is the most elaborate investigation of the subject yet undertaken, and it might be thought that other experiments were unnecessary ; but since then several government institutions and societies, and a

FIG. 14.

large number of individuals have investigated the question anew. The paper by Köppen, of the *Deutsche Seewarte* at Hamburg, bearing the title *Investigations on the Determination of Air Temperature (Studien über die Bestimmung der Luft-temperatur und des Luftdrucks. Erster Abhandlung. Untersuchungen über die Bestimmung der Luft-temperatur),* and published in 1888, gives the latest summary we have of what has been done by the various experimenters.

Even the bare results of these researches cannot be given here, but it may be said that each of the recent forms of shelter probably has some advantage over others; but it has not been possible to obtain an agreement among meteorologists to adopt any one form to the exclusion of others, and the question of a standard form for universal use is still an open one, and not likely to be settled in the near future. Köppen proposed a simple form of thermometer shelter, in which the thermometer bulbs are placed between two horizontally-placed boards, as shown in the accompanying figure, No. 15.

Normal Temperature. — It will be obviously impossible to arrive at

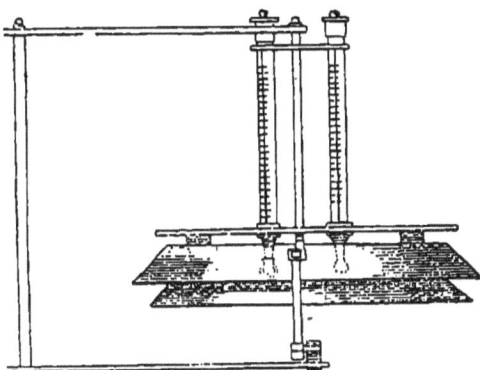

FIG. 15.

any definite decision as to what is the best mode of exposing a thermometer in order to obtain the true temperature, until some absolute standard for comparison has been adopted ; and yet this is what meteorologists have been trying to accomplish. It has been assumed, and is still so considered by many meteorologists, that if a thermometer is whirled quite rapidly in a shady place, so as not to be exposed to the direct rays of the sun, and so as not to radiate heat itself to the sky, then its readings after whirling will indicate the true air temperature. Usually in performing this operation the thermometer has a string attached to the upper end, and is whirled after the fashion of using a sling. The whirled wet bulb thermometer was first used by Saussure ; but the application of the method to ordinary thermometers was first made by Bravais in 1836, and later by Arago.

As a means of producing an artificial current of air which will flow past the thermometer bulbs, three methods are in use by meteorologists : the slinging, or whirling by hand at the end of a string, as just mentioned ; the ventilating, in which the current is caused by aspiration or some form of fanning ; the rotating, in which the thermometers are whirled by a machine of some kind, or, as in the case of the Renou method, the shelter is rapidly rotated.

Observations by Joule and Thomson have shown that if a wind blows against an obstacle the temperature of the air is higher on the windward than on the leeward side ; this is due to the dynamic heating by increase of pressure, and in exposing a thermometer care must be taken not to subject it to this source of error. Even for a thermometer exposed to a free current the reading is increased by $0°·1$ C. for a velocity of wind of ten meters per second, and by about $0°·5$ C. for a velocity of about thirty-five meters per second.

In 1817 Fourier proposed using two thermometers having bulbs of different degrees of radiation and absorption : viz., the plain glass bulb, and one covered with lamp-black ; and then from their readings he would compute the correction to be made for radiation.

In 1851 Liais proposed using three thermometers having bulbs of different absorptive powers. In

1865 Abbe (of Washington), then at Pulkowa, also proposed and applied the use of two thermometers. Fourier gave a simple formula for the reduction, depending on the absorption of glass and lamp-black; and quite recently Ferrel gave a more complete formula taking into account the velocity of the convective current of air to which the thermometers are exposed. Hazen made an independent proposal of the method of Liais, and his bringing the subject to public notice has resulted in some recent experiments in whirling of two thermometers, one with a gilded bulb, and the other with a bulb covered with lamp-black; but Abbe had recommended in 1883, and on earlier occasions, the use of the black and silvered or gilded bulb thermometers for determining the temperature even in open sunshine.

In 1857 Neumayer conducted experiments with white, red, and black bulb thermometers at the Flagstaff Observatory at Melbourne, for the purpose of determining the effects of radiation, but just what use he made of them has not been made public by publication; and Aitken independently proposed the use of thermometers having bulbs with different powers of radiation. It is seen, then, that so many different persons have independently suggested this general method that its discovery cannot be assigned to any one of the later proposers.

In the summers of 1885 and 1886 (as well as at other times), Wild made some very careful and complete experiments at Pawlowsk, Russia, with various kinds of thermometers, in order to arrive at some practical method for obtaining the absolute true air temperature. He used thermometers with plain-glass, gilded, and blacked bulbs, and compared them, both at rest and with strong current of wind produced by a rotary motion, with the thermometers in his form of shelter (Fig. 12).

These experiments show, in the first place, that in the present condition of our knowledge of the theory of heat it is impossible to obtain, under all circumstances, in bright sunshine and in the shade, even with rapidly moving air, the temperature of the air by means of two thermometers having bulbs of different coefficients of radiation, as, for instance, the blackened-bulb and gilded-bulb thermometers; however, if these thermometers are in the shade, and enveloped in moving air, the true temperature is probably very nearly obtained by subtracting from the observed temperature of the gilded thermometer 15 per cent. of the difference between the two thermometers. Some of the observations made with a free sun exposure gave as much as 1° C. too high a temperature at about 20° C. by this same formula. The plain thermometer whirled in the midday sunlight gave results about 0°·4 C. too high, and in the shade 0°·5 C. too low, and in the evening after sundown this depression below the air temperature increased to 0°·8 C. In the most unfavourable cases on clear, calm summer days, the Wild-shelter sometimes gave a temperature 0°·5 C. too high for the ordinary thermometers without an air current, but when, by means of the ventilator beneath the inner metal shelter, a current of two meters per second was forced through it, the reading of the thermometer was only 0°·1 too high ; so that if this shelter is used in the lower latitudes where such days frequently occur, the ventilating attachment must be used. On these calm days with strong radiation, the air temperature can vary 0°·5 C. or more, at distances from one to four meters above the ground, according to the method of sheltering the thermometer, whether it be in the shade of a house, of a small shelter, or of a large sail canvas.

Sprung has also recently published the results of some experiments made at Berlin, and his work shows, with that of others, the necessity of internationally conducted experiments.

§ 2.—*Barometers*.

Early History of Barometers.—The barometer was discovered by Torricelli in the year 1643, a year after the death of Galileo, who had shown that the air has an appreciable weight. The simplest form of this instrument is obtained by filling a glass tube, sealed at one end, with pure mercury, and placing a finger over the open end invert the tube and plunge this end, finger and all, into a cup containing mercury ; then remove the finger, and, if the tube is somewhat over thirty inches in length, the mercury will flow out until the surface of that in the tube is about thirty inches above that in the cup, if the experimenter is near the sea-level. This experiment was first actually performed by the mathematician Viviani at the suggestion of Torricelli, and his instrument only differed from the above description by there having been a glass bulb blown at the closed end of the tube, so that when the mercury ran out, this bulb, as well as a portion of the tube, was left empty.

If this tube is clamped in a vertical position to a rigid support, and a scale of inches is placed beside it, also vertical, with the lower end, marking the zero of scale, resting on the surface of the mercury in the cup, it will be observed that from day to day, or sometimes from hour to hour, the scale reading of the height of the mercury in the glass tube will vary considerably. This variation may amount to as much as a couple of inches in the course of a few days.

The instrument is, however, seldom used in this simple form, because it could furnish only very rough measures of the changes in the atmospheric pressure. Experience and experiment have caused many different forms of barometers to be constructed, in order to combine accuracy and convenience, and it is the gradual improvement of these instruments that we wish to follow out.

In Fig. 16 is shown an early form of the barometer as used by members of the Accademia del Cimento ; in this a nearly closed cistern was used ; the air being admitted by an orifice in a neck, and the gradations, which were arbitrary, were small protuberances on the glass tube.

In order that the glass tube, in the case of the simple cistern barometer already described, may not be easily injured, it is usually surrounded by one of metal, and as this tube is in all parallel to the glass one, it can be graduated and serve as a scale for reading the height of the mercury column. Then, too, that the instrument may be easily carried about, the mercury cup at the bottom is so arranged that it can be screwed on to the lower end of the metal baro-

FIG. 16.

meter tube. The instrument in this simple form is a cistern barometer with immovable bottom, and is much used at sea.

If, instead of having the cup or cistern of mercury at the bottom of the tube, the lower end of the tube is bent into a U-shaped form, we have the so-called syphon barometer, which was a very popular form until within recent years. In this case the difference in the height of the surface of mercury in the two legs of the tube gives the atmospheric pressure. (See Fig. 17, where the upper and lower scale readings are to be added together.)

Various Forms of Barometers.—With this general idea of the construction of the barometer in mind, it will not be difficult to understand the details concerning some of the improvements which have been made in the last few years, and which are described below. The most popular form of instrument is the Fortin barometer. This is a cistern barometer, in which the lower end of the reading scale terminates in a fixed ivory pencil (within the glass cistern), the lower point of which is at the zero of the scale. The lower part of the mercury cistern usually has a soft leather lining which is pressed upon by a screw from below. The large milled head of this screw is turned by the observer until the surface of the mercury in the cistern is raised so high as to just come in contact with the ivory point marking the zero of the scale;

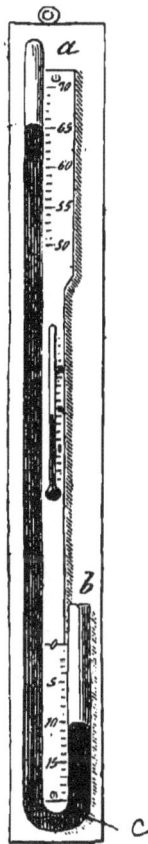

FIG. 17.

then the scale reading of the height of the mercury in the tube gives the "barometer height." There are numerous makers of this form of barometer, but they differ only slightly in details of construction, and the general form is to be found illustrated in most books on physics or meteorology. A cheap construction by Baudin, of Paris, is, however, of sufficient novelty to deserve notice. This instrument is shown in Fig. 18. The scale divisions are ruled directly on the glass tube, and the cistern is entirely of glass, but it has a metal rim which screws into another metal rim encircling the lower end of the barometer tube. The zero of the scale is at the end of a pencil of glass which is attached to the barometer tube, and is brought in contact with the mercury in the glass cistern when a reading is to be made.

FIG. 18.

About 1830 Kupfer invented the syphon - cistern barometer, various forms of which are now used extensively in Continental Europe, under the names of Turrettini, Köppen-Fuess, Wild-Fuess, &c. Many improvements have been made in the original design, and probably the best at present in use are those of the Wild pattern, manufactured by Fuess of Berlin. As the name suggests, this form of instrument is a combination of the cistern-syphon barometers in which the attempt is made to bring into one instrument the good points of both forms. The arrangement of the barometer tube is best realised if we imagine a syphon barometer tube with the shorter leg cut through at the beginning of the bend (at the point c shown in Fig. 17),and the bent portion straightened out; both of the cut ends are then inserted in a closed cistern filled with mercury, beneath which is an adjusting screw, as in the case of a Fortin barometer. In making a reading, the mercury is forced up into both legs of the barometer until the mercury surface in the short leg is brought up to a fixed index at the zero of the scale, when the reading of the latter, showing the height of the mercury in the longer leg, gives directly the barometer reading.

Fig. 19 shows the Köppen-Fuess form, in which the principle of construction can best be seen, as both tubes are straight, and two brass bars take the place of the usual encircling brass tube, thus permitting the glass tubes to be seen. In making an observation with this instrument the screw head s at the bottom is turned towards the right until the mercury in the short tube on the right is brought up to a fixed

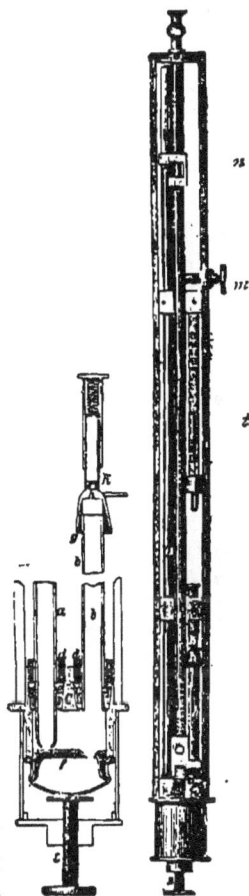

FIG. 19.

zero index at c: the upper reading sight at n is then adjusted to the top of the mercury column in the long tube on the left by means of the screw at m. The thermometer at *t*, it will be noticed, is unprotected in this case, to make the exposure uniform with that of the glass tube of the barometer.   This form of barometer is little used outside of Germany.

Fig. 20 shows the Wild-Fuess barometer, which is much more complex than the one just described, owing chiefly to the fact that in this form there is a bend in the longer leg, which permits the short leg to be located directly under the upper part of the longer leg, so that the two surfaces of mercury which are to be observed will be in the same vertical.   In the diagram the upper part of the barometer is shown at the left, and the lower part at the right ; a portion of the outer brass tubular covering being omitted in this last case in order that the construction of the glass tubes can be seen to better advantage.

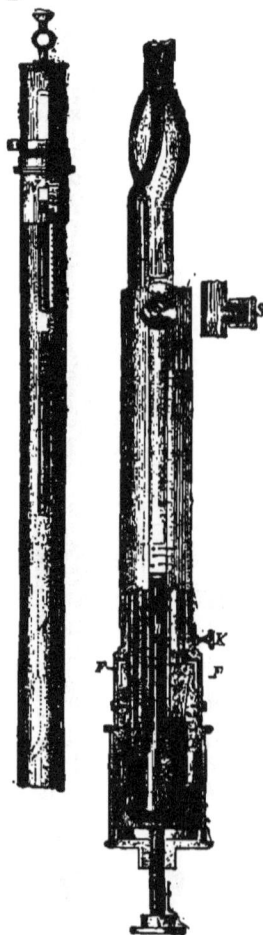

As this barometer has come into very extensive use on the Continent of Europe as a refined form of portable instrument, a more lengthy description is desirable.   The tubes A and B connect with the closed mercury cistern C. A is the longer tube and passes up, being considerably constricted, to a bulb blown in

FIG. 20.

the continuation of the tube B ; here a turn to the right is made, and A passes through this bulb O until it is directly over the short tube B. When A leaves the bulb O it expands into the full diameter which it has at the point of reading above. The short tube B is of the same diameter throughout up to the air hole S, and it is continued up to the place of meeting the tube A merely for the additional rigidity it gives the instrument. In making an observation with the instrument the bottom screw G is turned until the mercury surface in B comes in contact with the zero sight (which is adjustable to the zero of the scale by means of the screw K) which is shown just above it in the diagram.

The height of the surface of the mercury in the longer tube is read off by means of the sight at N, which is carefully adjusted by means of a screw motion. The framework of this sight, on which is also placed the vernier divisions, encircles the brass tube covering the glass tubes.

The outside air enters the short tube through the small opening in the arm covered by the screw cap S; this screw cap being left open during the observation, but kept closed at other times.

For this barometer the attached thermometer is within the encircling brass tube, but the scale is exposed to view through a slit; the thermometer bulb being well covered.

When the barometer is not in use, the mercury is screwed up into B until about all the air is expelled through the hole at S, when the cap at S is screwed on and any outflow of mercury is prevented. In this condition the barometer may safely be inverted for transportation, as the tube A becomes filled with mercury long before the mercury in B reaches S; but in inverting the barometer the screw cap S must be

6

turned towards the manipulator, and the cistern end
of the barometer moved upwards to the right (*i.e.*,
with the right hand with which it is grasped below) ;
and in restoring it again to its normal position, S
must still be towards the person and the exact
reverse directions be traversed by the right hand.
In this process of inversion the short tube is kept
uppermost, and there is thus no chance for the
bubble of air to enter the long tube A as it passes
from S along the short tube B to the highest point of
the cistern.

A great theoretical advantage of this form of instrument is that it admits of a determination of the
amount of air in the imperfect vacuum above the long
column of mercury in the tube A. This can be done
by a double measuring of the difference in the height
of the mercury in the two tubes; first when the vacuum
space in A is four or five inches long, and then again
after it has been reduced to a fraction of an inch by
forcing the mercury from the cistern up into both of
the tubes A and B by means of the lower adjusting
screw G. If any air is present in the vacuum space,
it will be more compressed in the second case than
in the first, and the second barometer height will
consequently be less than the first. The relation of
this difference to the space occupied by the air, in
the two readings, allows the amount of air in the
vacuum chamber to be computed. This is an application of what is known as the Arago method to
a portable barometer. In carrying out this process,
the screw cap S must be entirely removed from the
barometer, when the reading micrometer N, vernier
and all, can be slipped down the brass tubular covering, and be used for making the scale reading of the

mercury surfaces in the glass tube B as well as that in A.

I have now given some idea of the forms of portable barometer in common use and also of their manipulation, but in all of them they have been used as received from the maker, already filled with mercury. It has long been considered desirable to have a form of barometer which could be easily filled with mercury, so that it could be transported from place to place with the tubes emptied of the mercury, the weight of which makes it such a risky matter to move a barometer. A number of methods have been proposed by which this can be accomplished, but none of them have been subject to such tests as the barometer constructed by Professor Sundell of Helsingfors, which has been brought into very prominent notice by having been used in the splendid series of international barometer comparisons, carried out by the inventor, which will be spoken of a little later. This barometer is shown in Fig. 21 and Fig. 22. The former shows the system of tubes stripped of the mounting and the encircling

FIG. 21.　　FIG. 22.

brass tube which serves as a scale as well as for protection.

A duplicate glass tube of a Wild-Fuess barometer is used as a foundation for this barometer, the inner diameter of the tubes at the points of observation being 11·1 mm. At the upper end of the tube there is attached by melting a spherical bulb A with a content of about 110 c. cm. ; between this bulb and the capillary tube B (having a length of 43 cm. and surface aperature of 1·63 sq. mm.), there is inserted a small spherical bulb *a*, having a content of 5 c. cm. The reservoir C, having a length of 10 cm. and an inner diameter of 11 mm., which is joined to the tube B, is connected with the drying tube G by means of a narrow tube which prevents any foreign substance from passing from G into C. The tube G is stopped up at the top with wool, and is fastened into C by means of a cork, the joint being made air-tight by means of resin and bees' wax. As a means of drying, phosphoric anhydric acid is used. At the lower end of the contracted tube D, a T tube is melted on, and so bent that the axis of the tube LM coincides with the axis of the open leg of the barometer tube which terminates at K ; the two tube ends K and L, which are about 5 cm. apart, are then connected by means of a rubber tube, which latter can be somewhat compressed by means of the screw P (see Fig. 22). The end of the tube M is connected by means of the wrapped tube S (length 120 cm., inside diameter 5 mm.) with the glass reservoir Q which holds 200 c. mm. At the top of Q is a drying tube containing chloride of calcium. In Fig. 22, N is the upper sight having a micrometer motion, R is the lower fixed sight, and Y is the attached thermometer inside of the brass tube, but with scale visible through the slit. The method of manipulation is as follows:

We will start with the barometer emptied. The pure mercury is put into the reservoir Q (Fig. 21), after which the cork which is used to stop up the lower end of G is pierced with a needle, and the screw F, which controls admission of the air to the lower open leg of the barometer, is tightened to prevent the entrance of air into this lower leg. By slowly raising the reservoir Q, which can be easily done as the tube S is of rubber, and is many times larger than represented in the diagram, being in fact 120 cm. long, the mercury will gradually rise in the barometer tube until it has forced the air out of the spherical bulb A and the reservoir C. By slowly lowering Q the mercury flows back again, and the air flowing into the tubes again through G is dried in its passage. A repetition of this filling and empty- ing process, two or three times, letting the air fill the barometer tube each time, will suffice to dry the tube thoroughly. The barometer tube is now finally filled with mercury and the reservoir Q is allowed to rest in the upper holder g_3 (Fig. 22), which is so placed, and the amount of mercury in Q is so chosen, that when Q rests in g_3, the tube C is just filled with mercury. The small opening in the end of the drying tube G is now closed by means of a heated knitting needle, which is used to melt the wax which had previously been pierced to admit the air. Q is then again lowered, and C is only so far emptied of mercury that the air still remaining in it holds in equilibrium, by its tension, a mercury column of the height of the capillary tube B. Then the mercury will break above at the beginning of the capillary tube and draw itself back to the vertical part of the tube ; on the other side it sinks lower and lower in

the bulb A, leaving above it a Torricellian vacuum.
Q should now be lowered as far as possible in order
to set free any little air bubbles that may have
found lodgment between the mercury and the sides
of the tube D. By an additional raising of Q the
remaining small quantity of air is driven over into C.
Q is then returned to the holder g. The instrument
is now ready for making an observation, and the
distribution of the mercury is as shown by the
shaded lines in Fig. 21. There is always, however,
a measurable quantity of gas consisting mostly of
water vapour remaining in the vacuum chamber ; and
the amount of this gas must always be determined
by the method of varying pressures (Arago's method).
Sundell has found the gas tension to be somewhat
variable, being as high as 0·37 mm. and as low as
·07 mm. A full and very interesting account of the
manipulation of this barometer is to be found in
Sundell's rather inaccessible original paper in the
Acta soc. Fennicae, tome xvi.

In regard to the accuracy of the readings (which
must not be confused with the absolute accuracy of
the instruments as referred to a common standard) of
the various forms of portable barometers, it is difficult
to give positive values, and the following are somewhat, although not entirely, personal estimates. If
two ordinary Fortin barometers are compared with
each other, the readings differ by 0·25 mm. from one
comparison to another. In other words, if we know
the error of the barometer as compared with an
accurate standard instrument, we can be pretty sure
of obtaining the true atmospheric pressure to within
0·25 mm. by a single reading at any time. A Fortin
barometer of the highest class will give single read-

ings, with an accuracy of perhaps 0·12 mm. A Wild-Fuess (syphon-cistern) barometer of the best construction will give single readings with an accuracy of perhaps 0·06 mm. For the latter instruments, if a series of ten readings is made simultaneously on two barometers, and each difference is subtracted from the mean of the ten, the average deviation of the single readings from the mean should not be more than 0·03 mm. There have been constructed a very few Fortin instruments which could perhaps be read as closely as the Wild-Fuess barometers, but the personal equation between two observers is undoubtedly less for the latter form of instrument.

Standard or Normal Barometers.—All ordinary barometers must first be compared with a standard barometer before the results obtained by them can be used, as they are never perfect in construction and differ among themselves. In practice the following plan has been usually carried out. A barometer of unusually good construction, in the filling of the tube of which great care has been taken, is selected to be used as a standard, and mounted permanently at some observatory of repute. This is adopted as a standard for the observations officially made by the country or state, and the barometers used in making observations in this region are compared with it. This comparison is made either by bringing the barometer, to be compared, to the observatory, and making the comparison with the standard direct ; or else by comparing a barometer with the standard and then carrying this second barometer to the observing station, and comparing it with the observing barometer in the location and condition in which it is to be used for actual observations. This last method is

considered the best, although a slight error enters
into the comparison due to the use of the intervening
portable barometer.

With the gradual increase of accuracy in making
barometric observations, and especially in the more
frequent combination of the observations made in
different lands, it became evident that these so-called
standard barometers should be replaced by still more
accurate instruments which should have an absolute
accuracy, and which should agree with each other the
world over, *after* certain theoretical corrections had
been applied to their readings. This great reform,
which we owe more to Professor Wild (of St. Petersburg), than to any other single individual, has only
partially taken place, and is still in progress ; and it
is this progress which I propose to outline. It would
require a special treatise on barometers to contain a
complete description of all the normal instruments
now in use, and so only a brief mention can be made
here of the different constructions ; and, as these
differences occur mainly in the method of reading
the height of the mercury column, they are pointed
out in the text treating of the errors of this operation.
I have, however, given diagrams of many normal
barometers, descriptions of which have been published ; believing that such a complete series has not
been heretofore collected in any single publication.

Regnault's normal barometer at the Collége de
France, Paris, was the first instrument constructed
that could lay claim to this title. This barometer,
which was constructed for his own use in carrying
out the investigations described in his famous memoir
published in 1847, had its origin in his desire to do
accurate work rather than in any idea of establishing

a standard barometer, with which other and less accurate instruments might be compared.   Regnault's normal barometer (Fig. 23) was a combined ma-
nometer   and barometer.   The tube A in the illustration is the barometer   tube, and the tube B (towards which the reading telescope of the cathetometer points) is the ma-
nometer tube.

Experiments at the Kew Observatory during the years 1853-4 re-
sulted in the final establishment in 1855 of the Kew normal   barometer (Fig. 24), the cathetometer belonging   to which was somewhat similar to that shown in

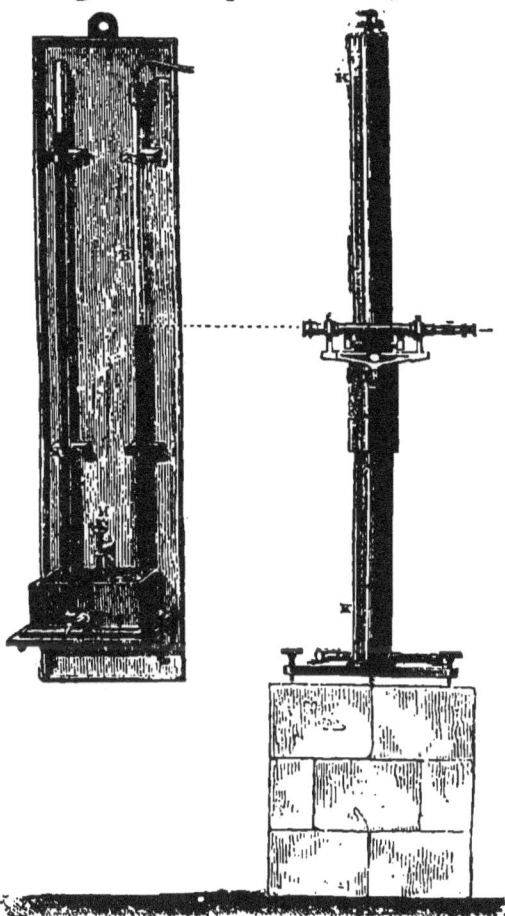

FIG. 23.

Fig. 23. This was really the first normal barometer set up as a standard for meteorological purposes. The Regnault and Kew normals are not very different in construction,

About ten years later, in 1866 or 1867, Wild
established at Berne a normal barometer which
embodied the main ideas of the
instrument established by him a
little later at the Central Physical
Observatory at St. Petersburg,
and which is shown in Fig. 25.
For outline plan see Fig. 31.

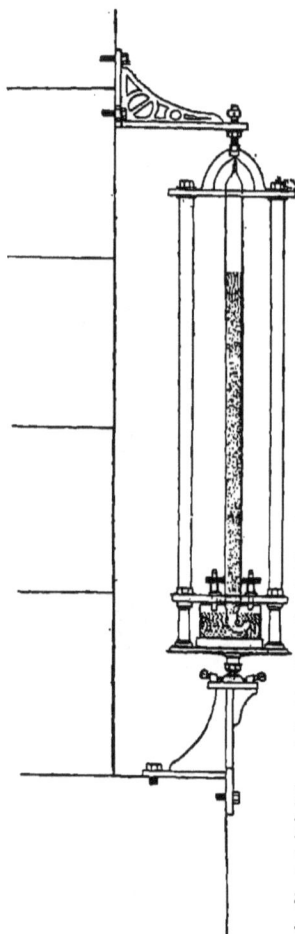

Another pause of ten years
followed, during which Wild,
Neumayer, Hellmann, and
others, strongly urged the establishment of other normals ; and
then, in 1881, the Fuess normal
barometer (Fig. 26) was mounted
at the laboratory of the Normal
Standards Commission in Berlin.
This and another barometer
mounted about the same time
by Pernet, at the International
Bureau of Weights and Measures
at Sèvres, followed out in the
main the method of construction
proposed by Wild.

About 1882, Marek's normal
barometer (Fig 27) was established at the International
Bureau of Weights and Measures
at Sèvres ; and in 1883 a Fuess
normal was obtained by Neumayer, for the Deutsche Seewarte

FIG. 24.

at Hamburg (Fig. 28). A new normal barometer
of novel construction was mounted, probably in
1886 (perhaps in 1885), by Sundell, at Helsingfors

(Fig. 29), according to a plan proposed by him somewhat earlier.

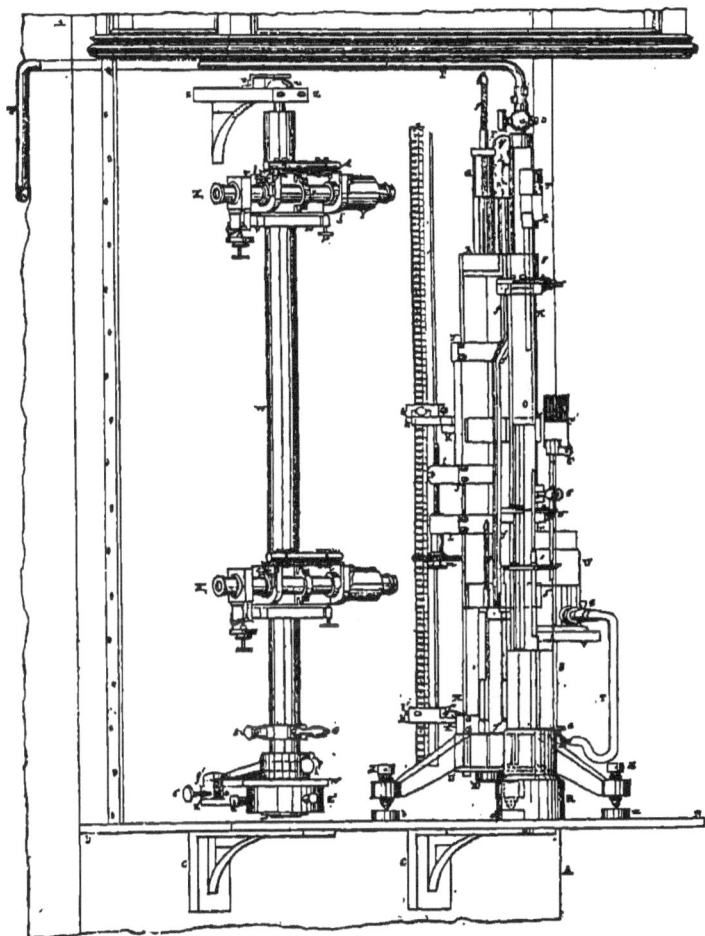

FIG. 25.

The latest addition to this list of normal barometers is the one mounted by Wild at the observatory at

Pawlowsk, Russia, in 1887 ; but the details of its construction have not yet been published.

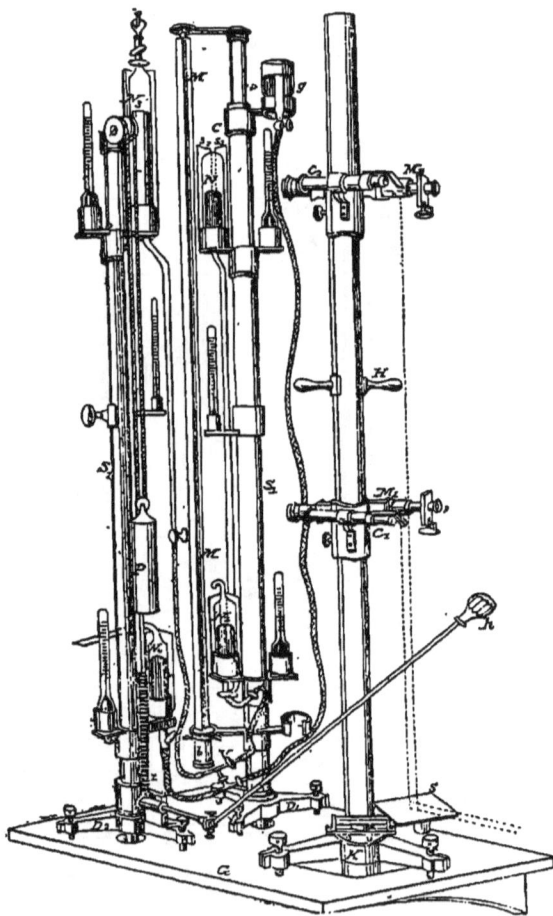

FIG. 26.

These various normal barometers may be divided into two classes. The Regnault and Kew normals form the first class, in which the practically straight barometer tube projects into a cistern of mercury, and the methods of reading the upper and lower

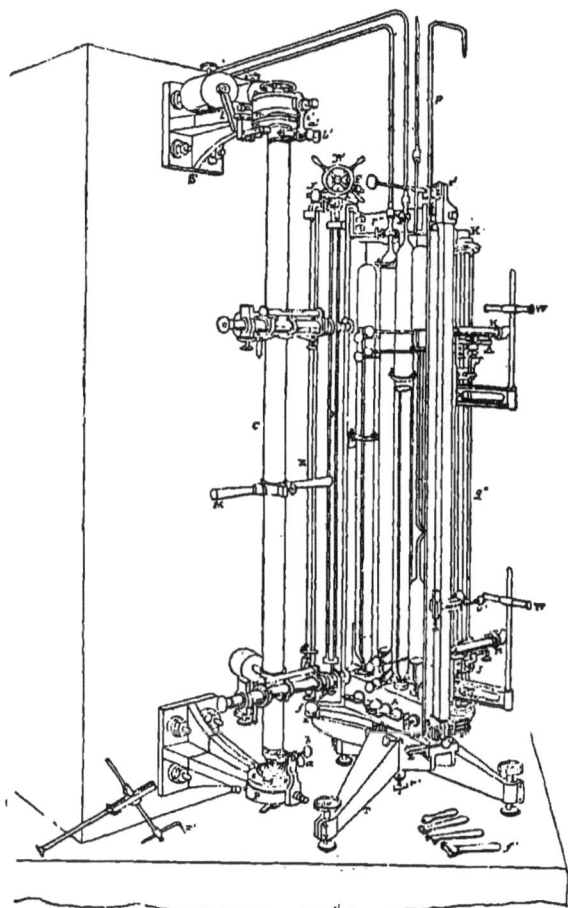

FIG. 27.

mercury surface are quite unlike. The remaining normals belong to the second class, inaugurated by Wild, in which the tube is of syphon form, thus permitting the same system of reading the mercury surfaces to be used above and below ; and an acces-

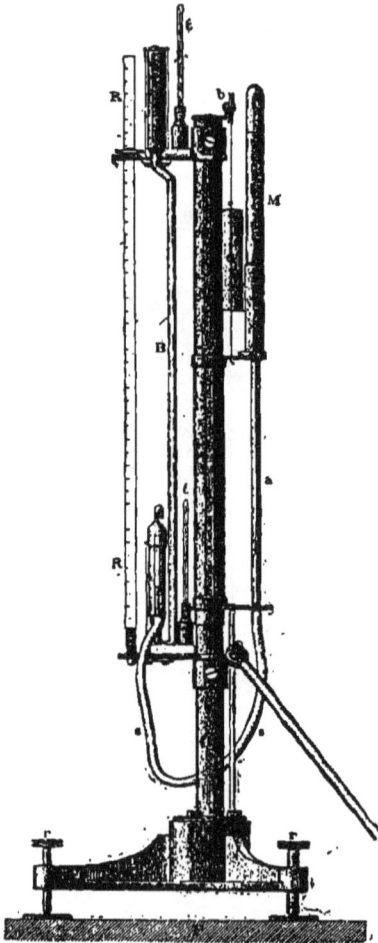

FIG. 28.

sory cistern of mercury to be used in adjusting the surfaces is connected with the barometer tube at the lower bend : this form of barometer is most distinctly shown in Fig. 28, as the usually accompanying mano-meter is not attached in this case : this is the normal barometer at the Deutsche Seewarte, Hamburg. The special differences in these barometers will be pointed out in appropriate places in the mention of the various errors pertaining to these instruments.

Sources of Errors in Barometers.—There are many sources of error in determining the true atmospheric pressure. Regnault claimed for his normal barometer, that the inaccuracy was not greater than 0·10 mm. Wild, however, set the admissible outside limit of any error at 0·01 mm. ; and he adopted this as a criterion in the investigation

of the St. Petersburg normal (Fig 25), which he
claimed, and I think justly, to be the first normal
barometer ever fully investi-
gated for its errors with this
precision.

I will now make a brief
mention of these special
sources of error, and give the
precautions necessary to keep
the errors within 0·01 mm.

1. Effect of the changes of
temperature of the mercury.

For an ordinary temperature,
of say 20° C., the mean temperature of the mercury
column, for its entire length,
must be determined with an
accuracy of ±0°·07 C. This
can only be done by mounting two, and perhaps three,
thermometers at different
points close to the barometer
tube, and they must have their
bulbs immersed in glass tubes
containing mercury, and with
diameters the same as for the
barometer tube.

2. Influence of temperatures
on the brass scale.

At about 20° C., the reading
of the scale temperature must
have an accuracy of ±0°·7 C.

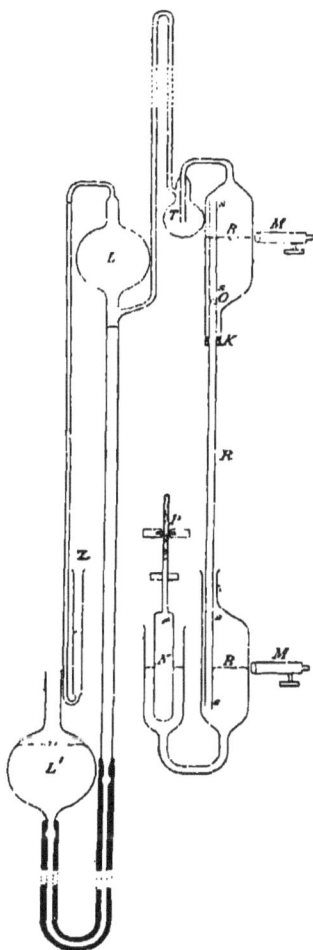

FIG. 29.

3. The correction to standard gravity.

The latitude of 45°, and sea level has been adopted

as the place to which the observations are to be reduced as a mean point ; and whenever the place of observation is more than 8' (two geographical miles), from the latitude 45°, or is more than 42 meters above the sea-level, the gravity correction must be applied.

4. The impurity of the mercury.

It is much easier to obtain pure mercury than to use that which is not pure, and then have to determine the specific gravity of it ; which it would be necessary to do in order to prevent an error of more than two units in the fourth decimal place of this constant.

Mercury of sufficient purity can be obtained by first washing commercial mercury in dilute acid ; shake the bottle containing the combination frequently during several days, and then pour off the acid, and wash the mercury in pure water. Now pour it into a clean glass vessel containing some chloride of iron, which being removed, and the mercury being filtered, it is ready for distillation ; and this can be done by means of a Weinhold distilling apparatus, or Wright's modification of the Weinhold form.

5. Inclination of the reading scale.

In mounting the graduated scale, great care must be taken to get it vertical, and it must be so firmly fixed that it cannot be easily moved by any jarring. A plumb line suspended from a steady support will do to show the verticality with sufficient accuracy.

6. Error of scale graduation.

The only way to find out this error is to compare the scale graduations with a normal scale of known accuracy. The comparisons must be made, not only for the extreme length of the scale, but also for perhaps a dozen points throughout the length.

Much of the accuracy of the scale reading depends upon the carefulness with which the graduation has been carried out.

7. Elasticity of the scale.

The scale having been compared with the normal in a horizontal position, and then hung vertically suspended from its upper end, the length will be found to have increased 0·01 mm. when the whole weight supported by the scale amounts to 0·8 kilograms for each square mm. of cross section.

8. Influence of gas in the vacuum chamber.

In the filling of a barometer tube with mercury, it is impossible to obtain an absolute vacuum, and the difficulty increases when the attempt is made to fill the complicated system of tubes of a modern normal barometer. It has been customary in the best investigated normal barometers to fill the tubes as carefully as possible, and then determine the amount of gas in the vacuum by means of the Arago method already referred to. That is, to measure the height of the mercury column when the mercury is low in the main tube and the length of the vacuum space is several inches, and then make a reading when the mercury has been raised in the tube until the vacuum space has been reduced to very small dimensions. The difference of the two readings compared with the dimensions of the vacuum space in the two instances, permits a determination to be made of the pressure of the gas which is to be applied to the actual readings of the instrument in order to reduce them to the case of an actual vacuum. For one of the normal barometers at the International Bureau of Weights and Measures at Sèvres, the reading of the normal barometer must be corrected by +0·13 mm.

to allow for the pressure of gas in the vacuum tube. The St. Petersburg normal requires a correction of about 0·06 mm. for the same purposes. The amount of gas in the vacuum tube depends principally on care in filling.

The best method for accomplishing this last is the one used by Wild at St. Petersburg, and also adopted at Sèvres. In the purification of mercury by the Weinhold process, it is distilled directly into the barometer tube, after the tube (and the system of tubes) has been exhausted of air as far as can be done by use of a mercurial air pump.[1]

Wild mentions the fact that in 1874 he used a Weinhold's distilling apparatus for filling barometer tubes. In 1883 I found this was still in use, and giving excellent results, with almost no danger to the barometer tube. The drawing given here (Fig. 30) is from a rough sketch I made of the apparatus, and is merely intended to give an idea of its shape and general appearance. It is drawn only approximately to scale, and shows the system of tubes emptied of air. All of that part to the left of the point P pertains to the mercurial air pump. A is the reservoir of mercury to supply the air pump, and the mercury flows into the tube C by way of the small hole E, which flow can be regulated by the handle B which moves the cover of the hole E. The mercury falls through the tube H into L and passes out through M into N, from which it is forced back into A again. For expediting matters, the tube G is used (G is shorter than thirty inches), and I and K are filled with mercury, which is drawn up through G and passes over into L by way of H.

[1] See Wild's monograph on barometers in the Russian *Repertorium für Meteorologie*, Band III. p. 145.

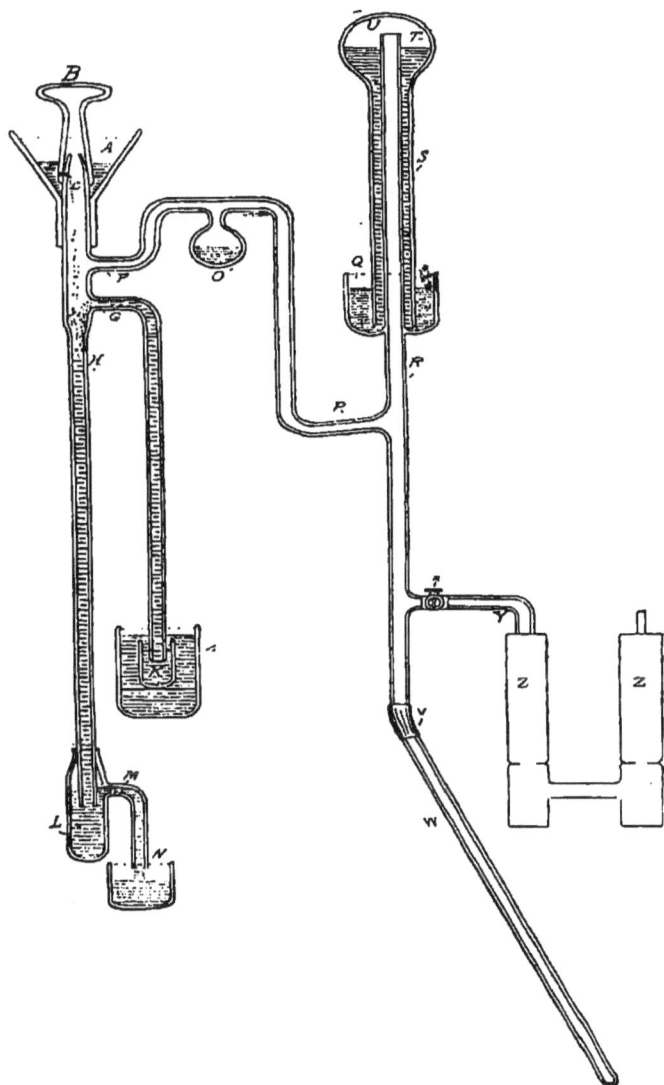

FIG. 30.

The mercury to be distilled and put into the barometer tube is poured into Q (having been first washed in acids, &c.). W is the barometer tube to be filled, and V is a short piece of unvulcanized rubber connecting W with the long glass tube R, the joints being sealed with sealing-wax. ZZ is an arrangement for drying the air, x is a stop-cock, R is a long tube running from x up through Q and inside of the tube S into T where it has an opening. NS is a tube passing into the mercury in Q, and has a large bulb (perhaps five or six inches in diameter) T at the top. The gas burner for heating the mercury is just below T, and encircles S. In operation, pour some mercury into A, K, and I, and the mercury to be distilled into Q. Open x, and by turning or raising B form a connection between A and C by E : the mercury now flows through C into L and carries air with it, causing more air to enter the system of tubes ZZ, where it is dried. After this has been continued until the original air has been expelled from the tubes, the cock x is closed (alternately opening and closing this during the operation just described is best). When x is closed the air of the system becomes rarified . by the action of the air pump, and the mercury in Q rises into S (around R) up into T, but not up to u. The mercury also flows from K up into G and over into H. When the air has been thoroughly exhausted from the system of tubes, then the burner under T is lighted, and the mercury in T will distill from T over into u, and fall through the cooling chamber R down into the barometer tube, where any vapour that may pass into P will condense into O. When W is filled with the pure mercury then the flame is extinguished and x is turned, and air let into the system. The barometer tube is then taken off by melting the wax, and is then ready for use. Paraffin is a good material for covering joints, and does not soil the mercury. During the operation of filling, the bubbles must be forced out of the tube by means of a Bunsen burner, that can be passed along the barometer tube and directed on to the place where any bubble is seen.

In filling a barometer of the Wild-Fuess control pattern having a tube of eleven millimetres diameter, about 375 grammes $\pm$ 25 grammes of mercury is used ; and it must not be forgotten to wash the mercury carefully in acid or chloride of iron before using. It is better to distil a little mercury into another tube first, so that any oxide of mercury that may be present can pass off, and then change .to the one to be filled. This is unnecessary when the apparatus is in constant use for distilling. It takes from fifteen to thirty minutes to get the air out of the tube, and the accomplishment of this can be recognised by the noisy action of the mercury in the air pump.

In the Helsingfors normal barometer, Sundell has for filling the tubes the arrangement shown in Fig. 29. The upper chamber B is connected with a drying vessel T, which contains phosphoric anhydric acid, and which in turn connects with the mercurial air pump LL'. The spherical bulbs LL' have a capacity of 193 cubic centimeters. Pure mercury is poured into the lower chamber B, and the mercury air pump is put into operation by filling the lower bulb L' with mercury and then raising it up until the mercury rises up into L, fills it (which necessitates the addition of more mercury to L'), and overflows into the tube shown on the left. Now

lower L' until the mercury in the tube connecting L
and L' is a little below the tube connecting it with
the barometer tube. Pour mercury into Z and it
will pass from the narrow tube into L and downwards. The circulation of the mercury is kept up by
adding mercury to Z, and, I suppose, by lowering L',
as there is no conveniently arranged overflow tube
joined to L'. The air is gradually exhausted from
the barometer tube by being carried out through L',
and as gradually the mercury rises from lower B up
through R and flows out through O into upper B until
the vacuum is as complete as possible; L' is then
lowered and the mercury assumes the position shown
in the Fig. 29. Sundell claims that the gas pressure
in the vacuum can be reduced to 0·005 mm. by this
process.

9. Errors of Capillarity. Very little reliance can
be placed on capillary corrections for small tubes,
and in order to reduce this capillary correction to less
than ·01 mm., the barometer tube must be at least
24 mm. (about 1 inch) in diameter, and just before
each observation the mercury must be made to rise
in the two legs of the tube. In the normals at
Helsingfors (Fig. 29) and at Pawlowsk the inner
diameter of the mercury chambers at the place of
reading is about 40 millimeters.

The mercury is made to rise in most of the
barometers of the Wild form by raising the accessory
connected reservoir M (Fig 28), but Sundell has a
plunger N (Fig. 29) which is lowered into the reservoir by means of a screw motion.

10. Errors of the methods of reading the height of
the mercury column. In order to keep the error of
reading down to the limit fixed upon (·01 mm.), the

utmost precautions must be taken in the whole detail of making the readings. Up to about 1873, there had never been a barometer that had been read with this desired accuracy. The various errors in reading may be arranged systematically as follows :—

a. The readings must take place in the same way for both mercury surfaces, that is, both surfaces must be as nearly alike as possible, both as regards form and method of reading ; and with any arrangement, such as the Fortin, where the top and bottom readings are made differently, the desired accuracy could not be attained. This form is shown in Regnault's barometer (Fig. 23) where the mercury surface at A is observed, and then by means of the cathetometer telescope the top of the adjustable screw M (of known length) is also observed.

The modern method of bending the tube of a syphon barometer, so that the lower mercury surface is directly beneath the upper one, gives two surfaces as nearly alike as can be obtained. This is best seen in Figs. 28 and 29, and also 31.

b. The mercury must be raised in both tubes just before the reading is made, and this is done by having another tube (of rubber) leading from the bend in the syphon to a small mercury reservoir (Fig. 31), that can be raised or lowered at will; thus changing the height of the mercury in the two legs of the syphon. This is also necessary for getting rid of capillary errors as already described.

c. With the ordinary form of cathetometer for reading the height of the mercury, only the most skilful observing and careful adjustment could give the desired accuracy, because the reading microscope must be at a little distance from the barometer tube

on account of changes due to heat from the body.
So, in all modern barometers the cathetometer scale
is suspended close to the barometer tube, and two
reading microscopes are mounted on a vertical stand
at a distance of several feet from the scale and
barometer.   When a reading is to be made, the cross
hairs of the telescopes are set on the upper and lower
mercury surfaces, and then the microscope
stand is turned slightly around the vertical
axis, and the telescopes pointed to the
scale and the divisions read off, the fractions
of the divisions being read by means of
micrometers on the microscopes.   In Fig.
27 the brass scale RR is shown at the
right, the old or less accurate form being
well shown in Fig. 23.

In the Helsingfors barometer (Fig. 29)
the scale divisions are
ruled directly on the
glass tube SSSS., and
are read by means of the
microscope MM.   A
similar method, pro-
posed by Thiesen, is (I
believe) adopted for the
new Pawlowsk normal

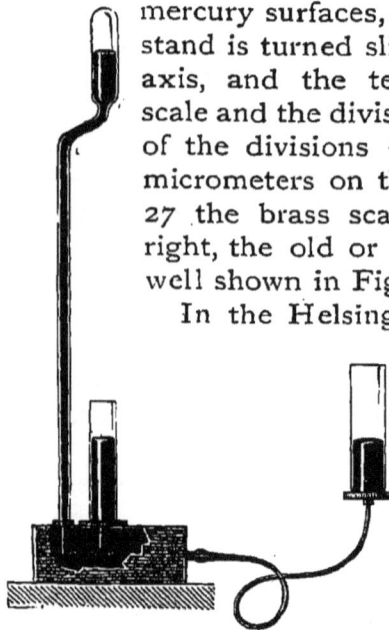

FIG. 31.

barometer.   This is
mainly, however, to aid in reducing the error next to
be mentioned.

d. It is important to be sure that the reading of
the true mercury surface is obtained.   In order to
accomplish this, various forms of illumination have
been proposed, and a steady improvement has been
made.   At first a piece of white paper, or other

light reflector, was put behind the barometer tube, and the dark mercury surface projected against this background was observed with considerable accuracy. In the Wild normal barometer (see g and g' at the right in Fig. 25) there is a plate, which is half black and half white, behind the mercury column. For a telescope reading the upper half is black and the lower half white, and the dividing line is brought just above the mercury surface, so that the top of it is projected against the white surface, and the black surface is just above, for contrast. For a microscope reading the white and black surfaces are reversed in position. In the Fuess normal barometer of 1881 (see Fig. 26) at the Normal Standards Office, Berlin, we find a different method of illumination by means of reflectors. Here the barometer is in a dark underground chamber, and gas illumination alone is used. The rays of light, which are rendered parallel by interposing a lens, strike the mirror S, and are reflected upwards to mirrors on the collimators C_1 C_2, and are by them directed through the collimator to those points to be read by the microscopes. Behind the mercury surface to be observed are double reflectors (see S_1 S_2 at N) with a space between. The left side of the front opening of the collimator is partially closed by a cover with sharp horizontal edge, and which has a fine vertical motion by means of a screw. The light from the right half of the collimator falls on the scale to be read, but that from the left half falls first on the reflector S_1, is reflected to S_2, and finally over the top of the mercury surface at N back to the microscope. The image of the sharp horizontal edge of the partial cover on the collimator is seen in the centre of the mercury surface after the double reflection, and is sharply defined in the reading microscope. This is essentially the same as the black and white background of the Wild barometer (Fig. 25), but is more convenient and probably more accurate. Perhaps the best method, however, is that proposed by Marek and used in reading one of the normal barometers at the International Bureau of Weights and Measures at Sèvres (Fig. 27). At least as early as 1873 Wild used a method of making an accurate reading of the mercury height in the open tube, by causing a needle to descend the middle of it, until the point almost touched the mercury. (I have recently seen a reference to this as Pernet's method.) In this way a sharp image of this needle point will be seen in the mercury, and if a setting is made of the microscope cross wires on this reflected image, and then on the point of the needle itself, the mean of the two readings will give the height of the mercury surface with great accuracy. (Any one can perform this experiment by taking a thin glass bottle an inch in diameter and pouring a little clean mercury into it ; then insert a needle from the top, and place the eye on a level with the mercury surface. Both the needle point and its reflected image can be seen with great distinctness.) This is not so easily done in the closed tube, but Marek has found and applied a method. By means of a collimator, as shown at HH on the right in Fig. 27, the image of the cross wires is formed in the middle of the barometer tube just above the mercury surface. This image and the reflection as seen in the mercury are both observed, and the mean of the two is taken as the surface of the mercury. This is done for both the upper and lower mercury surfaces. The error of a reading is about

·0015 mm., as found by Marek; this is far within the necessary limits, and the method is so simple that it will probably be used very extensively in the future. Thiesen has proposed another modification of this principle. He uses a wide barometer tube, so that the observing mercury surface will be about 40 mm. in diameter, and a pencil point or other sharply defined object is placed outside the tube, but near it, where this object observed directly and also its image reflected in the mercury can be used as points of observation; and the graduations can be on the barometer tube, thus greatly simplifying the manipulation of the normal barometer. This method has been applied by Sundell, and can be seen in Fig. 29, where the scale divisions are shown between SS and SS opposite the reading microscope MM. (It must be remembered that Fig. 29 is merely a sketch of the plan of the actual instrument.) Wild also mentions the application of Thiesen's method to the new normal barometer at Pawlowsk. For an illuminating power the electric light is undoubtedly the best.

In the case of the Kew normal barometer (Fig. 24) there are two steel pointers about two inches long, and one has the edge curved like the meniscus in the barometer tube. The lower end is in one case a knife-edge rounded like the upper meniscus, and in the other a point; both are brought in contact with their reflections. The mercury is agitated in the tube by rotating the barometer and so producing a slight vertical movement by centrifugal action in the cistern, and the result of this is assumed to be the equivalent of the Fortin cistern action.

ℓ. Considerable attention has also been given to

another error which may enter into a barometer
reading and which is due to the influence of the
wind.   If a barometer is placed to the leeward or
windward side of an obstacle, its reading will be
slightly decreased in the former case and increased in
the latter.   Theory gives unsatisfactory values of the
amounts of this error.   Abbe supplies the following
theoretical table of increase of pressure on the windward side :—

Wind. Miles per hour	20	40	60	80	120
Increase of barometric pressure in inches...	0·015	0·061	0·138	0·383	0·548

In order to determine the amount of this error
within a room, experiments were made on Mount
Washington by the United States Signal Service in
1886, and the following table shows the results; but
it must be remarked that a barometer could not be
read with the greatest accuracy in such a high wind
on account of the unavoidable jarring of the building,
and consequently of the instrument :—

Date.	WIND.			EFFECT ON BAROMETER OF OPENING THE—					
	Direction.	Average velocity (miles per hour).	Gusts (miles per hour).	Chimney flue.	Leeward window.		Windward window.		
				Effect.	Aspect of Window.	Effect.	Aspect of Window.	Effect.	
1886.				Inch.		Inch.		Inch.	
Sept. 29	W.N.W.	65	85	—0·005	E.N.E.	—0·014	W.S.W.	+0·029	
Sept. 29	W.N.W.	65	85	+0·003	E.N.E.	—0·008	W.S.W.	+0·045	
Oct. 15	W.S.W.	70	85	+0·006	E.N.E.	—0·011	W.S.W.	+0·045	
Oct. 21	W.S.W.	65	80	+0·003	E.N.E.	—0·018	W.S.W.	+0·062	
Oct. 22	N.N.W.	70	90	—0·006	E.N.E.	—0·025	W.S.W.	+0·032	

Comparison of Standard or Normal Barometers.—

Since the establishment of truly normal barometers, it has become very important that they should be compared with each other, and that the older standard barometers of various countries should be compared with them. Since the report of Professor Neumayer (of Hamburg), there have been not only numerous comparisons between isolated standards, made during the travels of individual meteorologists, but there have also been several series of comparisons of international standards made by persons who have carried portable barometers from place to place with the aim of determining as nearly as possible the differences existing between these standards and so-called normal barometers. The results obtained in this work are far from satisfactory, and show very clearly that Wild was right when he urged that each meteorological service must have its own thoroughly investigated normal barometer which will furnish normal readings at any time ; and that it will not do, as is commonly supposed, to use an ordinary barometer as a standard, on the strength of its having been compared either directly or indirectly with a true normal barometer located at some distant place.

Hellmann, of Berlin, has at various times compared many of the European standard barometers, but as his results have not all been collected and reduced to any one system, they have not been included in the table given here. In 1881 Chistoni, of Rome, carried a portable barometer of good construction to various cities of Europe, and made comparisons with standard barometers there, but we do not understand that the journey was made especially for the purpose of comparing these instruments. In 1883 F. Waldo, then of Washington, U.S., also made a comparison of various

standards, not only in Europe but later also in
America, in which from one to four portable barometers of the Fuess-Wild (Fig. 20) pattern were
used.   In 1886 Sundell, of Helsingfors, by means of a
single barometer (see Fig. 21), which was carried from
station to station with the tube emptied of mercury,
made the most extensive series of comparisons yet
undertaken.   I cannot refrain from expressing regret
that he did not also carry a barometer of the usual
construction, the tube of which should remain filled
during the journey. . In 1887 Brounow, of St. Petersburg, also made, by means of a single barometer, a
long series of comparisons of various European barometric standards.   The results of these four separate
series of comparisons, which may be said to give the
present state of our knowledge, are to be found in the
table on page 95.   The figure 1 after the name of the
barometer mentioned, shows that the comparison was
not made directly with this normal barometer, but
by means of a second instrument adjacent to it, the
relation of which to the normal was known.

The bracketed numbers after the results given
for Waldo's observations denote the number of the
portable barometers used in the comparison ; in
this series the observations at Berlin, and a portion
of those at St. Petersburg, were not made by Waldo,
but entirely by meteorologists at those places.   In
the last column are given the corrections to some of
the various barometers, as applied at the institutions
where they are located, in order to reduce their barometer readings to some adopted standard.   These
are as given by Sundell for the date of 1886 or 1887,
except in the case of Toronto, Canada, which was
obtained by Waldo in 1883 ; and in the results given

TABLE OF RESULTS OF INTERNATIONAL BAROMETER COMPARISONS.

Ref. No.	Place of Comparison.	Institution Where the Comparison Was Made.	Designation of the Barometer.	Correction to Reduce to the Normal Barometer at St. Petersburg according to — Chistoni 1881. (mm.)	Waldo 1883. (mm.)	Sundell 1886. (mm.)	Brounow 1887. (mm.)	Corrections applied to actual readings to obtain the local standard. (mm.)
1.	Paris.	Central Meteorological Bureau.	Regnault 1.	-0,04	-0,05 [2]	0,16	0,11	+,12
2.	"	Astronomical Observatory.	Fortin.		0,10 [2]			-,15
3.	"	International Bureau of Weights and Measures (Sèvres).	Wild-Pernet 1.		-0,24 [2]	0,21	0,10	
4.	Berlin.	Meteorological Institute.	Marek 1.			0,32		+,10
			Chappuis 1.	-1,16	-0,04 [1]	-0,02	-0,02	
5.	"	Normal Standards Commission.	Fuess, No. 76.			-0,05		-,08
			Greiner, No. 320.			0,13		
6.	Vienna.	Central Meteorological Institute.	Normal Barometer 1.	-0,17	-0,25 [3]	0,14	0,11	
			Pistor, No. 279.	0,11	-0,08 [3]	-0,05	0,07	
7.	Hamburg.	Deutsche Seewarte.	Normal Barometer 1.	-0,01	-0,04 [4]			
8.	Kew.	Observatory.	Standard Barometer.	-0,02	-0,10 [2]	0,06		
9.	Rome.	Observatory.	Deluil, No. 6.			-0,05		
10.	Stockholm.	Academy of Sciences.	Pistor and Martins, No. 579.	0,00		0,28		+,12
11.	Christiania.	Meteorological Institute.	Wild-Fuess, No. 214.	-0,11		0,17		
12.	Copenhagen.	Meteorological Institute.	Jünger and Co.			-0,44		
13.	Brussels.	Astronomical Observatory.	Wild-Fuess, No. 87.			-0,01		
14.	Utrecht.	Meteorological Institute.	Fortin-Tonnelot.			0,05		
15.	Zurich.	Swiss Central Office.	Becker.			0,08		
16.	Munich.	Meteorological Institute.	Wild-Fuess, No. 168.			0,00	0,18	
			Fuess, Syphon bar.			0,13	-0,32	
			Wild-Fuess, No. 43.			0,01	-0,16	
17.	Chemnitz.	Meteorological Institute.	Wild-Fuess, No. 4.			0,18		
18.	Helsingfors.	Laboratory of Professor Sundell.	Wild-Fuess, No. 163.			-0,08		
			Normal Barometer.			0,18		
19.	Washington.	Signal Service.	Wild-Fuess, No. 129.					-,18
			Casella, No. 1155.					-,60
			Girgensohn.					
			Standard.		-0,18 [4]			
20.	Toronto.	Canada Meteor. Observatory.	Newman 33.		-0,09 [1]			-,18

in the present table these corrections have been applied. This table is as compiled by Schönrock, of St. Petersburg, with some slight additions and a single correction. The negative sign indicates that the amount of the correction is to be subtracted. Where no sign is given a positive sign is to be understood.

It is seen from this table of recent intercomparisons of standard barometers that the results obtained are quite contradictory. In 1889 Schönrock attempted to find out the causes of these discrepancies, but was able to offer only very general possible explanations. In 1890 Köppen made a more critical study of this same material, with somewhat better success ; but even he was unable to find any definite reasons for the great differences existing for the individual barometers. He finds, however, that if a correction of −0·09 mm. is applied to Waldo's observations as referred to the St. Petersburg normal, and a correction of +0·10 mm. to Sundell's, and −0·06 mm. to Brounow's observations, then a much better agreement is found to exist between the comparisons. It may be added that Wild finds that the new normal at Pawlowsk practically agrees with the St. Petersburg normal ; and as late as 1889 Schönrock says that at that date these two normals and Sundell's normal at Helsingfors were the only ones whose errors had been thoroughly investigated. Speaking generally, the following causes may be assigned as accounting for some of the discrepancies of the results given in the previous table :

Sudden changes may occur in the barometer readings, due to displacements of some portions of the apparatus.

Gradual changes undoubtedly occur in all barometers with greater or less degree of slowness.

Where an intermediate sub-standard is used in the comparison, its readings may change before it is compared with the normal.

Insufficient number of days of comparison, and variation in the height of the barometer.

Differences in illumination undoubtedly exist for the various comparisons, due not only to the diurnal variations of light, but also to the fact that the portable barometers have not been mounted in exactly the same place and position in each comparison at any one observatory. A uniform artificial illumination is certainly to be recommended in all future comparisons of this kind.

Insufficient investigation of the normal barometers prevents an invariable normal reading being obtained at the different times.

Fluctuations due to capillary changes in all barometers with tubes having small bores would undoubtedly account for differences as large as those found in most of the comparisons that were made, and the careful experiments of Schönrock about to be mentioned will show the folly of placing any reliance on the usually applied tables of capillary corrections for tubes of various bores.

Capillary Action in Barometer Tubes.—In the case of barometer tubes in which there exists a capillary depression, there are changes occurring in the form and height of the meniscus, which materially change the readings of the instrument. Wiebe, Thiesen, and others have made some investigations on the absolute amount of these changes, but the question needs much more notice than has yet

been given to it.   While this change is due mainly
to the tube becoming fouled, yet other items
also must be taken into consideration.   Schönrock has made some very interesting comparative
experiments on tubes having different bores, using
the 25·2 mm. diameter barometer tube of the St.
Petersburg normal as a standard.   He finds that:

1. For the tube 12·8 mm. in diameter under different circumstances of pure and impure (oxidized)
mercury the depression varied from 0·23 mm. to 0·41
mm.   For comparatively dry air over the meniscus
the variation was considerable, and the greatest readings were obtained ; but when the air was about
saturated with moisture regular readings of about ·33
mm. were obtained.   With this same bore of tube,
and for clear mercury and dry air above, a depression
of 0·39 mm. gave a meniscus of 1·71 mm. height
when the pure mercury was slowly raised in the tube,
and of 1·44 when it was raised rapidly.

2. For the tube of 11·2 mm. bore, for pure mercury
and dry air above, the depression varied between 0·45
and 0·52 mm., with an average depression of 0·49 mm. ;
for moist air above the mercury the variation was
between 0·43 and 0·47 mm., average 0·45 mm.
The depression varied between 0·34 and 0·52 mm.
in all the experiments.   The height of the meniscus
was 1·36 mm. for a rapid rise, and 1·45 mm. for a slow
rise for pure mercury.

3. For a tube 8·4 mm. bore, the depression varied
between 0·69 and 0·83 mm., with average depression,
0·78 mm. for dry air above the mercury, and 0·71 mm.
for moist air ; the greatest extremes were 0·56 and
0·85 mm.   The height of the meniscus was 1·25 mm.
for a slow rise, and 1·05 mm. for a rapid rise of the
mercury.

4. For a tube 6˙4 mm. bore, the variation of depression for dry air above, was between 1˙13 and 1˙25 mm., with the average depression 1˙18 mm. ; for moist air, between 0˙91 and 1˙09 mm., with the average 1˙00 mm. The greatest extremes of depression found were 0˙91 and 1˙34 mm. The height of the meniscus was 1˙11 mm. for a slow rise, and 0˙80 mm. for a rapid rise.

Such differences as these show that no reliance whatever can be made on the usually published corrections for the capillarity of small tubes, if one wishes to obtain accurate results.

The Aneroid Barometer.—Concerning this instrument I will only briefly remark that it is chiefly useful as a house barometer, and should never be relied on for scientific purposes where great accuracy is required. It is a very delicate instrument, and, being subject to sudden changes, can only be used with confidence when it can frequently be compared with a mercurial barometer. Before an aneroid can be used at all, it must be carefully compared with a standard mercurial barometer at all of the various temperatures and pressures to which it may be subjected in its use. The effects of the changes of temperature vary for different instruments. Wild gives the following changes with time, in the corrections of an aneroid by Goldschmid: April, 1869, correction= −15˙13 mm. ; December, 1873, correction= −27˙16 mm. Another instrument by the same maker changed its correction only 2˙5 mm. in three years.

A thoroughly investigated aneroid of the best construction can be relied on to within 0˙2 mm., only when frequently compared, and when kept in one position. In travelling, the readings of an aneroid may be considerably altered, and it is impossible to fix a limit to its inaccuracy. The instrument must always be gently tapped with the finger before a reading is made, and it is also necessary to observe the temperature and apply a corresponding correction just as for a mercurial barometer, but the same table of corrections does not apply to all aneroids of the same construction, as is the case for mercurial barometers.

The Thermo - Barometer.—If a thermometer is carried to an elevation above the sea-level it will be found that water will boil at a lower temperature than at sea-level ; this is due to the decrease in atmospheric pressure. The law of this change in the temperature of free steam with the change of pressure is known, and if this temperature can be accurately observed the atmospheric pressure can be computed. In order to determine the pressure within $\pm 0 \cdot 1$ mm. the error of the temperature determination must not exceed $\pm \cdot 004^{\circ}$ C., and any such accuracy is simply impossible. However, thermometers are now constructed of Jena glass, by means of which the pressure can be determined within, say, $0 \cdot 3$ of a millimeter with portable apparatus in the laboratory, but it seems doubtful if this accuracy can be attained during a journey. This boiling-point apparatus is, however, far preferable to an aneroid barometer when no mercurial barometer is available for comparison. The general precautions and methods used for determining the boiling point on a standard mercurial thermometer are applicable to the manipulation of this instrument.

Barographs.—It has long been considered desirable to record the readings of a barometer automatically, and in accomplishing this many different forms of mechanical construction have come into use. These

can be divided into the following classes : 1. The
purely mechanical arrangement, which has a float in
the open leg of the barometer, and the motions of
this float are recorded. 2. The electrical method, in
which a platinum wire is made to descend the open
leg of the tube, and when this wire, which is a pole
of a battery, reaches the mercury surface, a circuit is
closed by having the other pole already touching the
mercury. The varying distance the wire moves from
a certain point in order to reach the mercury surface is
measured, and the successive measurements locate the
barometric curve. 3. The photographic method, by
which the height of the mercury in the closed leg of
the barometer is photographed continuously. 4. The
balance barograph, in which the variation of weight
of the mercury in the tube is measured when the latter
projects into a cistern containing mercury, but is not
in rigid contact with the vessel containing the mercury. The barometer tube is suspended from one
end of the balance arm, and is counterpoised by a
weight on the other end.

There are, perhaps, twenty different constructions
of barographs, but they nearly all belong to these four
classes, although there have been other forms proposed but not much used. These instruments are
of various degrees of accuracy ; perhaps the best one
being the Sprung balance barograph with sliding
weight, constructed by Fuess, of Berlin. The mean
deviation of a registration of this instrument from the
direct observations on a good ordinary barometer is
only $\pm$·04 mm. ; and this is about equal in accuracy
to the direct comparison of two portable barometers of
the highest degree of excellence.

Several other barographs are scarcely inferior in

accuracy to the Sprung form, especially the Wild-Hasler, as used at St. Petersburg, and the Photographic at Kew. Eylert has also succeeded in getting good results from the Fuess barographs of the *Deutsche Seewarte* model. Much depends on the method of reduction of the barogram, however, and the very common method of interpolating the direct recorded readings when compared with those from a mercurial barometer is not sufficient. The barogram is the record traced by the instrument on a sheet of paper.

The Sprung balance barograph is shown in Fig. 32, and with the photographic barograph has the advantage over some other forms that it gives a continuous record of the atmospheric pressure, while most others record it only at stated intervals, as, for instance, every ten minutes.

In the accompanying diagram of Sprung's balance barograph, the barometer tube B is suspended from the one arm (on the right) of the balance, and is counterpoised by the weight g and various minor accessories attached to the other arm. The barometer tube remains practically stationary, and when the changing pressure of the air varies the force acting downward on the barometer tube, the effect is at once counterbalanced by a slight shifting of the wheel R, which rides on the opposite arm of the balance, and whose motion is controlled by the carriage V below it ; this latter has attached to it the pencil S which marks on the tablet T its backwards and forwards motion. The tablet T has a uniform vertical slow motion imparted by the clockwork, and the fixed pencil S' marks on the tablet a straight line which is used as a base line of reference from which to measure distance of the points on the irregular line drawn by S, and which follows the fluctuations in the barometric pressure.

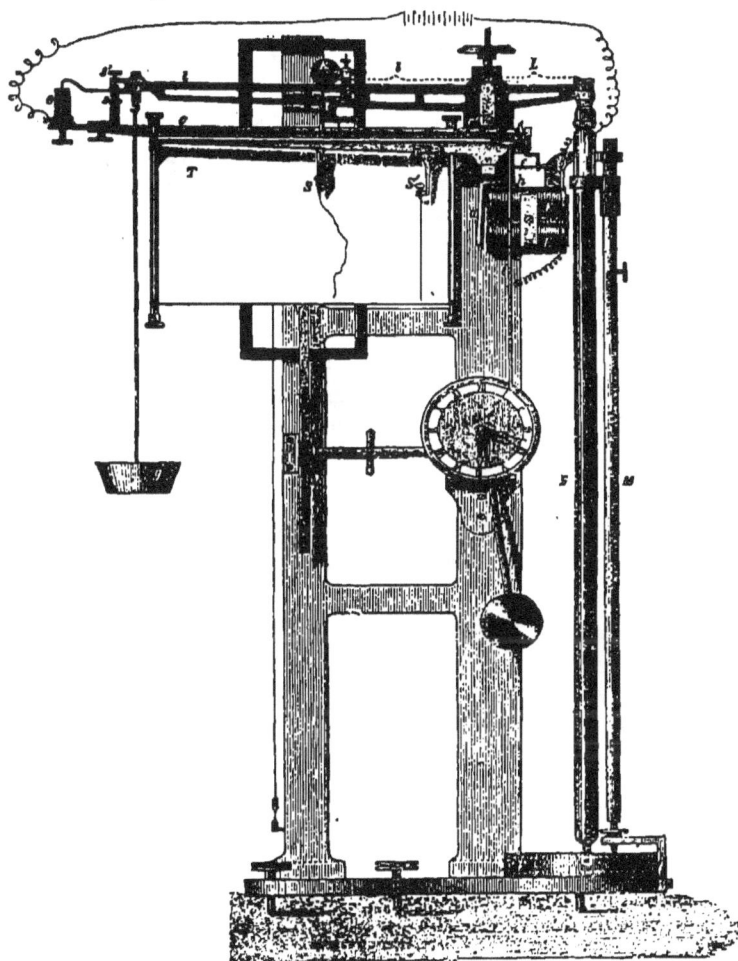

FIG. 32.

As Sprung has applied this same principle to making registrations of several of the meteorological elements, and as the instruments are fast growing in favour, a more detailed statement of the method of operation seems to be necessary here.

The lower end of the barometer B dips into the large mercury cistern F, but it is in contact with no rigid body below the point of suspension on the balance arm. M is merely a parallel reading scale to be used when required, by means of which the height of the mercury in the tube of the barograph may be read as for an ordinary barometer. The vertical rod t has a constant rotary motion imparted to it at its lower end by the clockwork, and at the upper end it has a horizontal bevelled wheel, which has a slight horizontal freedom, permitting it to press against either of two vertical levelled wheels which are rigidly attached to the screw CC' (on the right). When a current is passed through the coils EE, the rod t is drawn towards the left by means of the armature a and the connection h, and the rotary motion of t is imparted to the screw CC' by means of the bevelled wheel on the left; when the current is cut off from the coils EE the spring F draws t towards the right, and the contrary motion is imparted to the screw CC' by means of the bevelled wheel on the right. Let it be assumed that the atmospheric pressure is decreasing, then the left end of the balance which has a slight play between S' and S descends until a contact is made with the mercury at c, when the circuit is closed and the armature a draws t towards the left, and the screw CC' is then revolved in a right-hand direction; this causes a motion of the carriage V, which propels the running weight R along the balance beam until the connection at c is broken, when the spring f is allowed to draw t towards the right, which causes the motion of CC', and consequently V, R, and S, to be reversed until the current is closed again. By means of this constant, but alternating motion, the equilibrium of the two arms of the balance is maintained within the limit of play at SS', and yet the carriage V and pencil S are permitted to move laterally in proportion to the amount of change in the barometric pressure. In the Sprung barograph in use at Magdeburg the pencil S moves over a space five times as

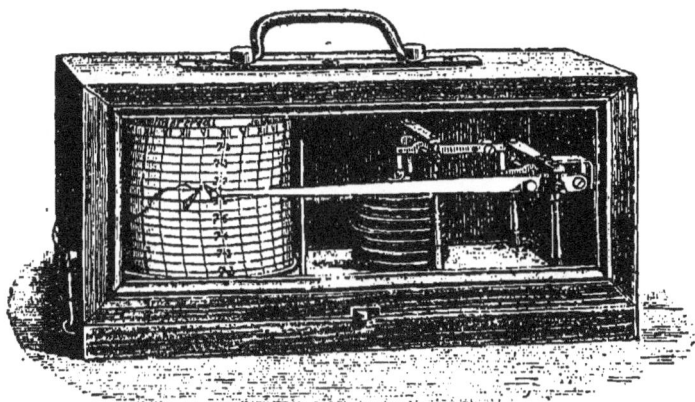

FIG. 33.

great as the change in the height of the mercury column.

The Aneroid Barograph.—The Hipp aneroid barograph is the most widely used at the older European meteorological stations, and the Hottinger form is also used, but at present the newly constructed Richard barograph (Fig. 33) is attracting the most attention from meteorologists. Any of these forms just mentioned will give results within, perhaps, 0·2 mm. when the instruments are compared tri-daily with the

mercurial barometer, and the records of the barogram
are submitted to an exhaustive treatment ; but at
most of the larger observatories where these instruments are in operation, their record is used merely for
the interpolation of omissions from the record of the
mercurial barograph.

§ 3.—*Apparatus for Measuring the Wind.*

*History of Apparatus for Measuring Wind Direc-
tion.*—The characteristics of the winds from the four
points of the compass were made the subject of
remark probably as early as 1000 B.C.; and 500 years
later the Grecians had added four additional intermediate directions, which were gradually increased
to twenty-four directions in still very early times.
Twelve points were in more common use, however,
and under different names, until the time of Charle-
magne, when the intermediate directions were compounded from the four cardinal points, somewhat as
in our present nomenclature.

The oldest arrangement, of which we have a knowledge, for accurately locating the directions of the
wind is the octagonal Tower of the Winds at Athens
(Fig. 34), which was constructed about 100 B.C. At
the centre of the roof was placed a wind vane in the
form of a triton, whose sceptre always pointed to the
wind octant, but was inclined downwards towards
the walls, where there were to be found not only the
Grecian names of the winds, but also allegorical
figures in relief, characterising the wind which came
from each direction.   M. Terentius Varro also relates
that he mounted a wind vane.

The first weather-cock on a church spire is accre-

FIG. 34.

dited to a church in Tyrol in the year 820, when
Bishop Rampertus von Brixen had it set up. But
the first wind vanes of the modern form were those
used by the Italian, Egnatio Danti, at Bologna and
Florence, in the middle of the sixteenth century,
and were of the construction shown in Fig. 35. The
regular observation of wind vanes for meteorological
purposes was not commenced in Italy until about
1650, however, and in England a few
years later. Hellmann has furnished a
very interesting account of these early
attempts at measuring the wind.

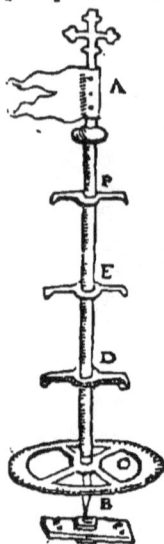

It would seem that for so simple an
instrument as the arrow vane, now almost
universally used, there could be very little
said concerning its errors and the theory
of its action ; and this is true to a certain
extent, but there are some points which
require to be considered.

Instead of the vane revolving around
the axis on which it is placed, as evenly
balanced as possible, it is the custom to
attach the vane rigidly to the axis, and
let the latter be supported between friction
wheels, so that it can turn with as little
resistance as possible. The lighter the
vane the truer will be its indications ; as, if the vane
is too heavy it is not sensitive enough to rapid changes
of direction in a light wind, and when once in motion
its momentum causes too great a change in a gusty
wind. It is desirable also to have attached to the
lower end of the axis to which the vane is fixed some
sort of damper, which shall prevent the vane from
following the quick unsteady changes of the wind

FIG. 35.

which have only local significance, and yet will allow it to follow the more permanent changes. This can be accomplished by attaching some light paddles to the lower end of the vane rod and immersing them in oil or glycerine, in which they will revolve with the rod.

It has also become the custom to construct the arrow vane with a double or spread tail, after the fashion set by Parrot in 1797. (A small vane with this arrangement is shown in Fig. 37.) By this means the tendency to continual oscillations, with slight sudden shifts of the winds, is much reduced. The best angle of divergence of these tails has not been a subject of agreement among meteorologists. Parrot made the angle of spread 45°, but in the English vanes the angle was made but $22\frac{1}{2}°$ (adopted about 1840); and this last has been quite universally adopted. Curtis has investigated theoretically the action of the single tail and spread tail vanes in steady winds, and finds that for the latter the proper angle of divergence is less than 30° in all cases, but that the best angle will depend on the amount of the shift of wind, and is consequently not assignable until the average amount of shift (in degrees) is known. Comparing, theoretically, the single and spread vanes, Curtis deduces the following relations for steady winds : " The oscillations of both vanes are smaller as the vanes are longer and larger ; the spread vane is always more stable than the straight vane, and this advantage in stability is greater for long vanes than for short vanes, and is independent of the wind velocity."

The arrow vane, as already described, shows the horizontal direction of air currents. It is also very desirable to know the vertical inclination of such currents, but no apparatus for this purpose has come into regular use among meteorologists. Benzenberg, Montigny, Hennessey, Descheverens, and others, have proposed various methods by which this has been attempted ; and Cassella makes a very ingenious self-registering double vane, one with a vertical and the other with a horizontal motion, but the difficulty in properly exposing any such instrument, so as to obtain trustworthy results, has not yet been overcome.

Apparatus for Measuring Wind Force or Velocity. —Instruments for measuring the velocity of the wind are usually called anemometers. These may be divided into three general classes : those measuring pressure, rotation, and suction.

The pressure anemometer is usually made according to one of two forms : either a metal plate of small dimensions, and pendulously suspended from its upper edge, is allowed to swing with the wind, and its departure from the vertical is noted ; or a plate of, say, a foot square is exposed, fronting the wind, and its pressure is received on springs, the movement of which permits the force to be noted. Many other forms have also been proposed, the earliest and simplest being that in which the air blows into the mouth of a tube which has some manometric attachment to show the variations in pressure.

In the rotation anemometer, a wheel is rotated by the force of the wind ; the wheel having some sort of a fan-shaped bladed surface, or consisting of merely four spokes terminating in cups presenting sides of different resistances.

The suction anemometer, which is as yet not much more than a toy, but which has great possibilities for future usefulness, consists of a tube placed vertically, so that the wind blows across the open mouth, by which action part of the air is drawn out of the tube: the amount of diminution of air pressure in the tube is measured by some delicate contrivances placed within the tube.

The pendulum anemometer, on account of its great simplicity and cheapness of construction, is at present used in the Russian Empire, and in some

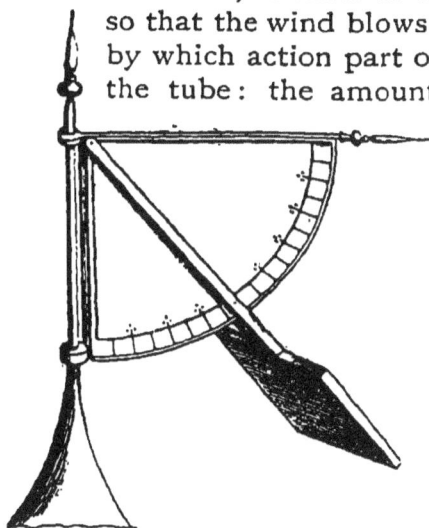

FIG. 36.

other, principally European, countries. This instrument consists essentially of a freely hanging metal plate suspended on a horizontal axis from above. The wind blowing against this plate causes it to swing out of the normal vertical position, the angle of displacement from the vertical varying with the force of the wind. This instrument was probably first proposed by Hooke, but on account of the uncertainty as to who was the inventor, Abbe recommends that it be

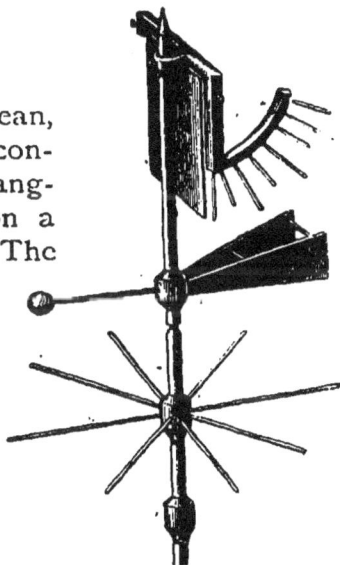

FIG. 37.

called the anemometer of the first meteorological committee of the Royal Society. The first description was published by Hooke in 1667, and its construction is shown in Fig. 36. In its most widely used modern form, as shown in Fig. 37, it is known as the Wild wind force tablet, and is mounted on the same vertical rod as the wind vane, which latter always keeps the plate facing the direction of approach of the wind.

Thiesen has given a theoretical treatment of the action of this instrument, and has introduced into his formula numerical values determined by Dohrandt and himself, by experiments with a wind force tablet mounted on the St. Petersburg Combes' whirling machine.

He deduces the following relation for the Wild tablets of the size 15 × 30 centimeters and a weight of 195 grammes ; when the barometer reads 758 mm. and the thermometer 15° C.

Wind Velocity. Meters per Second.	1.	2.	3.	4.	5.	6.	7.
Angle of Displacement from the Vertical, in Degrees.	1°·1	4°·1	9″·2	15″·7	23″·0	31″·0	38″·7

Wind Velocity. Meters per Second.	8.	9.	10.	11.	12.	13.	20.
Angle of Displacement from the Vertical, in Degrees.	45°·7	52″·1	58″·0	62″·5	66″·4	74″·0	80″·5

A change of weight of 1 gramme, in the tablet, will give a change of one-fourth of 1 per cent. in the velocities for the same angle.

At ordinary wind velocities Wild considers that this instrument will furnish results within 10 per cent. of the truth ; but it is not suitable for observing winds of over 20 meters per second velocity, on account of the slight increase in the angle (from the vertical) with a large increase of wind.

The invention of the cup-anemometer is attributed to Edgeworth, who suggested it to Robinson ; but in the Royal Ethnographical Museum at Berlin there is an apparatus very similar to it, only it is made of wood and the cups are oval and not complete hemispheres ; this came from Thibet, and probably illustrates a form of wind-mill, the small copy being used as a wind-mill god ; and in connection with this, Dr. Grünwedel (of the Museum) remarks that in Pallas's *Mongolische Völker*, 1770, there is

FIG. 38.

a diagram of a similar apparatus. The accompanying diagram (Fig. 38) is re-drawn from an illustrated article in the *Century Magazine*, January, 1891, giving an account of Rockhill's journey in Mongolia and Thibet. It is seen that this device, which is un-doubtedly very old, is almost the exact model of the modern cup anemometer. We find here also the suggestion of the double-cup anemometer. So that, although Edgeworth may have used it first as a scientific instrument, and probably re-invented it, the idea of construction is by no means new.

FIG. 39.

The Robinson cup anemometer (Fig. 39) [1] consists of four hemispherical cups of thin metal, attached to

[1] In this illustration the anemometer is shown as mounted on the highest point of the top of a building, as it is usually exposed to the

the extremities of two light metal rods crossing each other at right angles, so that all of the four arms (or spokes) formed by the rods are of the same length. The cups are attached to the arms (as the rods are usually called) in such a manner that they all face the same way—that is, either to the right or to the left. The arms are placed in a horizontal position, and are rigidly attached to a vertical spindle at their intersection. This spindle passes downwards through a hollow stand encircling it, and which supports it, but does not prevent its turning as freely as possible, except that the unavoidable friction will offer a slight resistance. The lower end of the spindle rests in a bearing (where also the friction has been reduced to a minimum), and carries, just above the end, an endless screw encircling it. This endless screw has a small toothed wheel geared into it, so that the wheel is turned by revolving the endless screw. This wheel is geared into a series of wheels, each of which may have an indicator and a dial to show the number of revolutions.

If, now, the instrument is exposed to an air current, the unequal pressure of the air against the windward convex and concave surfaces of the cups will cause the cups, and consequently the spindle, to rotate around the vertical spindle as an axis. This rotation is communicated to the train of wheels, and the number of revolutions of the anemometer cups around the vertical axis can be read off from the dials of the wheels, which are so arranged that a certain number of revolutions of the spindle causes one revolution of the

wind. The particular model shown here is the "Kew" anemometer, and beneath the cups are to be seen the bladed wheels of the Beckley wind vane.

first wheel, and a certain number of revolutions of this wheel causes one revolution of the second wheel, and so on. By means of this apparatus the number of revolutions of the anemometer cups is obtained ; and the distance passed over by the middle point of one of the hemispherical cups is, of course, 2 × 3·1416 × R, where R is the radius of the circle described by this middle point. This is not the true velocity of the wind, but Robinson found by experiment, however, that if the velocity of the cups is multiplied by 3, it will give approximately the velocity of the wind. And this factor 3 is used by all nations in the reduction of the anemometer observations. In order to save the labour of making this reduction each time, the wheels and dials of anemometers are so made that the miles and fractions, or kilometers and fractions, can be read off directly, the factor 3 being taken into account in the construction.

Robinson constructed a cup anemometer in 1846, and shortly afterwards gave a mathematical analysis of its action ; but his theory of the instrument, published in 1850, was not very satisfactory, although the results were approximately correct. The constants to be used in his formulæ he determined by experiments with a large anemometer on a whirling machine ; he found that the velocity of the centre of the cups must be multiplied by 3·00 (as just mentioned) in order to get the wind velocity. Robinson found that a small anemometer tested also on a whirling machine (Combes' apparatus) gave nearly the same results, and meteorologists, generally, accepted this value.

Other experiments of Cavallero in Turin, and Stowe in England (1870–72), showed, however, that this result was only approximate ; the latter's observations making it clear that the constant of multiplication (adopted as 3) is not the same for anemometers of different dimensions, but varies with the relation between the length of the arms and the radius of the cups themselves. Dohrandt at St. Petersburg was the first to show that the constant of multiplication could be computed if the dimensions of the radial arms and cups are known. These and other experiments have shown that anemometers of the size commonly used record wind velocities about 20 per cent. too great, and although this has been known for nearly twenty years, yet no change has been made in the erroneous value introduced by Robinson, and the wind observations published by the various meteorological institutions at the present time have only a relative but not absolute value.

For common sizes of anemometers, it is fairly accurate to deduct 20 per cent. of the anemometer record (given in meters per second) and then add 0·8 meter per second to the result, in order to obtain true wind velocities. Recent experiments by Dines show that for the large cup-anemometers of the Kew pattern the true constant multiple is from 2·0 to 2·2 instead of 3·0 for moderate velocities. It is very probable that many experiments on the relation of wind velocities to wind pressures have been made, in which this anemometer correction has not been properly applied.

Although the theory of the instrument has been given by several physicists, notably by Robinson and Thiesen, in the general terms of hydrodynamical analysis, yet they consider that the subject is one that can be best treated by experiment. And while the theory may serve to outline the general action of the instrument, still in actual practice it is better to investigate each separate instrument, as must be done in case of thermometers.

Standard Anemometry.—Numerous methods of testing not only the comparative but also the absolute accuracy of anemometers have been proposed ; but the method which has been usually adopted in practical meteorological work is that in which the instruments are mounted side by side and their readings noted for similar wind velocities ; and yet this can furnish only comparative results.   During the last twenty years, however, experiments with the Combes' whirling apparatus have become so frequent, that its use for testing anemometers will probably become general, since it can be used at any time and for any velocity up to its limits of operation.   This apparatus (Deutsche Seewarte model, Fig. 40) which, according to Thiesen, was used by Smeaton and Zeiher before Combes, consists of a horizontal arm, Z Q t, of at least ten feet in length ; this arm has a horizontal motion around a vertical axis, Z, at one end of it.   The free end, t, carries another vertical arm, to which is attached the vertical spindle of an anemometer D, the cups being horizontal.   The long horizontal arm is now made to rotate about the vertical axis at Z by causing this last to revolve ; which can be done by means of a small engine of one- or two-horse power (man power has sometimes been used).   The anemometer will consequently sweep horizontally through a circle, of which the revolving axis, Z, is the centre, and the long horizontal arm, Z t, the radius.   The motion of the anemometer causes a resistance from the air, which acts on the anemometer in nearly the same way as if the anemometer were at rest, and the wind were blowing with the same lineal velocity that the anemometer has in its revolution. Of course the space passed over by the anemometer

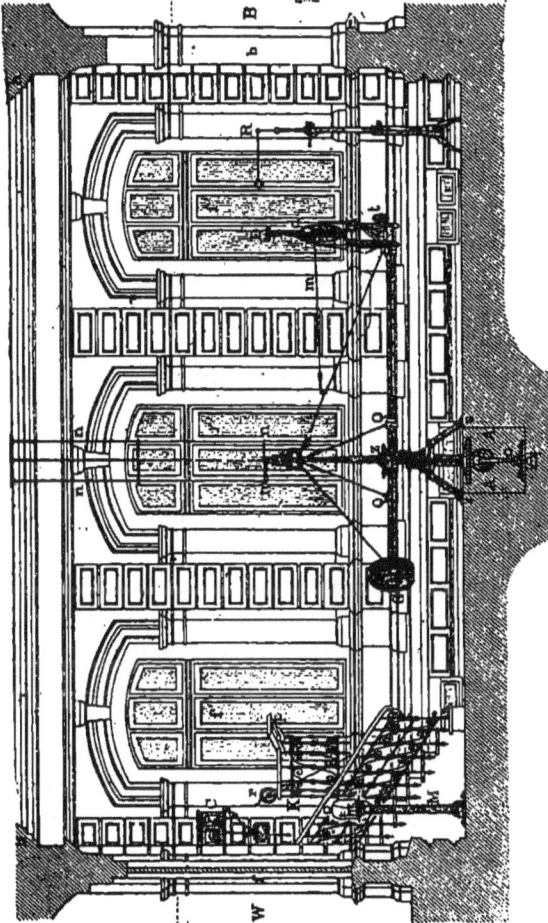

FIG. 40.

can be readily computed when we know the length of the horizontal arm, Z t ; and if the time is noted in which a certain number of revolutions are performed, and a reading of the anemometer dial is made at the beginning and end of this time, we can find out how many feet of motion of the anemometer cups around the spindle correspond to the number of feet passed over in a second by the whole anemometer at the end of the revolving arm. Knowing the actual velocity of the counter-wind caused by the rotation of the arm, the readings of the dial of the anemometer cups can be reduced to actual measure. The lineal velocity is deduced from that recorded by the anemometer by causing the latter to be whirled in the direction with the hands of a clock, then counter clock wise, and taking the mean between the two results.

The Deutsche Seewarte apparatus is located in the middle of an inner court, which is closed at the top by glass roofing, and is surrounded by corridors opening into the court, and the following additional details are given to show more clearly the method of its manipulation.

The long horizontal arm is not more than three feet from the smooth polished floor, but the anemometer D is perhaps eight feet above the floor, and three times that distance from the ceiling. The vertical axis Z of the revolving arm passes downward to a trap A just beneath the floor, where the rotary motion is communicated by a horizontal axis by means of a conical gearing; and this long horizontal axis is rotated by means of a gas motor in the basement. At one side of the court there is a frame R, supporting at the end of the horizontal rod a small anemometer, which can be placed just above the path of the centre of the anemometer to be tested. (In the Physical Observatory at St. Petersburg the small anemometer is placed just outside and in the same plane of the circle in which the large anemometer is whirled.) This small anemometer is for measuring the velocity of the weak current of air produced by the motion of the anemometer, and which has a motion *with* the anemometer (*Mit-wind*). The velocity of the current is about 5 per cent. of the velocity shown by the large anemometer, and this must be allowed for in determining the true reading of the latter.

In one corner of the court, at the left, there is a little balcony K, where the observer may place himself when the apparatus is in motion. Here there is a crank handle *r*, that has communication with two conical cylinders in the basement, near the engine, around which the driving belting passes. By means of the crank this belting is caused to slide along the cones, so that the velocity of the driving arm, and consequently of the whirling apparatus and the anemometer, can be varied by the operator in the balcony.

In the balcony is also mounted a chronograph, *p*, on which is automatically registered, electrically, the seconds, the number of revolutions of the long rotation arm, and the number of revolutions of the anemometer cups. This complex record allows of a more complete discussion of the observations than could be otherwise made.

Apparatus of this nature have been constructed by Robinson in Ireland; Stowe, Whipple, and others in England; Hagen, Rechnagle, and Neumayer in Germany; Wild in Russia; Marvin in America, &c.; but the most complete one is that of the Deutsche Seewarte, at Hamburg, constructed by Neumayer, and shown in Fig. 40.

When all of the meteorological services shall have adopted a whirling machine of some uniform size and exposure, on which to test anemometers used by them, then the observations of wind velocities in different countries can be made comparable; but even then the results will be purely relative. Experiments have still to be made which will show conclusively the relation of the velocities obtained by the whirling apparatus to true absolute wind velocities as actually observed in the open air.

Estimates of Wind Force.—It will be found that instruments for measuring wind velocities have been used by the majority of meteorological observers in but few countries, and at sea not at all. Where this is the case, the personal estimates of wind force are made by the observers according to some arbitrary scale. While it would probably be just as accurate to use the more readily understood expressions— calm, gentle, fresh, strong, gale, hurricane, to indicate the various grades of wind, yet in dealing with masses of such observations it is more convenient to replace these expressions by numbers. As early as 1780 what is known as the Mannheim Scale was introduced, and for many years widely used. This scale extends from 0 (calm) to 4 (strong gale). Since then several scales have been used, at various times and in different countries, in which the number of graduations, so to speak, have been from 0 (calm) to 6, 7, 8, 9, 10, and 12; these last numbers indicating the most violent winds that might occur. Of these, the most interesting and important is the scale 0–12, which was introduced by Admiral Beaufort in 1805 for use on board ship; and this is at present almost universally used at sea. This scale was based on the amount of sail a ship could carry in the different winds, but the system also came into use at land stations, especially those on the coast; although in applying it to land stations in general only half the number of divisions are used (0–6). In the past twenty years a 0–10 scale has also been widely used in the international work of the United States Weather Bureau; and Abbe ("Meteorological Apparatus and Methods") has given its graduations in terms of the other scales, which are brought together here in a little table.

U.S. Weather Bureau Scale 0–10.	Beaufort's Scale 0–12.	Jalinek's Austrian Scale 0–10.	Wild's Russian Scale 0–10.	Old Russian Scale.	French Meteor. Ass'n. Scale 1–7.	Paris Met. Bureau Scale 0–6.	English Land Scale 0 6.	Mannheim Scale 0–4.
0	0	0	0	0	0	0	0	0
1	0.5 and 1	—	1	0	1	1	0	—
2	1.5 and 2	1	2	1	2	—	1	1
3	2.5 and 3	—	3	1	3	2	—	1 to 2
4	3.5 and 4	2	4	2	4	3	2	2
5	4.5 and 5	3	5	2	—	—	—	—
6	5.5 and 6	4	6	3	5	4	3	3
7	7	5	7	3	—	—	4	—
8	7.5—9	6.7	8	4	6	5	5	4
9	10	8.9	9	4½	—	—	6	—
10	11.12	10	10	—	7	6	—	—

Abbe has also given the velocities and pressures of the winds for these different graduations of the 0–10 scale, but as they differ widely from some recent results in this same line of work, and as little information is given as to their derivation, they are omitted.

Very little has been done towards determining the value of these scales of estimation in terms of actual wind velocities, but Mohn has brought together

(*Meteor. Zeits.*, Feb., 1890) the results of a number of comparisons which have been made within the past twenty years, between the estimations on Beaufort's Scale and anemometer observations. As a result of such comparisons, made chiefly at lighthouses on the English, German, and Norwegian coasts, the following values, given in round numbers, were obtained:

Beaufort's Scale.	1	2	3	4	5	6	7	8	9	10
True velocities. Meters per Second.	2·	3·5	5·5	7·0	9·0	11·0	13·0	15·0	17·	19.

A few observations made on the open ocean give about the same values up to 5 of the Beaufort's scale, but somewhat larger velocities corresponding to the scale estimations of stronger winds.

Estimations of winds above 10 of this scale are scarcely to be relied on. In order to reduce these true wind velocities to those given by an ordinary anemometer, whose readings are obtained by means of the multiple 3, it is necessary to subtract from each velocity 1 meter per second, and then increase the remainder by one-fourth of its amount.

Self-Registering Anemometric Apparatus.—The recording apparatus for the ordinary pressure anemometer is so arranged, that the force of momentary gusts of wind is clearly brought out, but total wind movements for any definite time, as an hour or a day, are obtained with great inconvenience.

The registrations of the cup anemometer with electric contacts do not show the force of sudden gusts of wind, unless they continue for some time; but the record is well adapted for showing the hourly or daily amounts of wind. The former is more important for engineering purposes, and the latter for meteorological purposes. The registering instrument which will accomplish both purposes, as well as each of these does its part, has yet to be constructed.

According to Westphal's researches, the attempts at constructing apparatus for measuring and recording wind-force and wind-direction extend back to the year 1667, and the names of Hooke, Pickering, Christian Wolf, Lentmann and Leupold are given as having been early experimenters.

Count d'Ons-en-Bray, in the year 1734, constructed a registering apparatus for wind direction very similar to the form devised later by Beckley; the wind velocity being recorded by a hammer raised by means of a small wind wheel, and allowed to fall against a recording paper, having a uniform motion, at every 400 revolutions of the wheel.

The Robinson anemometer register, in use in many of the principal European observatories, is the Beckley form ; in which a long vertical rod is geared to the anemometer spindle by means of an endless screw. At the lower end of this rod is a gearing by which a small horizontal cylinder is made to revolve (once for every 100 kilometers of wind in the Fuess instrument). A spiral knife-blade-shaped strip of metal encircles the cylinder once in its whole length, and the edge of this metal blade touches a chronograph cylinder beneath it at one point only at a time ; but during a complete revolution of the small cylinder, each point of the edge of the spiral blade comes in contact with the chronograph at successive points from left to right. The chronograph sheet being of metallic register paper, the contacts of the metal blade are recorded.

A similar arrangement records the wind direction, the small cylinder and its spiral metal strip making a complete revolution for each revolution of the wind vane, and a particular point of the strip always coming in contact with the chronograph sheet for the same position of the wind vane.

Perhaps the best form of anemometer-register in common use is the electrical, in which a platinum contact is inserted in the wheel work showing the number of revolutions, in such a manner that for every kilometer or mile of wind a circuit will be closed and cause a registration to be made on a chronograph. There is less friction with this method than with others. The wind direction is registered by means of an electro-magnet and pencil for each direction—for instance, eight magnets and pencils for eight points of compass, the circuit being closed automatically by the vane entering that sector of the circle of revolution of which the direction is the middle point.

The anemometer record can be made on the same register as that used for showing the direction of the wind vane, and the apparatus so constructed that the miles or kilometers will appear opposite the wind direction at the time of registration ; thus enabling one to easily obtain by mere inspection of the anemogram the velocities for the various directions.

But a form of integrating anemometer, which has been independently proposed and its success proven by years of use in several countries, but first in Russia and France, in which the anemometer is included in the circuit of the conducting wires of the direction register, causes the miles or kilometers to be recorded by the pencils showing the direction at that time ;

thus recording the velocities under the headings of their proper directions. This is not only a great convenience to the person who takes the data from the anemogram, but it affords a ready material for the rapid

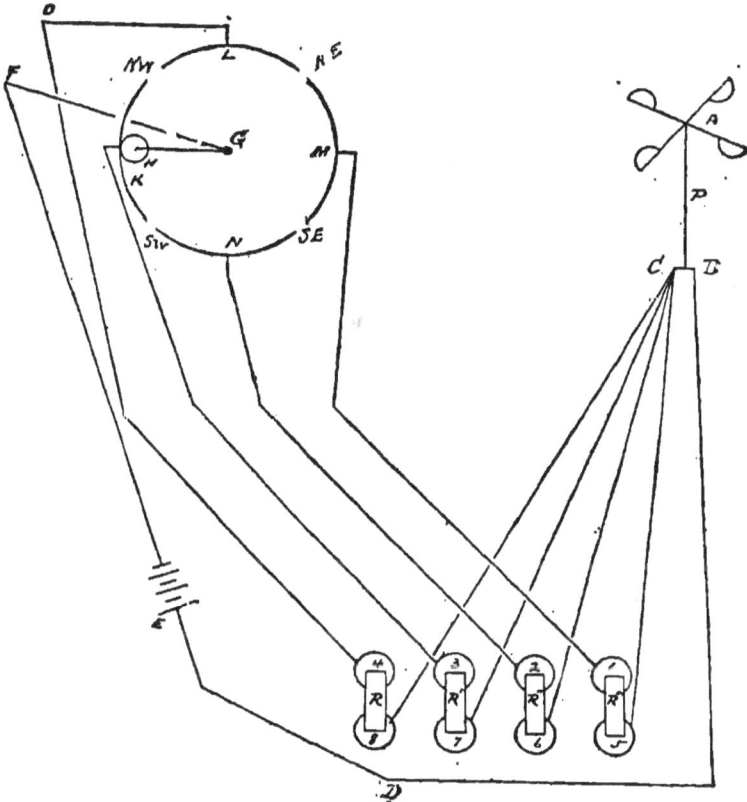

FIG. 41.

comparison of the directions and velocities of wind on successive days, as the record sheets for each day can be laid side by side.

The general principle upon which this device works is shown more plainly by the accompanying Fig. 41.

In this A represents the Robinson anemometer; P, the supporting frame; and C and B, the posts to which the conducting wires are attached as in the ordinary form of electrical self-registering anemometer, in which C and B have metallic connection through P at the completion of each mile or kilometer of wind as shown by the anemometer dial; at other times the circuit connection between C and B is broken. By means of a wire, B is connected with one pole of the battery E. Wires also pass from C to binding-screws on 5, 6, 7, 8, electro-magnets of the recording apparatus. The left-hand part of the diagram is shown in a horizontal plan. G is the lower end of a rod passing from the roof to the room beneath. This rod, being in rigid connection with the wind vane, will revolve with the latter. Near the lower end of the rod at G an arm is placed at right angles to the rod, and terminates in the small friction wheel H. This wheel H runs along a metallic rim encircling G; the rim not being continuous, but having small breaks at the points touched by the wheel H when the vane points N.E., S.E., S.W., N.W. These breaks are so short that when H, in its revolution around G, leaves one segment of the rim, it almost instantly rests against the next. The segment L is connected with the binding-screw of 4, by the wire passing through O. Similarly K is connected with 3, N is connected with 2, and M is connected with 1. The continual wire GF is in metallic connection with GH, and is also connected with the free pole of the battery E.

We will now suppose that the anemometer and vane are exposed to the wind, which is from the west. We shall then have the arrangement shown in the diagram. The circuit connection C 7 3 KHGFEDB will be complete (except at BC), and whenever the anemometer closes the circuit BC, the armature R' will be attracted by the double coil magnet 3 7. Thus, for any winds between S.W. and N.W., the action of the armature R' will indicate each mile or kilometer of wind. Similarly the armatures R, R'', R''' will indicate northerly, southerly, or easterly winds.

By attaching recording pencils to the armatures R, R', R'', R''', and allowing a chronograph sheet to pass beneath them, we can register in separate columns the amount of wind from the four points. By doubling the number of segments, electro-magnets, and recording pencils, the velocities can be recorded for eight points of compass; and by recording a combination of two directions at a time twice as many points of compass may be registered as there are recording pencils.

There are many more common methods of recording the wind direction. The simplest of these is to prolong the spindle, rotating with the wind vane, to a room below, and let this spindle terminate in a conical bearing, so that the friction will be small. Near the lower end of the spindle mount a cylinder encircling it and rotating with it. On this cylinder will be placed the recording sheet. Against this sheet there presses a recording pencil which has a slow motion in a vertical direction only, and which is regulated by clock-work. The cylinder rotating with the wind vane, and the pencil pressing against it continually, will cause the wind direction to be recorded for every instant of the time during which the pencil is moving in the vertical, which may be one day or several days for a single recording sheet, as has been allowed for in the construction. The motion of the

pencil can be imparted by the driving weight of an
ordinary clock of good construction.

Measurement of Air Movements at High Altitudes.—
Since the instruments in common use for denoting
the direction or measuring the velocity of the wind
must be placed in the currents to be observed, they
can only be used near the earth's surface unless they
are mounted on some mountain peak; but even this
gives observations for relatively low altitudes; and
besides, under the most favourable conditions of environment, the air which is to be measured cannot be
kept free from local vitiating influences. While it has
been proposed to observe the motion of toy balloons,
the smoke of exploded shells fired vertically upward,
the direction of smoke from chimneys of factories
and boats, and to mount anemometers on captive
balloons or kites, yet very little of this kind of work
has ever been done.

The movements of the clouds offer the readiest
means of observing the upper currents, and these
have been included in the routine of observation in
most meteorological services so far as *direction* of
motion is concerned, but the measuring of the velocities has been neglected by all but a few observers.

(1) Direction of motion. The direction of approach
of clouds is usually merely estimated by the observer,
but this method, according to which most observations
are made, gives only very approximate results unless
the clouds are at the zenith. In some few observatories a cloud mirror is used, in which a ruled reflecting
surface allows the motions of the image of the cloud
to be co-ordinated, and some such apparatus as that
of Fornioni's mentioned just below, adds greatly to
the value of the observations, and relieves them of

large personal errors which exist for the ordinary method.

Fornioni's nephoscope, shown in Fig. 42, consists of a circular compass box about fifteen centimeters in diameter, with a free moving magnetic needle mounted at the centre. Above this needle, which cannot be seen in the illustration, there is a fixed horizontal glass mirror or reflector BB. On this mirror are traced the lines showing the principal points of the compass. In the sector extending from north to north-west the amalgam is removed from a portion of the mirror, so that the needle can be seen, and degree divisions are also so fixed that the position of the needle can be read. A pointer T can be moved around the whole

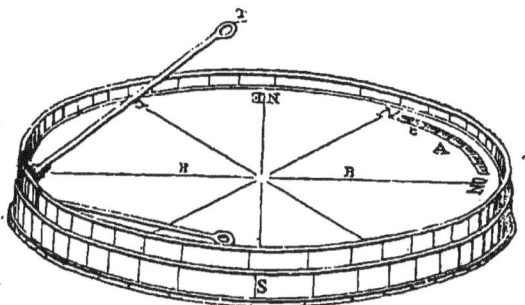

Fig. 42.

circumference of the instrument. In use, the instrument is placed horizontally and oriented with regard to the *true* meridian. T is now brought into such a position that the eye forms a third point in a visual line passing through the aperture of the pointer, the centre of the mirror, and the reflection of some chosen point of the cloud.

" The direction of the displacement that the reflected image will have undergone after a time, proportional to the velocity of the cloud, or inversely as its distance, will constitute the exact direction of the movement of the cloud."

In estimating the directions of clouds not in the zenith the following principle, as worded by Abbe, is found to be of great assistance : " All planes containing the observer's eye and the horizontal paths of the clouds moving parallel to each other, will intersect the horizon in a common vanishing point, whose bearing is the azimuthal direction of motion of the clouds."

(2) Velocity of motion.   If the position of a cloud can be determined at the beginning and end of an interval of time, its velocity is determined.   The following are some of the methods which have been proposed for accomplishing this.

The trigonometric method, proposed by Riccioli about 1650, is the one which has in its various forms promised the best success.   At the two extremities of a carefully measured base line two observers are placed, who measure simultaneously the altitude and azimuth of the cloud.   Simple computation will then fix the position of the cloud.   Two successive measures of this kind will give not only the velocity but also the direction of the wind at the elevation of the cloud.

If the observer is so located on an elevation that he can have a view of the surrounding country, he can observe the time of passage of the shadow of a cloud between two points of known distance apart, and by means of a map of the neighbourhood the directions as well as the velocities can be determined ; and if the angular velocity of the cloud is also observed, the elevation can be computed.

The nephoscopic apparatus, which was first devised by Aimé in 1845, and later other forms by Braun, Marie Davie, Linns, Cecchi, Fornioni, Finemann, and probably others, is very convenient for determining the apparent and relative motions of the clouds, which is a very desirable substitute, for the absolute motions if these cannot be observed. The principle on which this is planned is as follows :—

A cloud moves from K_1 to K_2 (in Fig. 43), and its reflection is found to move from m_1 to m_2 on a horizontal mirror, as seen by the observer's eye at o ; then "the distance m_1 m_2 is parallel to $K_1 K_2$, and is shorter than it, in the ratio of the vertical altitude O B to $K_1 K_1^1$, or in the ratio of the distances O m_1 to $m_2 K_2$,

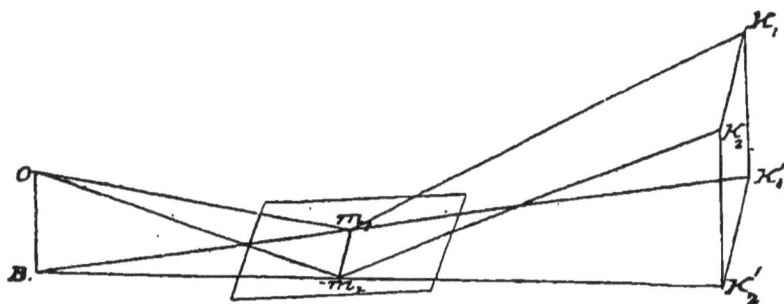

FIG. 43.

or in the ratio of the horizontal projection B m_1 to $m_1 K_1^1$." (Abbe).

Of the various forms of this class of nephoscopes which have been constructed, mention can be made of only one, that of Finemann, which is shown in Fig. 44. This instrument, like that of Fornioni, has a compass for orienting it when an observation is to be made, and also a horizontal mirror with lines drawn to the points of compass. This circular mirror has a circle of a certain radius drawn on it and concentric with it. A scale is placed at one side of the mirror and perpendicular to it ; this scale has a slow motion screw which moves it up or down, and the whole scale is so attached to a revolving band that it can be moved entirely around the edge of the mirror. In using it, the scale is so adjusted that the eye brings into the same range at the centre of the mirror the reflection of some chosen point of the cloud, and also that of the pointer. The selected point of the cloud will now move from the centre of the mirror along some one of the radial lines, and this gives the direction of the cloud motion ; and the time t is noted, which it takes to move from the centre of the mirror to the circumference of the concentric circle.

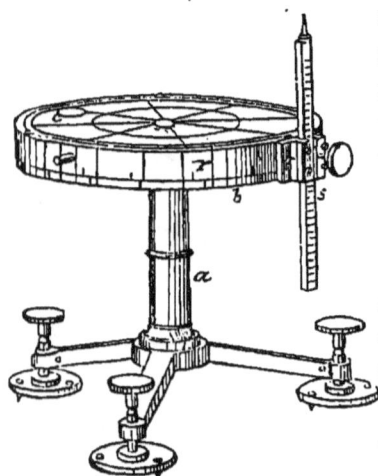

FIG. 44.

Now, as these measures are purely relative, it is convenient to use as a unit of reference the time T which it would take for the cloud to pass through a distance of 15° from the zenith, so that the observed time t must be reduced to this scale. This, trigonometry easily gives. If the radius of the circle on the mirror is chosen 26·8 millimeters, and the height b of the scale is also read in millimeters, then T is given by the simple formula, $T = 0.01\ bt$ (Abbe). These values of T can be compared, and furnish very useful relative values of the velocities of the upper air currents.

§ 4.—*Instruments for Measuring Atmospheric Moisture and Precipitation.*

In dealing with the continually repeated inter-change which goes on between the air and surface of the earth, as regards the water which is found first in the one and then in the other, and which the Germans so conveniently call the *Kreislauf* (circuit) of the water, meteorologists have found it desirable to measure, 1st, the amount of moisture in the atmosphere, 2nd, the rate at which it descends to the earth as precipitation, 3rd, the amount of this precipitation, and 4th, the rate at which it is taken up again into the air : the first is accomplished by means of hygrometers, the second and third by rain gauges, and the fourth by evaporimeters.

Moisture of the Atmosphere.—Moisture may exist in the atmosphere both as an invisible vapour (aqueous vapour), and as visible particles of water, which may or may not be frozen. In applying the methods of gases to its measurement, it frequently becomes necessary to first reduce it to a uniform gaseous condition, which may be done by changing the temperature or pressure of a limited portion of the air to be investigated ; after which the relations which laboratory experiments have shown to hold between the temperature, density, elastic pressure and volume may be applied. These relations for saturated vapour were thoroughly studied by Regnault, whose results were generally adopted ; but the rediscussion of Regnault's observations by Broch, of the International Bureau of Weights and Measures at Sèvres, has furnished the best tables showing these relations.

Early Hygrometers.—According to Hellmann, the earliest instrument of which we have record for mea-

suring atmospheric moisture is the balance hygrometer
described by Cardinal Nicolaus de Cusa in a post-
humous work published in 1472. He said that if a
considerable quantity of dry wool is placed on one
side of a balance, and stones are placed on the other
side until the balance is brought into equilibrium, experience shows that the weight of the wool increases
or decreases according as the air becomes moister or
drier. Reference is also made to the fact that the
instrument is useful for conjecturing changes of temperature. The exact date of the invention cannot be
determined ; but since the Cardinal died in 1464,
when Leonardo da Vinci was but twelve years old,
the usually received priority of the latter as the
suggestor of the hydroscope can no longer be accepted.

In 1554 Mizauld published the first notice of the
influence of humidity on the strings of musical instruments ; and a few years later Baptista Porta
called attention to the hygrometric qualities of the
" beard " of wild oats, which substance was widely
used, as we find that Torricelli applied it nearly a
century later.

The first condensing hygrometer was constructed
by the Grand Duke Ferdinand II. of Tuscany. This
instrument, as shown in Fig. 45, consisted of a reservoir, having the shape of the frustrum of a cone, and
made of cork, painted on the inside, and covered with
sheet metal within. The point of the cone was a
conical glass funnel without outlet, beneath which was
placed a glass measuring-cup. Upon filling the reservoir with snow or powdered ice, the moisture in
the air was condensed on the cooled outer surface of
the funnel, and the drops fell into the measuring-cup,

and the time was noted which was necessary to collect a certain amount of this moisture. In 1665 Ferdinand tried to establish regular observations with this form of apparatus, and sent a number of them to his friends. But the first regular continued hygrometric observations of which we have a record are those commenced by Boyle on June 30, 1666, at Oxford, the *Geranium Moschatum* being used as a hygrometric substance.

Modern Methods of Determining the Amount of Moisture in the Air.—The most important of the various modern methods of determining the amount of moisture in the atmosphere are the following : the weight of moisture for given volumes of air, or volumetric method ; temperature of saturation ; temperature of evaporation, or psychrometric method ; spectroscopic method ; the proportional saturation of certain animal and vegetable substances ; and, in a general way, the optical method.

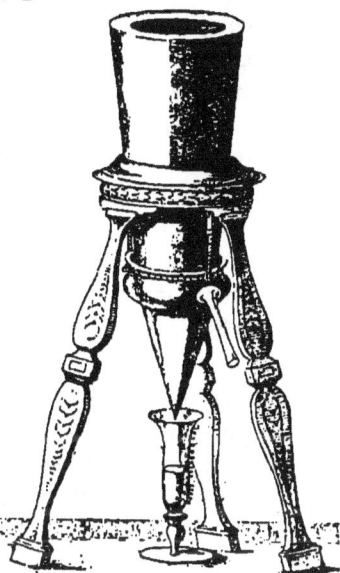

FIG. 45.

Of all these only two are in general use among meteorological observers ; viz., that in which the psychrometer, or wet and dry bulb thermometer, and the hair hygrometer are used. It is true that considerable use is made of the optical method in which the colours of the sky or clouds are noted, as an aid to weather forecasts, and the spectroscopic method is used at some few observatories ; but neither of these have been sufficiently studied to allow quantitative values to be assigned to the observations.

The most refined methods of obtaining the amount of moisture in any given amount of air, properly belong to the laboratory for chemical physics.   If air is allowed to pass through a glass drying tube, in which is placed a large surface of glass-wool, covered with strong sulphuric acid, or containing phosphorous pentoxide and glass-wool, and the feeding current passes through a small-bore tube at the rate of about two litres per hour, then a litre of the air thus dried does not contain over $0·0025$ of a milligram of water. This accuracy is of course far greater than any meteorological requirements necessitate.

The tension or pressure of the vapour in the atmosphere may be determined directly by the method of Renou (1858) and Mantern (1880), by adding as much moisture to a definite amount of enclosed air as will entirely saturate the latter.   The observed pressure of the air in the two conditions, as measured by a manometer, allows the vapour tension to be computed, as we know the pressure exerted by fully saturated air.

Edelmann (1879) invented an apparatus for measuring the pressure of a definite volume of air before and after its moisture had been absorbed by sulphuric acid ; but the apparatus which has been most successfully used by a number of meteorologists (as, for instance, by Sworykin, Pernter, and Marvin) is that devised by Schwackhöfer (1878), and which is shown in Fig. 46.   While this apparatus is somewhat cumbersome, yet it is sufficiently portable to allow of its use on scientific expeditions, being all contained in

a case easily carried by hand.   In this apparatus the
air from outside is drawn into the receiver B through

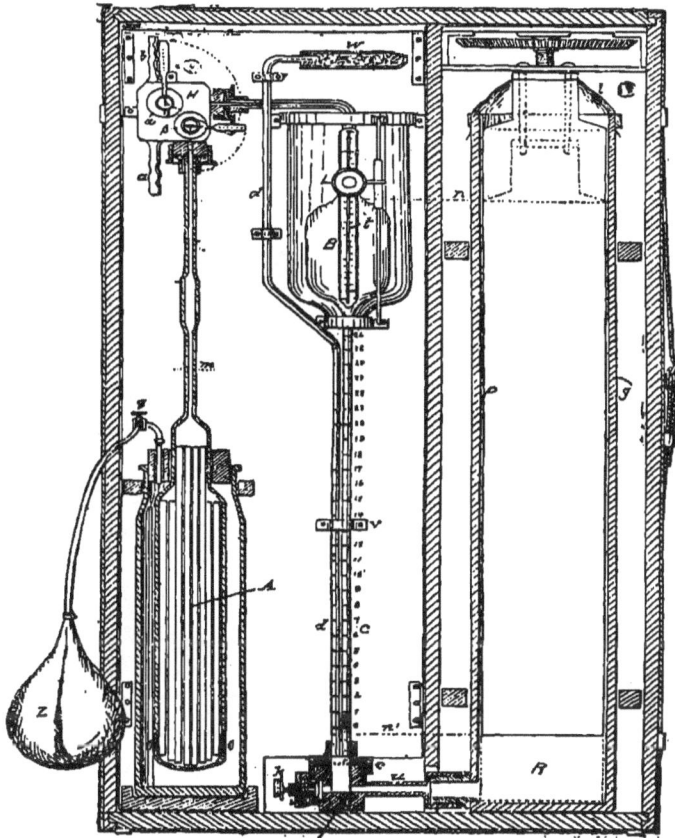

FIG. 46.

either of the glass tubes *a* or *b* (in the upper left-hand
corner).   When communication is cut off by means
of the three-way cock, the confined air fills the

upper tube connecting with the reservoir B and the graduated tube beneath to about the zero, where it is limited by the surface of the mercury which fills the lower part of this tube, the tube u and the lower part of the glass cylinder P. By means of a hand crank and screw motion at the top, on the right, the loose-fitting wooden piston F can be forced into the mercury in the lower part of P and cause it to rise in P and also to fill B to any desired height ; the suction power for filling the system B with air is obtained by starting with B full of mercury and raising the piston F. The reservoir B is surrounded by a second glass cistern containing glycerine for preserving a steady temperature, which is determined by means of the thermometer t. By means of the cocks a and β the confined air in B is placed in tubular communication with the drying chamber A, which contains glass rods and sulphuric acid. A communicates with the surrounding chamber by means of the small openings oo near the bottom. By squeezing the rubber bag Z, the air passes from it into the chamber containing the acid and forces this up to m ; and the cock g being closed, the acid remains at this level.

The air which is in B and its connecting tubes is now forced into A by pushing the mercury into B by operating the plunger F ; and the sulphuric acid is in turn forced from A into its enveloping receiver, and the cock q being opened again, the bag Z is once more filled with the air which it originally contained. As soon as the air in A has been deprived of its moisture by the action of the sulphuric acid, the bag Z is again squeezed simultaneously with the raising of the plunger F, and the acid flows again into A, and the dried air again fills B and its secondary tubes. It will be found, however, that the mercury in the graduated tube c does not stop at the same point which it did before, but reaches a little higher in the tube, as the pressure of the dried air has been lessened by the abstraction of the water vapour.

Free access of outside atmospheric pressure is maintained by means of the tube d, which has the mouth at w protected by cotton, and the lower open end inserted in the same mercury with c. In order, now, to find the volumetric change in the dried air, the mercury surfaces in the tubes c and d are raised by lowering F, until these two surfaces coincide and indicate that the pressure of the air above each is the same. A reading of the volumetric graduations on c now shows the change of volume in the air caused by abstracting its moisture. The whole operation requires perhaps ten minutes for its accomplishment, and many thousand observations can be made after once filling with sulphuric acid.

Of the hygrometric instruments in practical use, and the results regularly published, the psychrometer or wet and dry bulb thermometers, and the hair hygrometer, are the chief ones. The last of these was invented by Saussure, and as there has been very little improvement since his time, except in slight mechanical details of construction, it may be dismissed with a few words. In the best instruments human hair is used which has been freed from oil, and which has a breaking stretch of 100 grammes, and an elastic stretch of 33 per cent. The expansion in length of such a hair ten inches in length is about o·03 mm. for 1 per cent. increase in humidity. The one end is fixed, and the extension or contraction of the hair is communicated by the motion of the other end to an

index which magnifies this motion. The reading
scale being properly divided from o to 100 as representing dryness and saturation, the index is set to
100 by inclosing the apparatus in a box in which wet
cloths have been placed. It is frequently claimed
that this apparatus furnishes results within 1 or 2 per
cent. of the truth, and so it may in the laboratory, or
for an immense number of observations, but Berg-
mann has shown (Wild's *Repertorium*) that for
several years' observations the averages of the months
of lowest humidity are higher by about 2 per cent.,
and of the highest humidity lower by about the same
amounts, than when the psychrometer is used ; and
for single monthly averages differences of ± 8 per
cent. occurred. Experience also shows that slight
repairs and adjustments are necessary about once a
year, and a new instrument must be substituted, or
else the old one repaired by an instrument maker,
about once in five years.

The Psychrometer. — The employment of a wet
bulb mercurial thermometer for denoting the relative amount of moisture in the air by the amount
it is cooled by evaporation was early carried
out by Baume, Saussure, Hutton, and Leslie ;
but the modern psychrometer, composed of two
thermometers, one with a dry bulb, and the other
having its bulb covered with wet muslin, was the
idea of Ivory (1822), and August (1825), who
worked out about the same theory, independently of
each other. August, however, pursued his investigations to a much greater extent, and brought the
apparatus into practical use, and it is now known by
his name. These early theoretical developments,
although improved by Apjohn, Regnault, and others,

have now been displaced by the more thorough work of Stefan and Maxwell, the latter having given a beautiful treatment of the subject in his article on "Diffusion" in the *Encyclopædia Britannica*.

Additional necessary experimental work has been done by Sworykin in Russia, and by Marvin on Pikes Peak. Ferrel has combined the results of these with Stefan's theory, and has published, under the auspices of the United States Signal Service, what may be considered the most recent advance in the application of the theory of the apparatus to its practical use.

Method of using the Psychrometer.—In using the psychrometer the same precautions are necessary as for an ordinary thermometer, and, in addition, the wet bulb must in some way be subjected to the action of a current of air. As long ago as 1786 Saussure secured a good ventilation of the wet bulb thermometer by whirling it in a uniform but rather rapid manner, and this same or an analogous method has been independently proposed by a number of scientists, as, for instance, Belli (1830), Arago (1830), Bravais (1836), Espy (1840), &c. While it has long been recognised that psychrometer observations, and especially when the temperature is below freezing, may be very erroneous unless this ventilation does take place, yet the Italian meteorologists were the only ones who made a general practice of it until quite recently, when so much light has been thrown on the subject by Sworykin's investigations (Wild's *Repertorium*), that its general adoption is only a question of time. Within a few years it has been put into operation in the great network of stations controlled by the United States Weather Bureau.

Where this ventilation does not take place the hair hygrometer is the more accurate at temperatures below freezing.

It is necessary that the wet bulb be exposed to a current of two or three meters per second for about two minutes before an observation is made, and increasing the velocity above this amount does not make the accuracy appreciably greater.

Assmann has invented a portable apparatus (Fig. 47), in which the thermometer bulbs are screened by metal tubes through which an artificial current of air is made to pass by the employment of a pumping process. It is claimed that this will give accurate results when used in full sunshine, and it is therefore very useful for travelling observers.

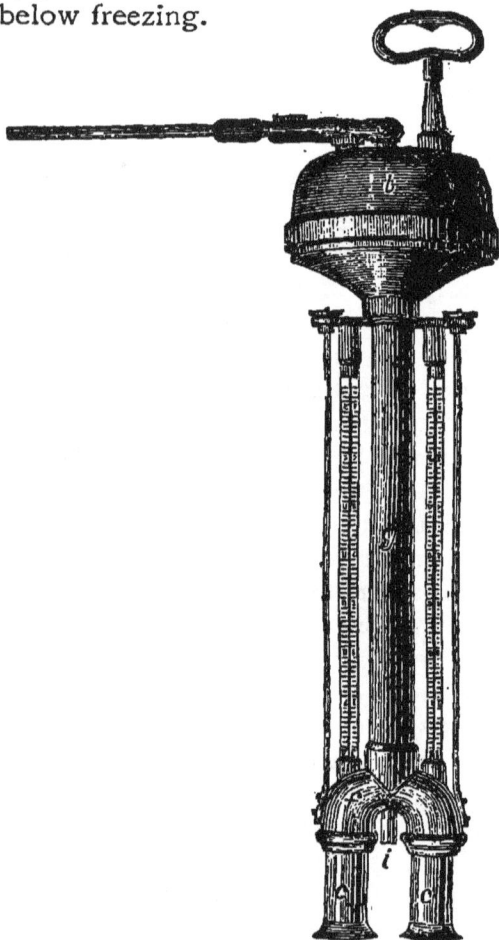

FIG. 47.

Self-Registering Hygrometers.—The hair hygrometer is easily adapted to a system of registration, but the most accurate method is that by photographing the readings of the dry and wet bulb thermometers, in which case the ventilation and wetting of the bulb must be continuous.

Measurement of Precipitation.—The measurement of precipitation, whether it be in the form of rain or snow, has been more widely carried on under the auspices of meteorologists than any other department of this science; for in most countries there are many volunteer observing stations where no other records are kept. Although a uniform idea prevails as to what is to be measured, viz., the amount of water, whether it be rain or snow, which falls upon any given horizontal area, and that this is best done by catching it in an open-mouthed receiver; yet so different have been the ways in which this receiver has been constructed and exposed, that the results obtained by meteorologists might differ for the same locality by half the amount of precipitation.

There are several distinct features which give rise to these discrepancies : the size and shape of the mouth of the catch basin, the means of preventing the escape of drops after they have once entered the mouth, and, most of all, the precautions which are taken in exposing the gauge so as to prevent the vitiating effects of other phenomena, and especially of the winds.

The question of shape and size of the rain-gauge is as much secondary to that of exposure, as the careful investigation of a thermometer is to its exposure.

So far as shape is concerned, it is now almost

11

universally the custom to use a cylindrical sheet-metal vessel with a circular mouth (of from eight to ten inches diameter is best for general use, but in tropical regions perhaps three inches is sufficient), and with a funnel-shaped false bottom, at some distance below the mouth, through which the water may pass into the lower part of the gauge, which does not necessarily retain the shape or dimensions of the catch basin of the gauge. The rim should have a clean-cut edge (not rounded), and the fall should not be straight down from the orifice, but, especially in regions of snow, there should be a double outward bevelling just below the edge which forms an encircling catch pocket to prevent the rain drops from spattering out and the snow from being blown out. Wild has also recommended the insertion of four metal strips at right angles to an axis concentric with the axis of the gauge, as a preventive against the snow being blown out after it has once entered the gauge.

There are various methods of measuring the amount of precipitation caught. It may be weighed by ordinary weighing scales, its depth may be measured by means of a graduated stick or rod, or it may be poured into a graduated glass volumetric vessel such as is used by chemists. In some of the best gauges there is a stop-cock at the bottom, through which the water may be drawn off without inverting the gauge. Snow should be melted before being measured.

In mounting a rain-gauge it must be so placed that the effect of the wind on the amount of " catch " will be least, and the loss of water, due to evaporation before the measurement is made, will be reduced to a minimum.

The effect of the wind can be in a great part annulled by protecting the mouth of the gauge by some arrangement like the encircling shield of wire-netting as proposed by Nipher in 1878 (see Fig. 48). The wind which would otherwise strike just below the mouth of the gauge is deflected downward by this netting, and the spatter of water into the gauge, which would result from the use of a solid shield, does not occur when the wire is used. It has long been noticed (perhaps first by Heberden, in 1766) that there is a de-crease in the catch of the rain-gauge with a slight elevation of exposure above the ground, and it was thought that for some unknown reason the amount of rainfall was actually less above, but it is now known that this deficiency is

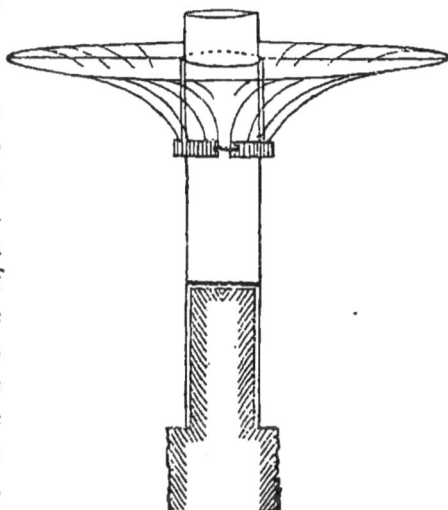

FIG. 48.

due to the action of the wind. A Nipher-shielded gauge will catch about the same amount of rain on the roof as on the ground, while an ordinary gauge may not catch more than three-fourths as much in the higher and more exposed position. It is worthy of note, however, that Wild, in planning the Pawlowsk Observatory (see Fig. 60), has so modelled the top of the tower as to deflect the wind downward and thus secure a normal condition on the roof.

The effects of evaporation can be reduced to a minimum by so placing the final receiver of the rain-gauge that the sun cannot strike it, and by making the mouth through which the water enters as small as possible.

Many series of experiments have been made to

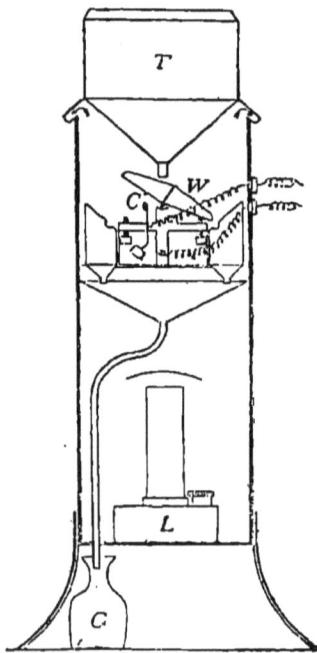

determine the best form of rain-gauge ; one of the latest being that of Hellmann (at Berlin), who wished to make the best selection for intro-duction among the stations of the recently reorganised Prussian Meteorological Bureau.

Self-Registering Rain-Gauges.—The self-registering instruments for recording the rainfall may be divided into three classes. There may be inserted in the gauge a float which will rise with the surface of the collected water, and the vertical motion of the float can be recorded by some one of the numerous methods of regis-tration ; another method, sug-

FIG. 49.

gested by Horner, is that in which the water passes from the gauge directly into a sort of small balance scoop which automatically shifts its position from one side to the other, and empties itself whenever it contains a certain amount of water, and the number and times of shifting can be recorded either electrically or mechanically (see Fig. 49) ; the third method is to make a continuous or

automatically intermittent record of the relative weight
of the water caught.

Evaporation.—While the observation of evaporation
from a free water surface exposed in a well-ventilated
thermometer shelter is to be accepted as the normal
plan for investigating this phenomenon, yet as there
are difficulties in carrying it out, the Piche evaporimeter (Fig. 50) has been proposed for ordinary
meteorological stations, since it is not
much more troublesome than a thermometer to manipulate. This instrument
consists of a graduated glass tube filled
with water and sealed at the upper end,
but with a porous paper cover at the
lower end. This porous paper constantly
presents a wet surface to the air, and the
moisture is evaporated; the renewal of
moisture in the paper being made at the
expense of the water in the tube. The
variation in the height of the water in the
tube shows the amount of water evaporated. It has been found that there exists
a fairly constant relation between the
water so evaporated, and that evaporated
from a free water surface; so that when

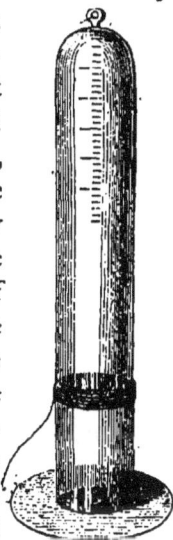

FIG. 50.

this relation has been accurately determined by
experiment, the records of the former may be readily
reduced to the standard terms of the latter. This
relative evaporation is, however, so dependent on the
wind velocity that this factor must also be carefully considered. Thomas Russell, of Washington,
United States, found that with a wind of fifteen
miles per hour five times as much water was
evaporated as in a calm. The rate of evaporation

will also depend on the nature of the evaporating disc, so that no numerical values are offered giving the factor of reduction for the Piche instrument.

The self-registering evaporimeter may be operated by means of the first or last principles just described for the rain-gauge; the water surface being artificially prepared, as in the case of the ordinary normal evaporimeter, and in case the rainfall has access to it, the amount of precipitation, as shown by the records of the ombrograph, must be taken into account. The loss of water from a Piche evaporimeter may be registered by a system of automatic weighing.

Wild has had constructed, and has used quite successfully, a self-registering combination weighing apparatus (shown in Fig. 51), which not only measures the rainfall and evaporation, but also gives the direct weight of snowfall. On the left is seen the registering apparatus, which it is not necessary to describe further than to say that B is a metal vessel open at the top, and by means of the connecting tube A is attached at G to the right arm of the balance, and is counterpoised by the weights C and D. The lower end of the pointer Z, which is in rigid connection with the balance beam, is automatically pressed against the recording sheet every ten minutes, and any shifting in the balance beam due to an increase or decrease in the water in B (and A) is therefore recorded at ten-minute intervals. The middle diagram shows on a very much larger scale the *summer* (time of no frosts) arrangement of the exposed portion of the apparatus. A passes through a hole in the roof of the small building devoted to the instrument, so that B is above the roof. A small louvred shelter, M M,

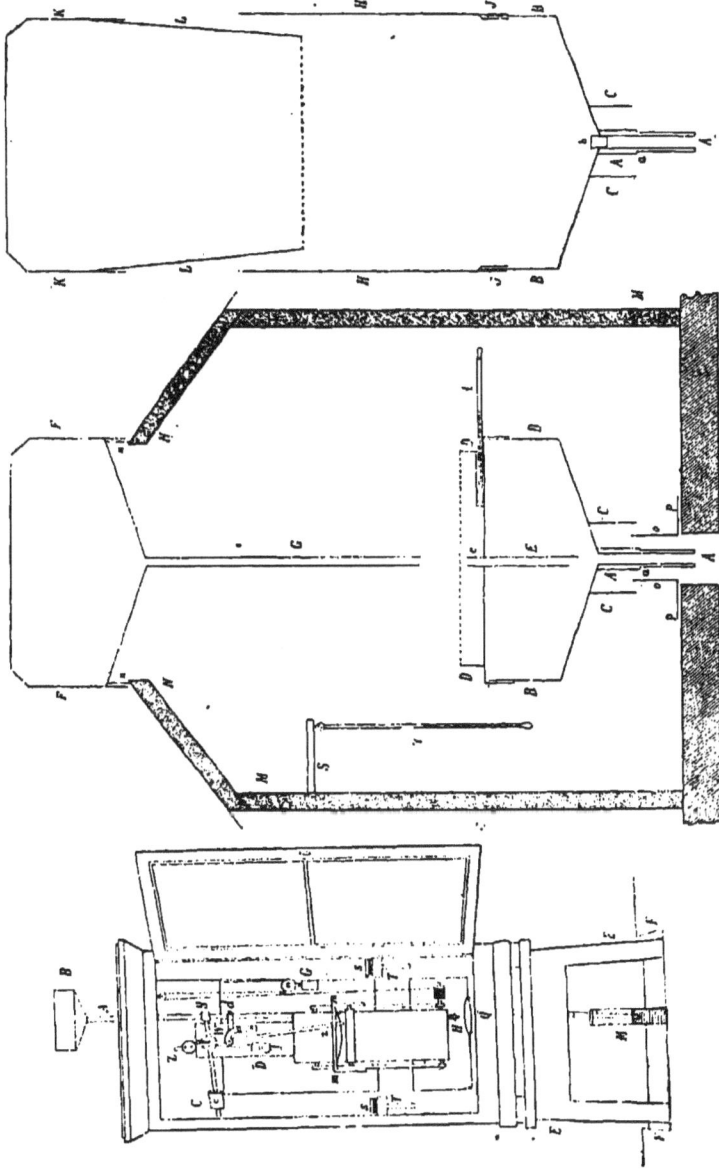

Fig. 51.

protects the vessel B.   There is fitted to the top of
B a shallow pan D, with a tube E passing through
the bottom, and with the mouth e just below the
edge of D, which is filled with water up to c.   This
water forms the evaporating surface.   At the top of
the shelter M is placed the rain collector F, from
which the water flows through G into D, and from
thence over into e, from which it is drawn off,
F through the cock H, at convenient intervals.   The
thermometer t shows the temperature of the water,
and the thermometer T the temperature of the air in
the shelter.   When rain is falling, the water in D, B,
and the secondary tubes e and A, increases the weight
on the right arm of the balance, which causes a
motion of the recording pointer.   When no rain is
falling and evaporation is taking place the weight is
less, and a retrograde motion of the pointer takes
place.   These differential motions are easily read off
from the recording sheet, and where the record is
made only once in ten minutes, in case of showers of
short duration, slight inaccuracies may enter, due to
the compensatory evaporation during a portion of the
interval of time, of the rain that has already fallen.
Sprung's method of making a continuous record by
the shifting-weight method would seem to be very
applicable in the arrangement of this instrument.
(See Barographs.)

In winter Wild replaces the metal pan D by the
close-fitting cylinder H (as shown on the right-hand
side of the illustration), and exchanges for the
collector F the bottomless collector KKLL.   The
snow now falling into the open mouth of K passes
downwards into B, and can become heaped up into
H as it accumulates, but before it reaches up to

where it can touch L it must be removed from the gauge. Also in winter the mouth of the tube A is stopped up by means of a plug, so that it may not become filled with ice and become troublesome.

§ 5.—*Measurement of Clouds and Sunshine.*

Clouds.—The measurements of degree of cloudiness which are made by meteorological observers in general are exceedingly rough personal estimates of the relative portion of the sky which happens to be cloud-covered at the time of observation. It is now almost universally the custom to consider the whole dome of the sky from the zenith to the horizon as represented by 10, and to represent the degree of cloudiness by a scale from o (clear) to 10 (entire sky overcast). The matter of estimating the relative portion of the sky which is covered is left almost entirely to the judgment of the observer, who, unaided by any apparatus, glances at the clouds and makes an estimate that a certain amount of cloud is visible ; and so he enters in his observation book, 1 for a small amount of cloud, 2 for a little larger amount, &c., up to 10.

In the accompanying diagram (Fig. 52), taken from a paper by Laurenty in Wild's *Repertorium*, vol. x., are shown the relative amounts for each tenth of the sky from the horizon to the zenith. For instance, a band extending around the whole horizon to a height of 5·7° would be 1 on the scale of o–10, and would be just equal to the portion of the sky overhead extending to 25·8° from the zenith.

It may generally be said that the difficulty of an estimation increases with the increase of the zenith distance ; and some meteorologists go so far as to take no account of the clouds below the altitude of

45°. And this last rule has the stronger claim, when it is considered that when a cloud is seen in the horizon we observe its vertical extent, while in the zenith only the base is to be seen.

Very few measures of thickness of clouds have been made, so that the great majority of observations merely denote in what proportions the various portions of the earth are screened by the clouds without regard to the density of this screen.

Perhaps the most efficient aid to judging of the

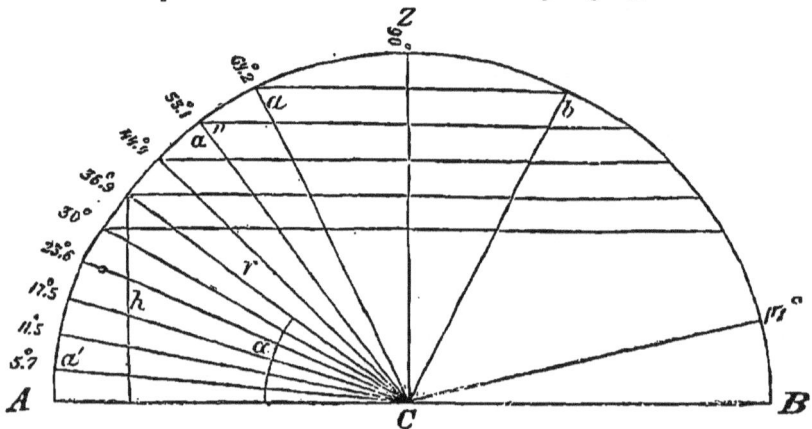

Fig. 52.

portion of sky covered by clouds is a cone-shaped framework of wire, such as is used at Pawlowsk. The eye of the observer being at the apex of the cone, the successive circles of wire as outlined against the sky divide it into zones, and the amount of cloud in each of these can be quite accurately estimated.

Altitude of Clouds.—It is a very important matter to obtain the elevation of clouds above the earth's surface, not only in order to measure the velocity of

the upper-air currents, but also to study the manner
of cloud-formation ; for each of the recognised types
has a distinctively physical cause, some of which are
known with considerable certainty, but others are as

FIG. 53.

yet mere speculation. A number of methods have
been proposed, the most successful of which promises
to be that of simultaneously photographing the same
cloud-region by means of two cameras mounted on
alt-azimuth stands at the ends of a carefully measured

base line. Knowing the altitudes and azimuths of any
common points on the two photographs, and the
length of the base line, a simple trigonometric computation (rapidly carried out by prepared tables or a

FIG. 54.

graphical process) gives the elevation, and the photograph shows exactly the kind of cloud which has
been observed. Such an instrument is the photo-
nephoscope (see Fig. 53), used at the Kew Observatory, and mounted by Abney about 1883.

The instrument in actual use is shown in Fig. 54, where one of the complete base stations is to be seen. The large tripod supports an overhead framework, which is necessary in adjusting the instrument. Communication is kept up between the observers at the two stations by means of signal flags and telephone connection. In fact, the telephone admits of such ready communication between the two stations that very useful observations can now be made directly by the observers without the application of photo-

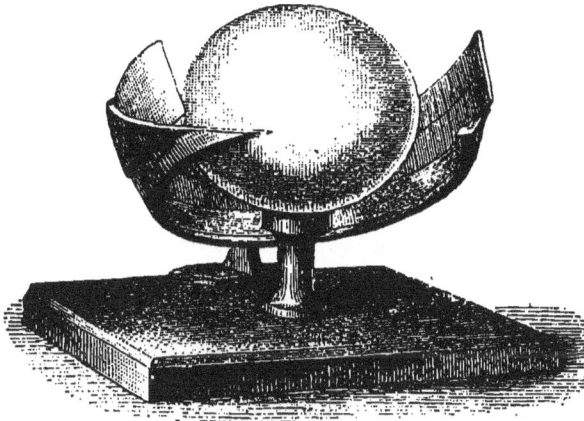

FIG. 55.

graphy, because the identification of some chosen point can be made with considerable certainty, which was an impossibility with slower methods of communication.

Sunshine.—The Campbell-Stokes sunshine recorder is the only instrument which has been used to any extent for the purpose of measuring the amount of sunshine ; and beyond its introduction at some of the stations controlled by the English, its use is almost entirely limited to some few large observatories. This instrument (see Fig. 55) consists of a spherical lens with the focus at a little distance from the surface. Partially encircling the lens is a strip of mill-board, so bent that its surface, held by means of a framework, coincides with the focal distance of the lens. The sun's rays passing through the lens burn a hole in the mill-board, and as the sun shifts its position the burning point on the board is likewise shifted, so that a narrow strip is burnt out. When clouds intervene this burnt strip becomes a broken line, so that its burnt part represents the time of sunshine and the unburnt part the time of cloud. By means of a graduation on the paper these times are converted into hours and minutes; but there is always a slight inaccuracy due to lack of definiteness in the edges of the burnt places, and the passage of clouds which obscure the sun for only three or four minutes is hardly noticeable.

A more accurate but much more troublesome method is the exposure of sensitised paper, which is affected by the action of sunlight.

Of the various other methods which have been proposed, and, as has been the case with many other devices for making meteorological observations, put into use by individual meteorologists, mention will only be made of Pickering's suggestion to apply in this case a variation of the idea which he has used so successfully with his meridian star photometer at Cambridge, U.S. The pole star has very little motion, and if a slowly-moving photographic plate is exposed toward it, a continuous record will be obtained on clear nights; but this will be broken by the passage of clouds which hide the star from view. Rotch has applied this method successfully at the Blue Hill

Observatory in Massachusetts. It will be seen that this method is supplementary to the Campbell-Stokes sunshine recorder, as the one can be used only by night and the other by day; but there would still remain the twilight hours unrecorded.

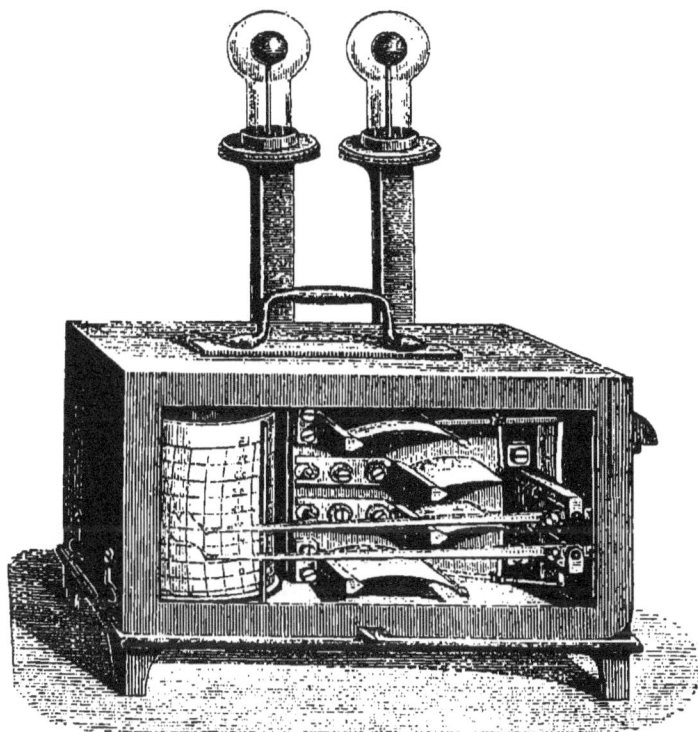

FIG. 56.

For obtaining a record of the times of cloud and sunshine the registering actimometer (Fig. 56) appears to be an especially valuable instrument. This consists essentially of two thermometers, one with a plain bulb and the other with blackened bulb, but each placed in a vacuum chamber of glass. The form of registration shown in Fig. 56 is that made by Richard, of Paris. The record of the blackened bulb thermometer would show almost instantly the changes from cloud to sunshine ; and in addition a computation would show the relative intensity of the sunshine on successive days. A good illustration of this is the use made of the actinometer by Upton on the Solar Eclipse Expedition to the Caroline Islands in 1883. By adopting Ferrel's formula for reducing actinometer observations he was able to show that on the day of the eclipse there must have been a slight haziness of the atmosphere, as the actinometer showed a less intensity of sunshine than was observed on other clear days, the amount being quite constant for some other days.

§ 6.—*Meteorological Observatories.*

History of their development.—For several centuries the word "observatory" conveyed only the idea of an astronomical observatory, and this meaning has become so deeply rooted that even now very few people would ask, "What kind of an observatory?" if reference is made to some institution under this name. In fact, in the earlier days of meteorology, the observations which were made and the apparatus used were of such a simple character that the place of observation was that which suited personal convenience and was freely changed at will, requiring no specially constructed building which distinctively marks the location of a telescope. The early observers were mainly physicians, teachers, cultivators of natural science, ecclesiastics, and others, who had no special training in the use of instruments of precision. But astronomers also made observations and kept careful record of the meteorological elements; this they could easily do, as it required but a few minutes' time at stated intervals, and would not interfere with their regular work, of which it gradually became a part.

But as astronomy became a more exact science, astronomers increased also the accuracy of meteorological methods; and it was to them, more than to the early experimenters in physics, that we owe the gradual development of the modern systems of meteorological observing. In later years the introduction of observations of temperature and pressure of the atmosphere into astronomical computation rendered imperative the making of accurate observations of these elements at the various observatories. The permanency of location of the astronomical observatories has greatly increased the value of the meteorological observations made at them, and they furnish reference or normal conditions for the reduction of irregular observations at different places in their neighbourhoods. The observations made at the Paris and Greenwich Observatories furnish excellent examples of these long-continued records. For instance, the rainfall observations at Paris extend back over two hundred years.

In the astronomical observatories, too, the instruments were of the best construction, and these came to be used as standards for those of inferior construction and stability used elsewhere. This was particularly true for barometers.

A long continuance of this condition made meteorology seem to be a part of astronomical work; and

so, when national meteorological undertakings were
commenced, the chief astronomical observatories were
deemed the proper places for such centralisation as
was necessary in carrying on this work : and when
delicate self-registering apparatus came into use, the
astronomers were about the only class of regular
observers who had acquired the skill necessary for
their manipulation. It was about the same with the
making of observations of the earth's magnetism.
Perhaps the most perfect development of this relation was that existing at the Paris Observatory up to
within about fifteen years.

The interest which was aroused in observations of
the phenomena of terrestrial physics by the work of
Humboldt, about the beginning of the century, and
which was continued a little later by Kupffer, Kaemtz,
Sabine, and others, resulted in the gradual divorce
of meteorology from astronomy ; and it has become
almost complete, so far as outside influence is concerned, although many astronomical observatories
still continue to publish meteorological observations.
While the independent organisation of many of the
meteorological services was effected before this time,
yet it was not until about the year 1850 that the
great meteorological institutions, which have now
become quite general, began to have an actual
existence. The Central Physical Observatory of St.
Petersburg and the Kew Observatory of the Royal
Society are about the best examples we have of the
newly-created institutions of that period.

Passing over the developing process which such
meteorological institutions have undergone, but noting
the fact that they early extended and still give to the
department of magnetism the same nurture which

meteorology formerly received from astronomical observatories, their present general features will be outlined.

Central Observatories. — In each country which possesses a network of meteorological stations, there is a central institution, from which the work of the stations is directed and controlled, and to which their observations are sent for publication and discussion, or use collectively. These institutions have in general the same plan, but each one presents some special

FIG. 57.

development of one or more features which gives it a distinct character of its own ; and in noting the various fields of usefulness, special attention will be called to the institution showing them most markedly, although this last must be to a great extent a matter of personal judgment.

Before proceeding with the description of these central observatories, attention is directed towards the Paris Municipal Observatory of Montsouris (Fig. 57). The work of this observatory includes not only the study of the air from the meteorologist's standpoint, but also from that of the chemist ; and it is an institution which should be duplicated in every large city in the world.   There are probably no more interesting reports of such work than the series of annual volumes devoted to " Meteorologie—Chemie —Micrographie. Applications à l'Hygiene," which is published (by Gauthier-Villars, Paris) under the title, *Annuaire de l'Observatoire Municipal de Montsouris.*

As these institutions were primarily needed as depositories for the collection and publication of observations from scattered observatories, they all perform this work, and in most cases it is done as well as could be wished.   As the extent of country to be covered by each, and the corresponding number of tributary stations is very different for the different countries, this work is carried on by various systems, ranging from the method of the United States Weather Bureau, Washington, in which a large force of workers must necessarily perform their prescribed duties in a more or less mechanical manner, to that of the Austrian Bureau at Vienna, where similar work is done by, or at least comes directly under the eye of, scientists, any of whom could creditably direct the affairs of such an institution.   In fact, this idea seems to be a governing one at the Vienna Central Meteorological Institution ; for it is the most conservative of all the great central offices ; its official work consisting of the making of local observations, the oversight of its secondary land stations, the preparation of the data thus collected for the international form of publication, and the issuing of weather forecasts from a city office totally removed from the main Observatory at Höhewarte ; and probably nowhere else can one find the unity and air of cosiness which is to be met with there, and which is due to the housing under one roof of about all that pertains immediately to the institution. At this observatory nearly all of the upper portion of the building is used as dwelling apartments for the chief scientists, thus allowing them to have easy access to the splendid library and to the work-rooms at all times. It must be remembered that this observatory has been a centre of international interest, and its influence widely felt through the agency of the meteorological journal which has been edited there for nearly twenty-five years, and although the official tabular publications contain little personal work of the director, Dr. Hann, yet his wonderful industry as a meteorological writer and critic, as shown in other channels, is unsurpassed.

While some of these Central Bureaus are so located as to combine the work of centralisation and that of a well-equipped and located observatory, such as is to be found at Höhewarte, Vienna, and at the Seewarte, Hamburg (Fig. 58), yet in most cases it has now come to be considered desirable to separate the two kinds of work by having a city office for administrative work, and an affiliate observatory at a sufficient distance in the country to ensure its best and most successful operation. Thus we find Kew Observatory to a great extent under the direction of the London Meteorological Office, Pawlowsk related to the Central Physical Observatory at St. Petersburg, and Parc St. Maur under the Central Meteorological Bureau at Paris. This is certainly the proper arrangement for securing the retirement and uninterrupted routine which are so necessary for the highest development of observatory work,

Special Work of some Observatories.—The work of verification of instruments to be used in making observations is a most important function of these central institutions, but such a department of work is much more fully developed in some than in others. No observatory has carried this on with a greater degree of success than that at Kew (Fig. 59), which is known all over the world by the certificates it has issued with instruments for use in nearly every land. While

FIG. 58.

this work at Kew is not of a higher order than that at some other institutions, as, for instance, the Central Physical Observatory at St. Petersburg, yet its influence has perhaps equalled that of all the others combined, especially in the extension of the matter to questions of practical life. About twenty years ago it was decided to take advantage of the reputation already acquired by the observatory, and establish a bureau for the testing of certain kinds of apparatus,

but principally those used in terrestrial physical
measurements, and to make a regular charge there-

FIG. 59.

for.   This work at Kew has grown to enormous
dimensions, and the number of instruments verified

in 1890 was almost 20,000, of which about 12,500 were clinical thermometers. The revenue derived from this department defrays almost, if not quite, all the expenses of the whole observatory. The success at Kew has emboldened other observatories to follow its lead in this partially commercial enterprise.

The work of weather forecasts and storm warnings is carried on with the highest degree of promptitude and efficiency by the Washington Central Office of the United States Weather Bureau. This is due to the following causes: the work of predictions has been hitherto the chief work of the service, to which all else was secondary; the entire time of the observers, on whose telegraphic reports the forecasts are based, is controlled by the service; a large amount of money is available for telegraphic purposes, and the right of precedence in the use of wires is maintained; and as most of the storms come from the westward, they cannot approach the Central and Eastern United States so unexpectedly as those arriving on the European coasts. Of the European weather forecasts and storm warnings, those by the English office at London, and the German office at Hamburg, are the most important and the most efficiently conducted.

But it is as meteorological observatories that these central offices deserve most attention; as it is to be expected that one will find there the model organisations of the country not only as regards methods, but also apparatus, and that the highest ideal of what constitutes observational meteorology will be realised. In many cases it has been impossible for the present meteorologists to carry out their own ideas very fully, as they have been obliged to make use of apparatus already provided, and an adequate sum of money to carry on the work, as it should be done, is not at their disposal.

The Observatory at Pawlowsk, Russia.—As a clearer idea of the work done at one of these observatories can be obtained by describing an actual case, the one which seems to me to be the nearest to the ideal observatory has been selected for this purpose.

The Pawlowsk Observatory came into existence under the most favourable circumstances. The director, H. Wild, whose ideas are embodied in it, had devoted a great amount of time during the fifteen years preceding its foundation to the study of an improvement in meteorological apparatus. His equipment of the astronomical observatory at Berne with meteorological apparatus (before and about 1865) marks an important epoch in the history of applied meteorology; and when, a few years later, he became director of the Central Physical Observatory at St. Petersburg, and consequently also of the Meteorological Service of Russia, he instituted equally important improvements in the methods of using meteorological apparatus, and especially in the manipulation of data furnished by self-registering instruments. But the facilities at his disposal were not just what were necessary to carry out fully his ideas: the opportunity for this came with the building of the Pawlowsk Observatory in 1876. Situated in a corner of the park of the Grand Duke Constantine, about twenty miles south-east of St. Petersburg, and one and a half miles from the village and summer resort, Pawlowsk, the observatory is located in a partially cleared grove of fir trees near the juncture of the wooded and open country. The land around is level or slightly rolling, and the buildings are so situated that the exposition of various instruments is good, except in the case of the anemometer, which may not be at a sufficient distance above the level of the tree-tops to feel the full force of the wind.

In order to obtain proper drainage, advantage was taken of the natural slope of the land to draw into an excavation the surface water and thus form a pond, which could also serve for evaporation measurements on a large scale.

The method of detached buildings which has grown in favour with astronomers, was adopted as the plan of arrangement. The grounds lie between two highways, and on the side of the main entrance are placed the half dozen economic buildings, residences for the staff, &c., and at a distance of about 150 feet from the road is placed the main observatory building (rear view, Fig. 60). At some distance to the rear of this building are various minor buildings for the exposure or housing of single instruments, while still further back and near the centre of the grounds are the main magnetic buildings, one of which, above ground and of wood, is for absolute magnetic measures, and the other is an under-ground (or rather, earth-covered) chamber, and contains the self-registering magnetic apparatus. The buildings and equipment cost originally about £15,000.

The principal building, most of which, as is usual in such institutions, is mainly given up to the accessory work rather than direct observation, has the floor of the lower story slightly elevated above the ground, and in this story are the porter's dwelling room, a mechanic's work-room fitted up for metal and wood working, a physical laboratory room with the

FIG. 60.

necessary comparator for measurements of length,
scales, arrangements for magnetising steel magnets
and investigating them, a chemical laboratory, a
photographic laboratory and dark room, a battery
room, and at the centre there is placed a central heating apparatus. Above this last, in the centre of the
second story, is a small vaulted rotunda, in which are
kept those instruments which require a constant temperature, viz., self-registering barometers, clocks, chro-
nometers, &c. ; and surrounding this central room is
a room for the superintendent, one for the physicist
or senior observer, two large ones for the younger
observers and library, and one for the current observer;
and this last room contains a number of meteorological
instruments and self-registers, and is connected with
the magnetic observatory by telephone. The roof of
the tower is 23.5 metres above the ground, and serves
as a place for observation of clouds, as well as
for the exposition of a sunshine recorder, rain-gauge,
registering wind-vane, and anemometers ; these last
being mounted from 3 to 5 metres above the platform.
Special attention is called to the peculiar construction
of the tower near its roof, that shape having been
chosen which was considered as best adapted to prevent vitiating effects of winds deflected by such an
obstacle as the tower.

Attached to the north side of the building, near the
ground, is a large instrument-shelter open towards the
north (see Fig. 60). This shelter is 5 metres high,
and 5.3 metres long and wide, and the roof is double,
and slightly inclined. The side nearest the building
is made up of glass windows and doors, which allows
space for a corridor between this and the wall of the
building, and in case of extra refinement of observa-

tions the instruments can be read through these windows by means of telescopes. The sides of the shelter are open to a height of 1·5 metres, but above this are closed by adjustable louvred blinds. The various instruments are approached by means of wooden grating walks ; otherwise, the shelter is floorless to the sod beneath.

The instruments are mounted on separate piers or piles. In this large shelter are placed : the special thermometer shelter for direct observation of the temperature and humidity of the air by means of the psychrometer, hair hygrometer, maximum and minimum thermometer, self-registering thermograph and hydrograph ; the inner shelter for the direct observations having a ventilating pipe reaching above the roof of the main shelter. A photographic psychrometer of the Adie (Kew) pattern, and a scale evaporimeter for direct observation complete the list of instruments exposed here. To the north of this main building is a large cleared space (partly shown in the foreground of Fig. 60), in which are located various isolated instrument shelters and other smaller buildings in which instruments are mounted.

I am indebted to Director Wild and the Superintendent, Dr. Leyst, for the following details concerning the Pawlowsk Observatory. They show the methods of a well-appointed meteorological observatory with a completeness not heretofore published.

Personnel of the Pawlowsk Observatory.—To conduct a completely equipped observatory the time of at least three persons is necessary, in order to carry on the meteorological observations and reduce them for publication ; the addition of another person will permit the parallel work to be done for observations

of the earth's magnetism. At Pawlowsk the staff
consists of a superintendent, a senior observer, three
junior observers, a skilled mechanic, and one or two
persons to perform the duties of messenger, &c.

The superintendent, in addition to his responsi-
bilities as such, personally supervises the working up
of the magnetic observations, while the senior observer does the same for the meteorological observations.
These two positions are filled by men of decided
scientific talents and attainments, who have had a
careful preliminary training at the Central Physical
Observatory ; the secondary position having in fact
served as a stepping-stone to higher positions of
scientific trust for a number of men. The younger
observers have been men of more moderate attainments, and their work is entirely a routine ; and consists of the making of prescribed observations, and
doing the laborious part of putting these in a form
for use in the discussions made by their senior
officers. There is also a large amount of observing,
such as the determination of instrumental constants,
investigations of new instruments and methods, which
devolves upon the superintendent and senior observer.
During the summer months, when the Director resides
at Pawlowsk, a great amount of such work is carried
out, or planned for future development.

METEOROLOGICAL INSTRUMENTS IN USE AT THE
PAWLOWSK OBSERVATORY IN MAY, 1891.

I. AIR PRESSURE.

For Direct Observation.

1. Mercurial barometer, Fuess No. 247, 1st class,
   for controlling observations and manipulation
   of the barograph.

2. Mercurial barometer, Turettini No. 58, 2nd class, for observations as at stations of second order.

Self-Registering.

3. Mercurial balance barograph, Wild-Hasler, with temperature compensation and arrangement for overcoming the adhesion of mercury to the glass; for hourly values of pressure to be published. Registers electrically every ten minutes.

4. Mercurial balance barograph, Sprung-Fuess, with sliding weight, tapping apparatus for overcoming the adhesion of the mercury to the glass, without temperature compensation. Registers continuously. For control.

5. Aneroid-barograph of Richard Bros. Registers continuously. For control.

Nos. 1 to 4 stand in the round central vault-like hall with almost constant temperature; No. 5 stands in the room for day service. All at the same altitude.

II. AIR TEMPERATURE.

For Direct Observation.

Mercurial enclosed scale (Fuess) thermometers of Jena glass, spherical bulb, graduated to $\frac{1}{5}°$ C.

1. In the sheet-zinc small shelter in the normal shelter (Wild's normal exposure), for stations of second order. Ventilation current $2\cdot5$ m. per second.

2. In the sheet-copper shelter with lattice floor
   in Wild's new shelter. Ventilation 5m. per
   second. For control and comparison.

3. In the case of the thermograph B, Wild-Hasler,
   on the grass plot; for the manipulation of
   the thermograph registers.

4. In the shelter of the thermograph C, Wild-
   Hasler, near the house; for the manipulation
   of this thermograph.

5. In the sheet-copper shelter, near the house,
   ventilation 2½m. per second; for comparison
   with the exposure on the house.

Self-Registering.

6. Thermograph B of the form Wild - Hasler,
   bimetallic spiral (platinum-silver), electrical
   control register every 10 minutes, natural
   ventilation through walls made of louvred
   blinds and a 3m. high black painted chimney,
   in Wild's shelter, on the open grass plot; for
   the publication of hourly temperature values.

7. Thermograph C, Wild-Hasler, same as 6, but
   brass-steel spiral and shorter ventilation
   chimney, in the thermometer pavilion, with
   walls of louvred blinds, open towards the
   north, on the north side of the main building. For control.

8. Thermograph by Richard Bros., in Wild's normal shelter, with a Bourdon tube, registering
   continuously for 7 days on a single chronograph sheet.

Extreme Thermometers.

9. Maximum : Mercurial ther- ⎫ In the shelter
   mometer with long de- ⎪ with
   tached column of mercury. ⎬ thermograph B,
   Minimum : Spirit thermo- ⎪ Wild-Hasler.
   meter with index ⎭

10. Maximum and minimum thermometer, same
    as 9, in the shelter of thermograph C, Wild-
    Hasler.

> Nos. 1, 2, 3, 6, 8, 9, in shelters, distant
> 40–50 metres from trees and buildings ;
> Nos. 4, 5, and 10 in the same as No. 7.

III. EARTH TEMPERATURE.

1. Under pure sand surface of the fineness of
   quartz sand, and free from sod and snow.

   > 1 thermometer (common) on the surface.
   > 4 thermometers (common) in vertical glass
   > tubes, the bulbs in brass hoods, at the
   > depths of 0·05m., 0·10m., 0·20m., 0·40m.
   > 4 thermometers in ebony tubes at the
   > depths 0·40m., 0·80m., 1·60m., 3·20m.
   > 4 thermometers in clay tubes at the depths
   > 0·40m., 0·80m., 1·60m., 3·20m.
   > 1 maximum and 1 minimum thermometer
   > (like II., 9) on the surface.

2. Under the natural surface (in summer under
   sod, in winter under snow). The ground of
   the fineness of quartz sand.

   > 5 thermometers in ebony tubes, which pro-
   > ject one metre out of the ground and are
   > painted white, at the depths of 0·00m.,
   > 0·20m., 0·40m., 0·80m., 1·60m.

3. On the natural surface (in summer on close-cut
   sod, and in winter on snow).
   1 common thermometer.
   1 maximum and 1 minimum thermometer,
   like II., 9.

IV. RADIATION.

1 radiation thermometer by Hicks (black bulb in a
   vacuum).

V. MOISTURE.

For Direct Observation.

1. Psychrometer   } in Wild's normal shelter
2. Hair hygrometer   }   (see II., 1).
3. Psychrometer in Wild's new shelter (see II., 2).
4. Psychrometer in sheet-copper shelter near the
   building (see II., 5).
5. Psychrometer   } In the shelter of the thermo-
6. Hair hygrometer   } hygrograph B (see II., 6).
7. Psychrometer   } In the shelter of the thermo-
8. Hair hygrometer   } hygrograph C (see II., 7).

Self-Registering.

9. Hygrograph B, Wild-Hasler, 1 hair, electrical
   contact register every 10 minutes (see II., 6) ;
   for hourly values of absolute and relative
   moisture.
10. Hygrograph C, Wild-Hasler, 1 hair (see II., 7) ;
    for control.
11. Hygrograph, Richard Bros., fitted with hair
    registering continuously for 7 days, in the
    Wild normal shelter. For control.

12. Hygrograph, Richard Bros., fitted with horn, registering continuously for 7 days. For comparisons.

VI. Precipitation and Evaporation.

For Direct Observation.

1. 1 pair Wild's rain-gauges for observations as at stations of second order.
2. 1 pair Wild's rain-gauges with Nipher's shield, for special observations (published in the *Annalen*) and control.
3. Wild's balance evaporimeter in a louvred shelter on the open lawn.

Self-Registering.

4. Ombro and atmograph, Wild-Hasler, with separate winter and summer collectors, a balance like that for Wild's balance-barograph, electrical contact registers every 10 minutes, in a shelter on the open lawn, evaporation basin in a louvred shelter on the roof. Used for the hourly record of the evaporation and precipitation. Is controlled by the rain-gauge, and balance evaporimeter.

All these instruments stand near each other on the open grass plot.

VII. Direction and Force of the Wind.

For Direct Observation.

Anemograph, Oettingen-Schultz. Direction in degrees of the vane on the tower. For control on the wind registration by electrical transmission of 8 recording pencils (16 direc-

tions) in the observing room below. Force,
read on the apparatus in the tower, for
control on the indicator (electrical transmission) in observing room below.

Self-Registering.

Anemograph, Munro, registers continuously on
metallic paper the force and direction of the
wind.

VIII. CLOUDINESS.

For direct observation, a cone formed of wires,
and including 120°, in which the field of
observation is divided into four parts by cross
wires directed towards the cardinal points of
the compass. For observations of the whole
sky from the tower, as at stations of the
second order, no instrumental means are
employed.

An instrument for determining the angular
velocity and direction of the clouds, by Fine-
man's method, and a cloud camera with an
angular opening of 90°, for direct observation
and photography (by Hesekiel, Berlin) are
not in regular use.

IX. SUNSHINE.

1. Heliograph by Campbell (sunshine recorder) ;
   for the publication of hourly values.
2. Heliograph by Jordon, photographic, for control.
   Both instruments stand on the platform of the
   tower, with free exposure.

X. Electricity of the Air.

Electrometer, Mascart-Carpentier, Wild's water
collector with constant water-pressure, and
Mascart's form of foot receiver.
Wild-Edelmann scale telescope which is, at the
same time, arranged for direct readings and
for photographic registration. A transport-
able electrometer by Exner, Vienna, is not
in regular use.

XI. Height of Snow.

6 metal strips divided into centimeters ; one
stands in a wooded exposure, the other 5 in
an open exposure.

DAILY ROUTINE OF THE METEOROLOGICAL OBSERVERS AT THE PAWLOWSK OBSERVATORY (IN MAY, 1891).

The three younger observers have, alternately, 24
hours of observing service, another 24 hours of sub-
service, and the third 24 hours free. The observer in
sub-service is regularly on duty from 9 A.M. to 4 P.M.,
except from 12 to 1. The one in service is on duty
from 6.20 A.M. to 9 A.M., after being awakened shortly
before 6 A.M. by a night watchman observer, and is
released at 9 A.M. by his successor, who remains on
duty till 2.10 P.M., and then after dinner at 4 P.M. is
again on duty till 10.10 P.M. ; if there is a thunder-
storm in the night he is awakened by the night
watchman. From 2.10 to 4 P.M. his place is supplied
by the observer in sub-service.

The daily order of making the regular observations
is as follows :—

6.20–6.38 A.M. Revision of instruments: in winter the cleaning of frost from the thermometers; moistening the psychrometers; comparison of the pocket-watch of the observer with the standard clock of the observatory.

6.38 A.M. First reading of Oettingen's anemometer.

6.39 A.M. Reading of the electrometer at rest, and connecting it with the water collector.

6.43 A.M. Observation from the tower of the quantity, form, and direction of clouds for the whole heavens.

6.44 A.M. The wind direction read in degrees from Oettingen's anemograph.

6.47 A.M. Observation of atmospheric electricity.

6.48 A.M. Second reading of Oettingen's anemometer.

6.50 A.M. Reading the attached thermometer and control barometer (Fuess) of the 1st class.

6.52 A.M. Reading attached thermometer and station barometer (Turettini) of the 2nd class.

6.55 A.M. Reading the psychrometer before ventilation in the copper shelter on the north side of observatory.

6.56 A.M. Reading radiation thermometer, and thermometer on the sod.

6.57 A.M. Reading psychrometer after two minutes of ventilation.

6.58 A.M. Reading psychrometer and hair hygrometer before ventilation, in the zinc shelter within the isolated normal wooden shelter.

7.00 A.M. Reading same as preceding, after ventilation.

7.01 A.M. Reading psychrometer, before ventilation, in the sheet copper shelter with lattice floor, in the new shelter.

7.02 A.M. Clouds observed through a cone of 120°; amount, form and direction of motion.

7.03 A.M. Reading psychrometer, after ventilation, same as at 7.01.

7.04 A.M. Reading earth thermometers in a sand surface free from snow and sod, at the depths 0·00m., 0·05m., 0·20m., 0·40m., 0·80m.

7.06 A.M. Reading 4 ground thermometers under the natural surface (sod or snow) at the depths 0·00m., 0·20m., 0·40m., 0·80m.; measuring the height of the snow in winter.

7.10 A.M. Reading psychrometer and hair hygrometer in the shelter of the thermo-hygrograph B (Wild-Hasler) on the lawn.

These are used for the mark-
ings, at 7.10, on the thermo-
hygraph and Ombro-atmo-
graph.

7.12 A.M. Temperature, of the water in the
evaporation vessel of the
evaporimeter, in the evaporation vessel of the Ombro-
Atmograph, and the air in
the shelter with these.

7.13 A.M. Exchange of the rain-gauge, and
measure of the precipitation ;
if snowy, after the melting of
the snow.

7.19 A.M. Wind direction (to 16 directions), and the first reading
of the indicator of the wheel-
work on the anemometer
Oettingen-Schultz.

7.20 A.M. Second reading on this wheel-
work, and repetition of the
wind direction.   After this
the reference marking is
made on the balance baro-
graph Wild-Hasler.

7.22–7.50 A.M. Reduction of the observations
just made and entering them
in the proper columns.

7.50–8.10 A.M. Magnetic service.

8.10–8.20 A.M. Arranging and telegraphing the
meteorological telegrams to
the Central Observatory at St.
Petersburg, by means of a
telegraphic apparatus, in the

service room, used by the observer in service.

8.20–9.00 A.M. Completing the reductions, and service in the photographic laboratory.

9.00 A.M. At 9 o'clock change of observers for 24 hours.

9.00–9.58 A.M. Control of all reductions and computations of the observations of the previous observer in the last 24 hours. Service in the photographic and chemical laboratory.

10.00 A.M. Reading the psychrometer and hair hygrometer in the shelter of the (second) thermo-hygrograph C, by Wild-Hasler, and marking the registration at 10 A.M.

10.02 A.M.–12.20 P.M. Continuation of the work of 9.0–9.58, and afterwards work on the records of the self-registering instruments.

12.20–1.03 P.M. The same as from 6.20–7.03, except that from 12.39–12.48 the anemograph Munro is to be attended to (paper changed, clock wound, &c.).

1.04 P.M. Reading of 13 ground thermometers under a sand surface at the depths 0·00m. to 3·20m.

1.06 P.M. Reading of 5 ground thermometers under the natural sur-

face at the depth 0·00m. to
1·60m.

1.10–1.12 P.M. Same as 7.10–7.12.

1.13 P.M. Reading of Wild balance evaporimeter.

1.19–2.10 P.M. Same as 7.19–8.10.

4.00–8.20 P.M. Working up the observations
and registrations of the instruments.

8.20–9.12 P.M. Same as 6.20–7.12, except that
the two heliographs of the
form Campbell and Jordan
are to be attended to be-
tween 8.39-8.43 P.M.   In
summer, as long as the sun
sets after 8.40, this is done
after sunset.

9.04 P.M. Reading and setting the two
extreme thermometers on the
sand surface.

9.13 P.M. Reading and setting the extreme thermometers on the
natural surface (sod or snow)
and of 4 extreme thermo-
meters for air temperature.

9.19–10.10 P.M. Same as 7.19-8.10 A.M., except
that in the evening the daily
averages for all the observations are to be computed.

The extreme care taken in the matter of observing
is shown by the fact that the standard clock of the
observatory is regulated by astronomical observations
made at Pawlowsk, although the great observatory of
Pulkowa is only a few miles distant.   In order to

insure greater accuracy an electric time signal is automatically transmitted to a local gong, in some cases, so that the exact time of observation may be made known to the observer, who stands in readiness to accomplish his duty at the sounding of the gong.

On certain days of the week and month there is special work to be carried out. Every Monday at 11 A.M. four Richard self-registers are to be attended to, and on this day water has to be distilled in the chemical laboratory. On every first day of the month the paper for the preceding month is taken from the Wild-Hasler instruments and replaced. On the last day of every month the books and sheets for meteorological and magnetic work for the next month are prepared. The observer in sub-service, on 16 prescribed days of the month, reads for 2 or 3 hours the absolute magnetic variations apparatus. In the remaining time he works on observation and registration records.

Numerous series of experimental observations are carried on at the observatory, which require additional service on the part of all the observers.

Every day fifty-eight thermometers for the meteorological and magnetic service are regularly read by the observers. The working up of the meteorological results is carried on by two or three of the younger observers, and supervised by the senior observer. Before the close of the month, the results of the preceding month must be completely worked up. The first proof sheets of the publications are read by the one who has worked up the results; the second, by the one who has supervised them. All constants of the instruments, correction tables, reduction scales, &c., are obtained by an observer, and supervised by the senior observer, or by the superintendent. The three younger observers work up under the supervision of the superintendent the magnetic observations and registrations.

§ 7.—*Mountain Observatories.*

Early History of Mountain Meteorology.—It became evident, very early in the history of instrumental meteorology, that if satisfactory explanations were to be given of the observed phenomena, something must be known of the conditions existing at some distance above the earth's surface at ordinary levels. Pascal recognised this when he caused the Torricellian tube to be carried to an elevated position, in order to note the change in atmospheric pressure.

But the originators or instigators of what is now termed high-level meteorology were Saussure, Gay-Lussac, and Humboldt ; and each had a separate field of action. Saussure's remarkable work in the Alpine region, published in his *Voyages dans les Alpes* (Neuchatel, 1787–1796), was a revelation of what could be accomplished by such methods ; and his results were at once applied to the forming of new theories. Gay-Lussac, about this same time and a little later, made his wonderful balloon ascents which for boldness and importance have never been surpassed, except, perhaps, by those of Glaisher ; and his observations furnished the necessary data for extending the atmospheric theories to regions far above those which had been previously visited by man. Humboldt made an immense number of observations during his travels in the very elevated regions of the New World ; and such was the care exercised by him, that even now his results have much more than historic interest, although the more complete work of recent years gives them at present mainly a local value.

But it was not enough to gather together the results obtained during irregularly conducted excursions to high altitudes. It was necessary to establish permanent places of observation; and this could be done only by locating them high up on the mountains.

Of the mountain observatories, that on the Schnee-koppe, Germany (lat. 50° 44′ N., long. 15° 44′ E.) has probably the earliest observations, since they date from 1786, but the records are fragmentary, and have attracted little attention. Most of these observatories have been established within recent years.

Use of Observations from Mountain Observatories. —In order to understand the conditions necessary for the correct classification of elevated observatories, it must be made clear what they are expected to accomplish.

Although much has been said on this subject in connection with the establishment of the older mountain observatories, yet the Vienna Meteorological Congress of 1873, in advocating a more universal adoption of this kind of work, considered it advisable to have a clear statement of the question, and Dr. Hann was appointed to make a report, which was submitted to the Meteorological Congress of 1879 at Rome. In general, these observations are expected to give the daily, seasonal, and annual march of the meteorological elements at that elevation above the sea; the decrease of temperature, variation of changes in pressure, increase of wind velocity, decrease of vapour capacity, increase of rainfall (to certain limits and then decrease), &c., with the elevation above the sea-level; and to show what relations exist at dif-

ferent elevations during the passage of areas of barometric maxima and minima, in order to be able to
formulate a system of atmospheric mechanics which
shall take account of the existing phenomena.

Some few of these questions can be investigated, at
least partially, at elevated plateau stations, still more
at mountain stations, where the land gradients have
merely the necessary steepness to cause a rapid rise
in elevation ; but for investigating all of them it is
necessary for the observatory to be located on a
mountain peak above any neighbouring mountain
tops, and with a companion observatory at the greatest
possible distance below it (if possible near the sea
level), but with the horizontal distance between the
two as small as possible; and for the best opportunity
to study the vertical relations in cyclones and anti-
cyclones the chosen site for the observatory must lie
in the path of their greatest frequency.

Partial List of Mountain Observatories.—The very
great expense attending the construction and maintenance of such observatories, has limited their
number, and they are for the most part confined to
Central Europe, where the great Alpine range and its
far-reaching spurs have offered favourable locations
in a number of countries. In other regions we find
that Scotland has furnished one, the United States
three, and perhaps India some additional ones ; but
these last are probably to be classed among plateau
stations. Mention must also be made of the occasional temporary observing stations which have been
maintained by expeditions, especially in North and
South America.

The following table shows the principal mountain
observatories, the plateau stations not being given :—

NAME.	CLASS, AS REGARDS EQUIPMENT.	ALTITUDE ABOVE SEA-LEVEL.
Germany.		Metres.
Wendelstein	II	1,728
Schneehoppe	II	1,603
Schneeberg	II	1,215
Brocken	II	1,141
Hohe Peissenberg	II	994
Inselsberg	II	915
Austria.		
Sonnblick	I	3,103
Hoch Obir	I	2,043
Schmittenhöhe	II	1,935
Schafberg	II	1,776
Gaisberg	II	1,286
Switzerland.		
Santis	I	2,467
Rigi-Kulm	II	1,800
Gabris	II	1,250
Chaumont	II	1,152
Italy.		
Etna	I (?)	2,990
Mte. Cimone	I (?)	2,168
Mte. Cavo	II	966
France.		
Pic du Midi	I	2,877
Mount Ventoux	I (?)	1,912
Mount Aigougal	I (?)	1,567
Puy de Dôme	I	1,463
Eiffel Tower (Paris)	I	300
Portugal.		
Serra da Estrella	II	1,441
Great Britain.		
Ben Nevis	I	1,343
United States.		
Pike's Peak	II	4,308
Mount Washington	II	1,914
Mount Hamilton (Cal.)	II (?)	

Probably the Cinchona Plantation in Jamaica, elevation 4,850 feet, and perhaps some of the high stations in India, should be included in this list.

Fig. 61 shows the observatory at the extreme top of the Eiffel Tower ; while Fig. 62 gives an excellent idea of the Pic du Midi Observatory and its surroundings.

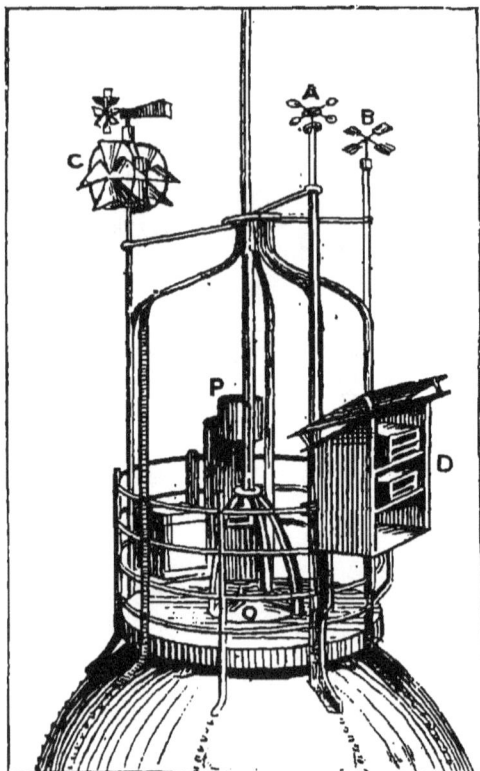

Fig. 61.

Difficulty in establishing Mountain Observatories. —These mountain observatories each have individual characteristics, due not only to difference in location, but also instrumental equipment and working force. Some of these can be pointed out, but for complete details it is necessary to go to the various publications in which descriptions are to be found.

Rotch's papers on the Mountain Observatories of Europe, in the *American Meteorological Journal*, are the best summary we have of these, as he visited and made a careful study of almost all of these observatories.

The greater portion of the large sums of money expended in the establishment of mountain observatories was used in overcoming the natural disadvan-

FIG. 62.

tages which had to be dealt with. In some cases it was necessary to construct the very roadway over which to transport the necessary building material, fuel, and food to the mountain top; and on account

14

of the large distances and steep gradients, the transportation itself has been exceedingly costly ; in many cases man power being all that was available, and a few pounds a load.

In the establishment of these observatories a very different problem from that for low-level observatories had to be worked out.

In the first place, various economic matters needed to be planned. Suitable habitation must be provided for the observers who were to isolate themselves from their fellow men for months at a time, and in a climate where but little out-of-door life during the greater part of the year is possible on account of the continued severe cold and strong winds. The living rooms must be directly connected with the observatory proper, as many times it is impossible for the observers to venture away from the building. Storage must be provided for fuel and provisions sufficient to last through an entire winter ; and telegraphic communication must be kept up between the observatory and some village, whence aid can perhaps be obtained in case of emergencies.

In the planning of the observatory itself, many matters connected with the placing of the various instruments have to be considered, in order to overcome difficulties which either do not occur for lower stations, or are greatly magnified by the exposed location of the elevated station. These difficulties are not the same, or rather not equally great, for all stations. For instance, at stations where the air is very moist, frost-work forms on the cups of the Robinson anemometer so frequently and so rapidly that ordinary records are valueless during the seasons when this occurs. On the peak stations, where the

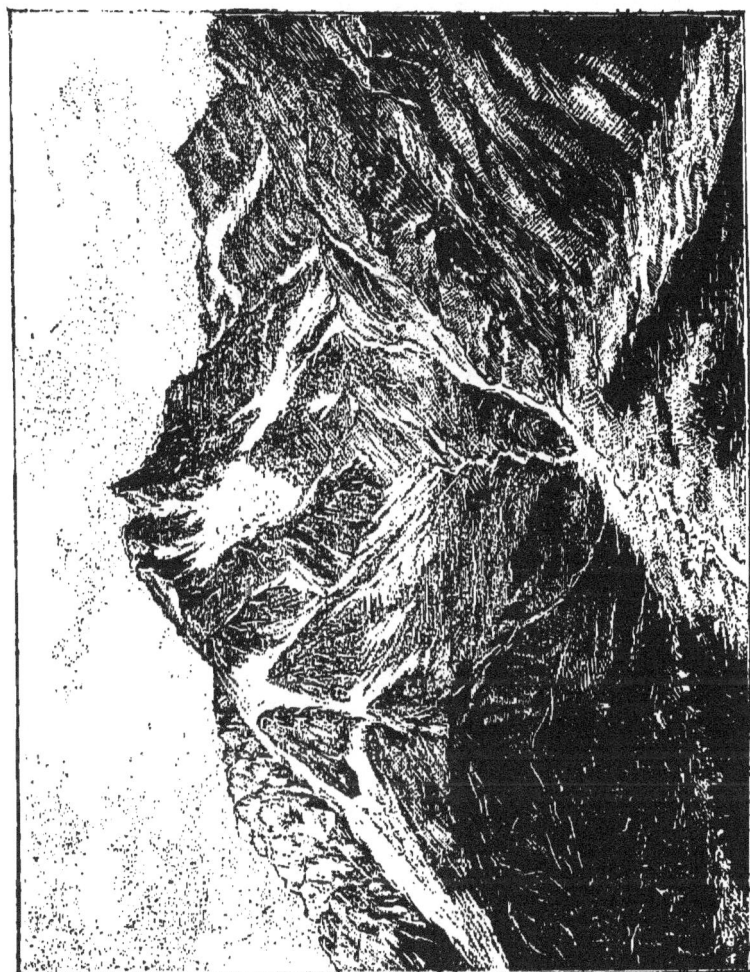

FIG. 63.

full force of the wind is felt, the difficulty in making any accurate measures of precipitation in winter is very great ; the snow is blown away from the gauge or drifts over it, or the mouth of the gauge becomes so clogged with frost work that little precipitation enters it.

Some Special Mountain Observatories.—The highest meteorological observatory in Europe is that on the Sonnblick mountain, in Austria. The general surroundings can be well seen in Fig. 63, in which the observatory is placed at the highest point shown ; and the character of the building and the immediate exposure is shown in Fig. 64. It is located at an elevation above sea-level of 3,103 metres, with the immediately accessible neighbouring inhabited valleys only half that elevation, and has adjacent base stations at Lienz (2,420m.), Zell am See (2,340m.), Kolm Saigurn (1,480m.), Schmitten-höhe (1,150m.), Obir (1,050m.), the numbers in parenthesis signifying the vertical distance of the stations *below* the level of Sonnblick. The relation of the stations at Sonnblick Peak and at Kolm-Saigurn are particularly favourable for the determination of the decrease of temperature with elevation, since the vertical distance between the two amounts to 1,480 metres, and the horizontal distance to only about 2,500 metres ; and the lower station has a remarkably free exposure, which prevents the stagnation of air which so frequently introduces anomalous data into such determinations. Hann remarks that no such favourable dual exposure can be duplicated in the Alps, and perhaps not elsewhere on the globe.

The Sonnblick observatory and the Pike's Peak observatory (Fig. 65) in Colorado, U.S., which is the

FIG. 64.

highest observatory in the world, are the best situated
of the mountain stations having a purely continental
location. But neither of these stations lies in the
path of greatest frequency of cyclones and anti-
cyclones. There are two, however, which could
hardly be better placed for the study of cyclonic
phenomena and the comparison of the general conditions of the atmosphere at about sea-level and at a
moderate elevation of perhaps a mile, viz., the Mount
Washington, New Hampshire, U.S., and the Ben
Nevis, Scotland, observatories. The former, as also

FIG. 65.

is the case with the Pike's Peak observatory, being a
second-class station with scarcely any self-registering
apparatus, and no very near permanent base stations,
has had its possible usefulness greatly lessened by
the lack of completeness in its organisation.

The Ben Nevis observatory (Fig. 66), with its companion base station at Fort William, only four miles
distant, and located at sea-level, while one of the
latest to be established, promises to offer a most
fruitful field for investigation. Its oceanic exposure
is on the side of approach of cyclonic disturbances,

which, by reason of the closeness of the mountain to the coast-line, reach it in their most perfect form ; because in their previous course they have been subject to the symmetrical oceanic conditions. How-

FIG. 66.

ever, the fact that the more or less distant land stations cannot entirely encircle Ben Nevis places it at somewhat of a disadvantage compared with the truly continental stations.

Management of Mountain Observatories. — The observers at these mountain stations are deserving of the deepest gratitude of those interested in meteorology. Isolated from all companionship during a great portion of the year, and kept indoors most of the time by the severe weather, their lives can only be compared with that of a lighthouse-keeper in an arctic climate ; and where a single observer must winter alone, on account of lack of means for furnishing additional observers, it must require a great amount of fortitude on his part to undertake the task. Fig. 67 shows an interior of the Ben Nevis observatory ; and its equipment and furnishings certainly indicate that utility and not comfort have been considered.

A few circumstances concerning the management of some individual observatories may be detailed. On the Brocken a hotel waiter makes the observations in winter, and the post-master in summer ; on the Wendelstein the watchman is the observer ; on the Hoch Obir there is a regular meteorological observer, who receives thirty florins a month, and part of the time an assistant at twenty florins per month, and the whole expense of the station is but nine hundred florins per year ; on the Sentis there is one observer in summer, receiving two thousand francs per year, and in winter he has an assistant ; on the Puy de Dôme there is a single observer with his family, who have three thousand francs per year income. On Ben Nevis there are three observers, and in this respect it is better fitted for undertaking necessary work than any other of the mountain observatories ; and since in this case an observer can always be on duty, direct hourly observations can be made, and the results properly tabulated and discussed by those who make them.

It must be impossible for one not having actually experienced mountain weather to know the disagree-

FIG. 67.

able features of the life of observers there. At most of the stations exposed instruments and everything else become frequently covered with frost-work which forms on the windward side when there is a fog and the temperature is below freezing ; and since this sometimes accumulates at the rate of an inch an hour, its removal from instruments must keep an observer very busy.　On the Wendelstein there is

FIG. 68.

little of this frost-work, but on the Brocken and Ben Nevis it is very frequent.　Fig. 68 shows the observatory at the latter place when the frost-work is present.

Exposure of Instruments at Mountain Observatories.
—An additional mention must be made of the difficulties in properly exposing instruments, and the
means of overcoming them.

On the Brocken, Assmann has successfully used a
movable precipitation gauge, which he places on the
lee side of the observatory. The exposed receiver of
this gauge has a lamp, doubly encased, burning continually beneath it, and thus keeping it warm and
melting the snow at once, so that it runs downward
into a cold, covered, collector, where the heat of the
lamp does not cause it to evaporate again. When
the exposed gauge is kept warm, frost-work does not
form on it. It has also been suggested to make the
anemometer cups double, so that a heated current of
air passing through between the outer and inner sheet
metal would keep it warm and melt ice or prevent
frost-work from forming. It would, however, seem
simpler to use an aspiration anemometer, which could
easily be kept heated. The thermometer shelters
and other places of exposure for apparatus must be
kept free from drifting snow and icy accumulations by
the labours of the observer, and no substitute has been
found for this work. The barometers must be attached
to an isolated pier built up from the solid rock, otherwise they would be jarred by the high winds, and an
accurate reading at such times would be impossible.
The dynamic effect of high winds is also to be
guarded against.

Taking into account these and other difficulties
which could be mentioned, it must be admitted that
mountain meteorology is still a very inexact branch
of science.

CHAPTER III.

THERMODYNAMICS OF THE ATMOSPHERE.

§ 1.—*Preliminary Ideas.*

THE temperature of the air near the ground, as measured by the thermometer, has been pretty thoroughly studied for a large portion of the earth's surface; and in this study the laws of direct radiation, conduction, and convection of the heat derived from the sun have long been considered. While meteorologists had to be contented for a long time with this foundation work, as we may term it, of a knowledge of the actual temperature of the atmosphere, yet the few temperature observations made in balloon ascents and on mountain peaks, showed that unknown laws were yet to be discovered which might be applied to the data which could be collected at the bottom of the ocean of air, in order to obtain an idea of what condition prevailed in elevated regions. The demand for a knowledge of these unknown laws was made imperative when the mechanics of the atmosphere began to be studied with mathematical rigour. Fortunately the study of the behaviour of gases under varying circumstances of physical condition has received the continued attention of very eminent physicists, who have provided meteorologists with a means of developing many of the special laws which are operative in the great gaseous ocean which we call our atmosphere.

While the static conditions of the air, which were at first studied on account of their simplicity, sufficiently explain certain classes of problems, yet the study of the atmospheric motions which are so ceaselessly transferring masses of air from one latitude and longitude to another, and from one altitude to another, has necessitated the inclusion of the more complex dynamic changes which must of necessity result from these ever-continued shiftings of bodies of air.

The laws of the dynamical theory of heat, as developed by Clausius and others, are directly applicable to the changes in the physical condition of these displaced air masses, and it is my intention to devote the present chapter of this work to following out some of the recent applications of these principles to atmospheric events.

While it is necessary to take for granted the reader's familiarity with the changes which occur in the condition of gases when they are subjected to various pressures, volumes, and temperatures, as developed in elementary text-books of physics, yet space has been taken to render clear the most important ideas which are introduced in these pages. As an elementary and easily-read introduction to this subject there is no better work than Maxwell's little book on Heat.

Evaporation of Moisture.—In investigating the physical conditions of the atmosphere we have generally to deal, not with dry air simply, but with a mixture of dry air and vapour of water. This vapour gets into the air by the process of evaporation, and is distributed by subsequent diffusion and convective air currents. In any limited space the evaporation will go on until the vapour has acquired a certain density which is dependent on the temperature. The amount of vapour which can exist for any given temperature is the same whether it alone fills the space, or whether the space has been previously supplied with dry air ; but in this latter case the time required for the vapour to permeate the whole space is greater than in the case of no air, for it must be diffused according to a kind of percolation. The density, pressure, and temperature are so connected that when we know any two the other can be ascertained, and for each temperature of the vapour there is a maximum pressure and maximum density which can exist without the vapour condensing and becoming water. When the vapour reaches this condition it is said to be saturated, and below this condition is called superheated vapour.

If dry air is first allowed to fill a confined space it will exert a certain pressure, and if vapour is introduced the pressure will be increased by the same amount that the vapour would have exerted if acting alone, and each gas exerts a pressure just as if the space occupied by the other gas were a vacuum. This very important principle was discovered by Dalton, who also found that the maximum density of vapour, when suspended above a water surface, is the same, whether the air is present or not ; but the more air there is present the slower will be the process of evaporation.

It must be remembered that the volume of a gas when under constant pressure expands by a uniform fraction of itself when the temperature is raised :

and that air, when raised from the temperature of
0° C. to 100° C., will change its volume from 1·000 to
1·3665, if the pressure remains constant. Also, if the
gases, air and vapour of water be taken in equal
volumes at any given temperature and pressure, then
for any other common temperature and pressure the
volumes will also be equal.

Indicator Diagrams.—The atmosphere being a
fluid, or rather a mixture of the two independent
gaseous fluids, air and vapour of water, of variable
volume, the study of its changes is so complex that
the analytical method is ill adapted for an elementary
explanation of these changes. A method of studying
such changes for steam, in tracing the work done by
a steam engine, was first invented by James Watt;
but great improvements in the general method were
made by subsequent investigators, among whom was
Clapeyron, who applied the use of the indicator
diagram, as it was called, to the study of fluids in
general; and, later, Rankine made great use of the
same method in his study of the steam engine. We
must try now to obtain a clear idea of the principles
on which this method is based, on account of its
recent introduction into the science of meteorology
by von Bezold of Berlin; for we shall neglect the
older ways of procedure in studying what are known
as the thermodynamic changes in the atmospheric air,
in order to present more fully the new application of
a method which has received the praise of such eminent physicists as Maxwell, Rankine, and others, and
which consequently we may feel great confidence in
using.

In using this method to determine the properties
of gases when acted upon by heat, the accompanying

diagram (Fig. 69) which has the same framework as
the ordinary rectangular co-ordinates in geometry,
serves as base of representation. In the plane OPV
lay off the rectangular axes OP and OV. The
volumes are represented by the distances along OV,
and the pressures by distances along OP. So that
the co-ordinates of the point N will represent these
conditions. The condition of the gas for any point
N, is then represented by the volume which has a
magnitude OM, and a pressure of the magnitude MN

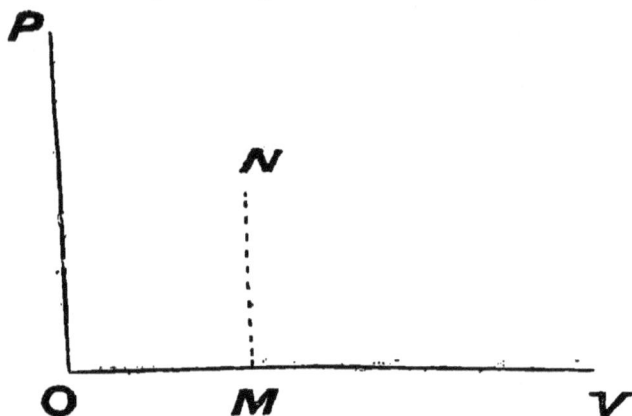

FIG. 69.

If now the temperature remains constant and the
pressure is increased, then the point N will assume
the position N', which shows the consequent diminution of volume as seen in Fig. 70. If the pressure
had been decreased, then N would have assumed
the position N'', which would have shown an increase
of volume.

If the volume diminishes as the pressure increases,
as it will in the case of constant temperature, the line
traced out by the point N is called an isothermal line.

If we take any gas, like air, for instance, a series of isothermals can be drawn for successive temperatures which will express the relations of pressure, volume, and temperature for that gas ; and if the pressure and volume are given, the temperature is easily determined by interpolating on the graphical table.

Since by Boyle's law the product of the volume and pressure is constant, when the temperature remains constant, the temperature curves traced out by the point N must be hyperbolas, of which OP and OV

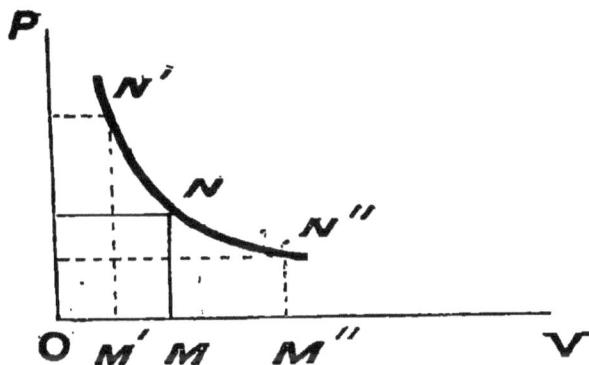

FIG. 70.

are the asymptotes. These temperature curves are called isothermal lines, or merely isothermals.

In the case of isothermal lines the temperature, as we have seen, is kept constant, and we will now consider the relations of pressure, volume, and temperature when no heat is communicated to or taken from the air. The curved line which would now be drawn on the diagram has been termed by Rankine the adiabatic line, and it is quite different from the isothermal line, although it has the same tendency and general direction of the curvature, on account of the

15

inverse changes of pressure and volume which govern
the location of points along the lines in both cases.

The differences of isothermal and adiabatic changes
are shown roughly in the accompanying diagrams,
Fig. 71, and Fig. 72. The two lines, if drawn on
the same scale on a single diagram, would intersect at
a certain point, since a greater pressure is required
to diminish the volume in what may be termed the
constrained adiabatic changes, and therefore the
inclination of the adiabatic line to the horizontal or

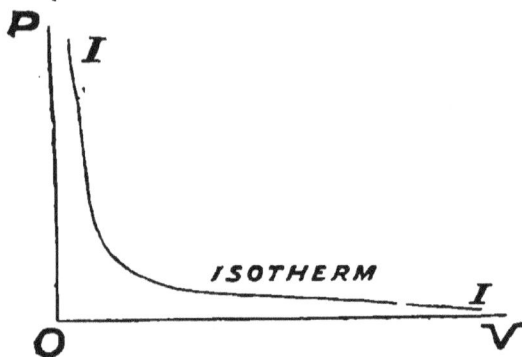

FIG. 71.

volume line is always greater than that of the isothermal line.

Vertical Circulation.—It has been customary in the
application of the mechanical theory of heat to the
theory of atmospheric motions and changes of condition, to consider the adiabatic changes merely ;
that is, those in which there was neither gain nor loss
of heat during the expansion or contraction of the air
mass.

It has been unfortunate that the limitation has
been made, for it has probably retarded the progress
which has been made in certain directions in explain-

ing observed phenomena. Take, for instance, the Föhn wind, which has been so thoroughly studied in the Alps, but which is also known under other names in other regions. The most acceptable explanation has been based on dynamical principles, as explained in another place, yet there is a very evident lack of completeness when no mention is made of any actual change in the amount of heat, although the far

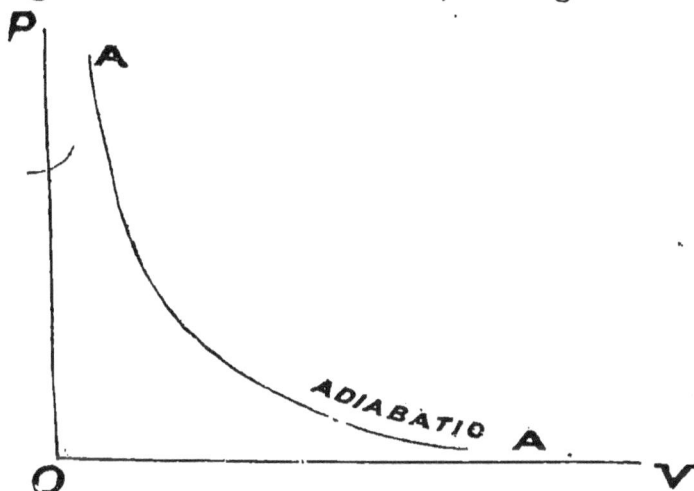

FIG. 72.

greater effects of expansion and compression so mask smaller thermal changes that the criticism of omission is not as forcible as in cases where the whole change is less. In the investigation of convective equilibrium, and especially in the interchanging motions in and between cyclones and anti-cyclones, the question of the addition and abstraction of heat can no longer be neglected, although the already very difficult hydrodynamical analysis will be ren-

dered still more complicated by its consideration.
It is somewhat remarkable that the method of Carnot,
so well known in thermodynamics, and applied more
practically by Clapeyron,[1] should not have earlier
attracted the attention of meteorologists. The work
of von Bezold in this direction is the most important, as he was the first to fully show its advantages. As long ago as 1884 Hertz developed
a graphical process (to be explained later) for
showing the condition of moist air subject to adiabatic change, and Sprung incorporated this in his
Meteorologie (1885); which shows that this last
mentioned meteorologist manifested considerable
keenness in recognising so quickly that the suggestions made by Hertz were of unusual importance ;
but several years elapsed before the real significance
of the process was fully brought out and insisted on
by von Bezold. Hertz used the graphical process,
and dealt only with the limited problem of adiabatic
changes of condition, and while the graphical method
is well adapted to illustrate a process, and to obtain,
easily, numerical results, it must be replaced by
analytical forms in the consideration of the problem
as a continuation of the hydrodynamical analysis
which has of late years been applied to atmospheric
motions. It is this difficult process which von
Bezold has attempted to carry out on the lines pursued by Clapeyron, which latter require, however,
considerable extension in the new investigation, and
without the narrowing limitation of adiabatic condition as adopted by Hertz. I shall attempt to follow
portions of von Bezold's work.

The experimental researches of Espy in the early

[1] *Poggendorf's Annalen*, Bd. 59.

part of this century, and the later theoretical deductions of Sir W. Thomson, Peslin, Reye, Hann, Ferrel, and others, showed to meteorologists the importance of the dynamic heating and cooling of vertically moving air masses as a consequence of the compression or expansion of the air. In the very important memoirs of Guldberg and Mohn, these authors have, in their treatment of this question, considered that the heat lost in the ascent of air currents is converted into the energy necessary for causing and maintaining vertical currents. This whole matter cannot be understood, however, without considering the main cause of the vertical circulation as being found in the differences of specific gravity in the air at different places, by which, in the place where the air is denser, a downward motion results, while where the air is lighter an upward motion ensues. Thus the energy necessary to raise the air in the one case is obtained by the fall of the air in the other case; and this motion once entered upon would continue indefinitely without the addition of externally derived force, were it not for the effect of friction. This compensatory flow, as it may be termed, is compared by Sprung (*Meteorologie*), von Bezold, and others, to the flow of water in a closed system of tubes, and Ferrel has illustrated it by the case of an open canal in which the two currents are separated by a neutral plane. In a water-heating apparatus we do not consider that the heat in the ascending water is lost in doing the work of raising the water, and that the cooling is the result of this work. Just so the changes in temperature of ascending or descending air masses cannot be connected with the work necessary for producing this motion, but it is entirely due to the expansion or

compression of the air, and the same changes of temperature would take place if the same changes of
pressure and volume took place in a horizontal
cylinder or a vertical cylinder. It cannot be too
strongly insisted upon, nor made too plain, that the
cooling of the ascending current is not brought about
by loss of heat in the work of raising the air ; nor is
the heating up of the descending air due to the gain
of energy by the fall. But the work of compression
and expansion cannot be neglected, as has been done
by some investigators.

As the atmosphere is never made up of pure dry
air, but is a mixture of air and water, it devolves on
the theoretical meteorologists to follow the course of
this mixture in its changes, and not be content with
the more easily treated hypothetical problem of dry
air alone.

It has been customary, as shown by the work of
Hertz and others, to adopt a unit of mass of the
mixture which shall remain constant during the
entire process, but this is not even very approximately correct, because, as a matter of fact, much of
the moisture is lost by condensation and subsequent
precipitation in the form of rain, hail, or snow.
Now, while the moisture is a changeable quantity, we
can consider the dry air as constant, and it is easier
to treat the two quantities separately as having these
characteristics than to treat the whole mixture as a
variable one.

Four different mixtures are to be considered, and
as these exist in the conditions which obtain in
layers of the air taken progressively they are aptly
termed stadia.

We have first the dry stadium, in which the air is

not saturated with moisture, and consequently in which there is no condensation and precipitation. The amount of the unsaturated or super-heated vapour in this stadium of the free air may be taken in general as constant ; for there is no precipitation, and the further accumulation of any considerable amount of water vapour only happens at the earth's surface, and the intermixture of layers of different degrees of humidity are now unconsidered.

The rain stadium is reached by the air of the dry stadium being cooled, or subject to adiabatic expansion. In this case the mass of the mixture is less than in the dry stadium, except in the extreme limiting case, where they would be equal. The amount of fluid water present is small, unless there is an ascending current of great strength which supports the raindrops, and, in fact carries them along upward with the air. Such upward currents exist in the interior of tornadoes according to Ferrel's views. The vapour of water in the mixture is in the saturated state.

In the hail stadium the mass of the mixture is less than that in the rain stadium, but they approach equality in the limiting case. The mixture is made up of the unit of dry air, the saturated vapour, the drops of water which are present, and the drops of ice or hail. This condition can prevail only when raindrops are present to be converted into ice, and when the temperature has fallen to 0° C. to make this conversion possible.

The snow stadium is usually attained by the gradual cooling of the ascending air, which passes through the rain and hail stadium, but in which the rain and hail precipitation is so small that the latter

stadium at least, is practically jumped over, and so in
this case there exists but the dry air, the saturated
vapour and the snowflakes. The mass in this layer
is still less than in the hail stadium.

Von Bezold points out that there may exist here
an application of the mechanical theory of heat
which has not yet been considered, viz., the recon-
version of a portion of the condensed water, or drops,
back into the vapour condition, by the process of
warming up or compression ; and if this is carried
out to a sufficient degree, the mixture will again
enter the dry stadium, but the amount of vapour will
not be the same as the original amount.

§ 2. — *Von Bezold's Graphical Analysis.*

The analytical discussion which von Bezold has
made of these conditions cannot be reproduced here
on account of the mathematical forms which are
necessarily used by him, but the importance of the
investigation demands that some notice should be
taken of such portions as can be easily put into an
elementary form. The number of variables which
have to be considered is sufficient to cause great
complexity in the equations of condition; so that a
graphical, but, it must be distinctly understood,
schematic, representation is all that can be produced.

The volume v, the pressure p, the temperature t, and
the water vapour x, are the quantities with which we
have principally to deal. The precipitation, whether
occurring as rain or as hail or snow, is so small that
its influence on the pressure and volume can be
neglected.

In the diagram, Fig. 73, let the ordinate OP

represent p, the pressure, and the abscissa OV represent v, the volume. A third co-ordinate OX is imagined drawn from O towards the reader, that is, perpendicular to the plane OPV, and this ordinate, or one of its parallels, represents x, the water vapour. By this choice of co-ordination the representation of all conditions having the self-same values of x may be confined to one plane which does not deviate much from the plane OPV (shown in the diagram), when we take an atmospheric pressure as the unit of pressure and lines of equal

FIG. 73.

length along the axes OV and OX as unit of volume (1 cubic metre) and mass. If we imagine successive (transparent) planes, representing the progressive conditions step by step, in, say, thousandths of the chosen unit, or from gramme to gramme, superimposed like the leaves of a book, then we can follow out the changes of condition by means of the curve on the plane OPV, which is the projection on this plane of the actual curve connecting the points in the successive layers. This simplification renders possible the schematic conception of a problem which

must otherwise remain unknown to the reader un-
skilled in mathematical reasoning.

Dry Stadium.—Mathematical considerations show
that the line representing the isothermal for moist
unsaturated air and also for dry air, is an equilateral
hyperbola, or a portion of one.

The main condition for the existence of the dry
stadium is that the existing pressure be less than
the pressure necessary for the forming of condensed

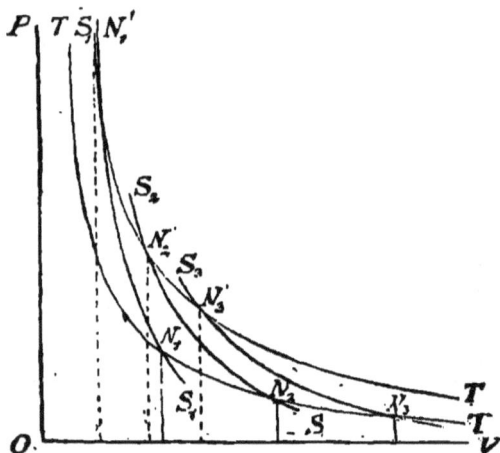

FIG. 74.

vapour at the same temperature ; this last pressure
is dependent on the temperature and increases with
the latter, but more rapidly than it. The characteristic
curves to be investigated are the isotherms, adiabatics
and curves of constant amount of saturation.

In Fig. 74, the curve S_1S_1 represents the
" dew-point curve " or " curve of saturation " ; and
this curve bounds the plane of dry stadium with
any constant amount of vapour. Consequently when
the curve of condition starting in the dry stadium

intersects the "curve of saturation," the dry stadium is departed from and the stadium of condensation, the rain, or snow stadium, entered. So that for the dry stadium, the isothermals (for any constant amount of vapour) must terminate in the "curve of saturation." Therefore the course of this latter curve is determined by the hyperbolic isotherms and the origin of the abscissas.

In the diagram Fig. 74 let N_1 be the point on the "curve of saturation," which is the starting-point of the isothermal line T, which is drawn for a definite temperature T and for a certain vapour quantity x. The "curve of saturation" $S_1 S_1$ corresponds to this vapour quantity x; and for any other vapour quantities x_2 and x_3 we should have the corresponding "curves of saturation" $S_2 S_2$ and $S_3 S_3$; and for a constant T the starting-points of the abscissas of the isotherms corresponding to the same temperature, but different vapour quantities, are proportional to these vapour quantities. The dew-point curves S_2, S_3, as drawn in the Fig. 74, show about how these curves run for vapour quantities of two and three times the amount for which the curve S_1 was drawn.

The isotherms, corresponding to the different amounts of vapour which really exist, would run so close together that they can roughly be taken as lying within the single line T, which is shown in the diagram; only it must be remembered that each has a starting-point on the hyperbolic curve, and the position of this starting-point depends on the amount of vapour of water in the atmosphere. By further expansion of the air, the curves of saturation S_2 and S_3 will replace S_t.

In the dry stadium, which lies to the right of the

curve of saturation in the diagram, the characteristic curves will nearly obey the same laws as for a perfect gas. We should have consequently, adiabatically, the products of the volumes and pressures a constant quantity, for any probable quantities of water vapour which might be present. Applying the proper numerical constant we should have $(pv)^{1.41} = (p_1 v_1)^{1.41}$, as expressing this more exactly.

Von Bezold now uses the so-called line of constant entropy as synonymous with the adiabatic curve ; and he is thus able to show that by the decrease of entropy by isothermal expansion from one volume to another, if (see the diagram) we proceed along the isotherms which cut an adiabatic S_1 (line of constant entropy) a second adiabatic S_2 will be reached ; the relations of expansion remaining constant.

As soon as we know the course of an adiabatic, it is easy, by a process similar to that for dew-point curves, to construct such additional adiabatics that each shall have a constant difference of entropy with the next adjoining.

Rain Stadium.—In the *rain stadium* the changes which occur are seldom reversible, that is, capable of retaining the mixture to the original condition, on account of the abstraction of water by rainfall. The line which indicates the changes of condition will seldom stretch out into the concave side of the dew-point curve, and it is only possible to go back to the old condition in case the condition of super-saturation exists and the line is above the curve of saturation. This is more fully explained later.

In the rain stadium the isotherm is an equilateral hyperbola as in the dry stadium, since the vapour pressure remains at a constant value, and the one assymptote coincides with the axis of p (the pressures) as in the dry stadium, while the other is slightly inclined to the axis of v in the direction of p. There is, however, very little difference between the isothermal curves for any given temperature in the dry and wet stadia, and the one joins on to the other in passing from one stadium to the other with only a slight notching of the curve, with point of the notch turned towards the right and above. This is noticeable in the curves showing the changes of condition in the air in the vertical air currents mentioned later, and it is due to the fact that the isotherm for the rain stadium contains the starting-point of all the isotherms in the dry stadium which corresponds to a certain amount of moisture, which is, however, less than the original amount of moisture which we started with in the dry stadium.

In investigating the adiabatic curve we must know the amount of heat which is added during any change both to the dry air and the water with which the air is mixed. The former is easily computed, but the latter is a variable quantity depending on the amount of water which exists in the form of vapour, and the amount of water which is lost in the form of rain ; and we find here two extreme limits, which have to serve as guides in the absence of any accurate knowledge as to the relative or absolute amounts of water lost out or retained. All of the waterdrops may fall out as rain, or all of them may be carried along by the motions of the air, so that no water is lost. While it is probable that in nature neither of these limiting cases occurs, or at least with any frequency, yet we must examine them to see the extreme conditions which are possible. Graphically represented, these

two conditions would mean that the same plane is
passed along ; that is, there is a constant quantity of
water, or else the path is along the dew-point curve
in which all the water immediately falls away and the
condition of saturation is just maintained. Von Bezold
thinks the latter condition is the nearest that which
we find in nature, although the former is that which
is usually assumed as a basis for investigation.   Still,
in the practical case of tornado action, where there is
a strong upward current assumed, this case probably
does occur.   These two cases von Bezold also distinguishes as having in the one case a maximum
super-saturation (*Übersättigung*), and in the other
case a normal saturation.   It will be seen that the
amount of heat added (algebraically considered) would
have to be zero, in order to have an adiabatic condition of change ; and this could occur only in the
case of the maximal excess of saturation, if the term
adiabatic is strictly limited in meaning.

   If the adiabatic condition is supposed to depend
simply on the assumption that, in the change, heat
is neither added nor taken away, then the two cases
just cited are certainly adiabatic.   However, if we
define the adiabatic condition as one in which not
only all external work is effected at the cost of
energy, but also the whole of the lost energy is used
in the external work ; then the definition does not
apply to the second limit mentioned, as well as to
the intermediate changes in condition when any of
the condensed water drops are present, but in which
some have fallen out, and even when the condition
of no gain or loss of heat is fulfilled.   When the
waterdrops fall out, we have a diminution of energy
by the amount necessary for the external work, and

also by the amount corresponding to the definite temperature of the water which falls out.

Von Bezold carefully specialises those cases in which the water all falls out by calling the change *pseudo-adiabate* (pseudo-adiabatic); while the general cases in which the water entirely or partly falls out, he calls simply *pseudoadiabatische* (pseudo-adiabatical).

The analytical statements of the conditions controlling adiabatic and pseudo-adiabatic curves, and which are called the equations of these curves, can not be reproduced here, but they show that the pseudo-adiabatic curve sinks more rapidly than the adiabatic, or, in other words, the temperature falls more rapidly when all of the condensed water falls out at once than when some of the raindrops are retained and carried along. Comparing these two curves with the dew-point curve, or curve of saturation, it is found that they sink more rapidly than the latter, as is evident from the fact that heat must be added to make the expansion take place along the dew-point curve ; and in the graphical representation adopted here, curves showing changes of condition when heat is added are always inclined less steeply towards the abscissa axis than when the change of condition is adiabatic. In the expansion, the adiabatics depart from the dew-point curve, towards the abscissa axis, and the quantity of moisture decreases.

The course of the adiabatic curve cannot as yet be explicitly determined so as to give directly the relation between the volume and temperature, or the temperature and pressure, or between the pressure and the volume ; and so an indirect method must be pursued in which use is made of the decrease of entropy in the passage from the initial condition to

the final condition. It is possible to compute the
values of the entropy for a number of selected values
of volume and pressure, as these are independent of
the individual constants of the whole system ; and by
interpolation, the entropy for a sufficient number of
the intermediate values of the volume and pressure
may be obtained to allow the construction of the
lines of equal entropy, which are the adiabatics. Von
Bezold finds it possible, and very convenient, to
select for computation those points which lie con-
secutively on isotherms, so that in passing from one
point to another it will be along the same isotherm,
and the entropy will be constant if this condition is
maintained. A consideration of these relations shows
that the isentropic curves cut the isotherms at a
sharper angle in the rain stadium than in the dry
stadium ; and any definite change of entropy is the
result of a greater change of volume in the case of
the dry stadium than would be necessary in the rain
stadium.

Limiting our view to a small portion of the plane
of co-ordinates, the isotherms will appear as parallel
straight lines running almost the same course in the
two stadia, the dry and the rain stadia ; and to
represent any chosen change of entropy, it would
be necessary to pass along the isotherms a longer
distance in the dry stadium than in the rain stadium.
The dew-point curve sinks in further towards the
positive side of the abscissa axis than the isotherms,
and the adiabatics show a bend at the dew-point
curve, as shown in Fig. 75, where S S represents
a portion of the dew-point curve ; TT, T'T' are
isotherms, and AA, A'A' are adiabatics.

The " pseudo-adiabatic " curve is still more difficult

to investigate, and still less only approximately correct solutions can be obtained even for the cases within the limits of actual use, which are fortunately more determinate than the general solution. But even the derivations for special cases are too complex to be explained here. It may be mentioned, however, that a consideration of the equation shows that the pseudo-adiabatic curve sinks more rapidly than the adiabatic. That is, the successive volumes for corresponding temperatures must be less for the pseudo-adiabatic changes than if the adiabatic had been followed.

Hail Stadium.—When the temperature descends below 273°C. on the absolute scale, or the freezing point of water, the conditions which have been outlined for the rain sta-
dium must be modified. When the temperature is just at the freezing point,

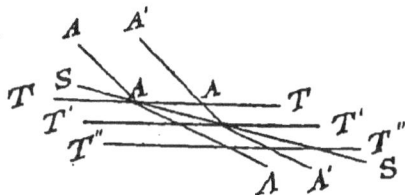

FIG. 75.

water and hail may exist side by side, and the change would be by isothermal expansion as long as this condition is maintained ; the adiabatic in this case coinciding with the isothermal line. Von Bezold holds that the adiabatic will in reality be a pseudo-adiabatic, since the hail falls out under *any* circumstances (although Ferrel's theory of the vertical circulation in a tornado shows that the upward current may be so strong as not only to support the hail stones, but to actually carry them upwards), and in this condition, any increase of vapour requires addition of heat, and the formation of more hail means a loss of heat.

16

In the case of continued expansion, but without gain or loss of heat, through the hail stadium, there will be a continued increase in the amount of hail formed ; but in this freezing process there is a simultaneous evaporation, so that at the upper limit of the hail stadium there is actually a greater amount of vapour present than at the entrance to this stadium.

The whole process and change in this stadium may be followed thus.—The determinite plane above the OPV plane, which corresponds to the amount of waterdrops *plus* the amount of saturated vapour, will be that one showing the condition of entrance into the hail stadium ; and it is somewhere above the dew-point surface. The line which indicates the distance from the assumed plane OVP will cut the dew-point surface at the plane (above the OPV plane) which corresponds to the amount of vapour just necessary for complete saturation. If now the mixture expands isothermally, the just-mentioned point of intersection with the dew-point curve will gradually be raised, and the projection of its path on the OPV plane will be an equilateral hyperbola. It is plainly evident that while the point of intersection with the dew-point surface is thus slowly rising, the plane of the initial condition (depending on the waterdrops and saturated vapour) is being receded from as the frozen water drops out, and finally the two will come together at the plane where the hail stadium ceases.

The question naturally arises as to the relative or absolute amount of hail which is or may be formed. At the entrance of the hail stadium, the mixture contained water in the form of raindrops and saturated vapour, and all of this water is turned into hail : so that the amount of hail will depend on the amount of water on entering the hail stadium, and the amount of change in the volume of the mixture by expansion. The hail stadium cannot be entered until there is at least an appreciable amount of water present in the form of drops ; and the more of these there are, the greater will be the hail fall.

Snow Stadium.—When the air is saturated with water vapour and becomes cooled below freezing without the formation of raindrops, then the condition of the snow stadium exists. In this stadium the main conditions are somewhat similar to those of the rain stadium, but here the heat of melting takes the place of the heat of condensation. The amount of vapour which may be present depends on the temperature. The lower the temperature the less the vapour, and so, in this stadium, the less will be the amount of snow ; and the lower the temperature becomes, the more nearly will the adiabatic curve in this stadium coincide with that of the dry stadium. We thus see why it is that in very cold weather the snow fall is comparatively light.

§ 3.—*Application to Föhn Winds and Cyclonic Circulation.*

In the widely-distributed and strongly-marked winds known as the Föhn, the moist air expands in passing upwards along the side of a mountain, and becomes compressed again in the descent on the opposite side, where it is much drier. It is interesting to follow von Bezold's application of the preceding ideas to this process. In Fig. 76 let a represent the original condition and Sa the corresponding dew-point curve, then the air expands according to the adiabatic of

the dry stadium until it cuts the curve Sa at the
point b; and the curve ab lies in one of the imagined
planes parallel to the OPV plane at a distance of xa.

The temperature curve corresponds to the initial
temperature shown by the line Ta, and we see that
in the change from a to b the temperature falls
rapidly. As soon as the condition b is reached, the
representative point glides down the dew-point
surface; the adiabatic of the dry stadium goes over
into the rain stadium bc, and forms with the curve

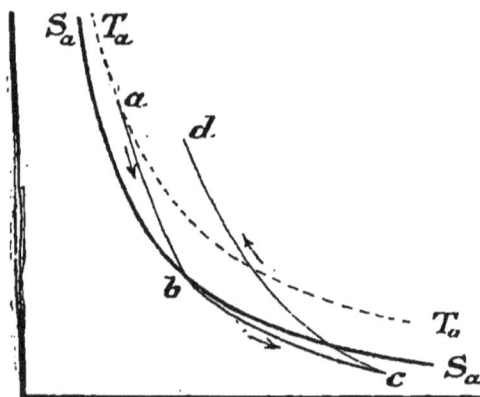

FIG. 76.

of saturation, a low angle. The temperature now
sinks by a uniformly continued expansion much
slower than before (as shown by the curve), and
water is condensed as the line bc cuts the dew-point
lines of lower amounts of vapour. The condensed
water falls as rain, and if the process is continued
far enough, later as snow; so that the line bc becomes,
in fact, only pseudo-adiabatic, or rather it is the
projection of a pseudo-adiabatic line. The hail

stadium is wholly passed over ; or even in an extreme case would be scarcely perceptible.

We will suppose the expansion to cease at c, and that compression or a sinking takes place. In this reverse process, if no water has been abstracted by precipitation, the path will be a retracing of the same track which has been already followed, viz., from c to b, and then from b to a : and the greater the amount of water retained the more nearly will this path be followed. So that when the air does not leave the dry stadium there is no noticeable change in the air when it reaches its starting-point. If, however, water is lost by precipitation, then in the process of compression the fall will be that shown by the projected line cd. In this case, as is shown in the diagram by the intersection of cd with Ta, the original temperature of the air is reached at a pressure lower than the original pressure, and in the farther descent which is necessary before the air reaches the original pressure the temperature is of course raised above the original temperature. So that now the amount of vapour is less, and the temperature greater than in the original condition, and we thus have the characteristics of the Föhn ; which are, moist cool air on the preceding side of the mountain, and dry warm air on the following side. And, as may be seen from a consideration of the diagram, this peculiarity is the more strongly marked, the nearer the starting-point a lies to the curve or saturation, and the longer the path bc within the rain stadium

Cyclones—Thermodynamics.—A further more important application is to the great vertical currents of air in the cyclonic and anti-cyclonic areas,

According to the usually accepted theory we have in cyclones an ascending current and in anti-cyclones a descending current; the current from the anti-cyclone acting as a feeder to the cyclone. In this interchange of air-masses between the two, the air is subject to very much such changes as it experiences in rising on one side of the mountain and falling on the other side, as in the case of the Föhn Wind. In the cyclonic systems, however, the distances and the times are so great that an adiabatic condition (that is, without addition or abstraction of heat), cannot be assumed. Since the ascending current in the cyclone may be hundreds of miles from the descending current of the anti-cyclone, due allowance must be made for the changes in heat just mentioned, according to the conditions which are liable to exist. In the summer the addition of heat would be the preponderating condition, and in winter the abstraction of heat; the daytime and the night-time would also have this same relative connection.

The conditions of the interchanging circulation between the cyclones and anti-cyclones in summer is shown by the following diagram (Fig. 77). The air starts from the condition a (of the cyclone) and by expansion, but at the same time receiving heat, passes along ab, which is within the adiabatic curve as the decrease in temperature would be slower on account of the added heat. When the rain stadium is reached at the point of cloud condensation, the path would be bc close to the curve of saturation; but at greater elevation, where we find a much stronger addition of heat at the time of insolation at high elevations where the condensation is less and the density of the clouds decreases, the path will be as shown by cd, and the condition will be within the dry stadium again, and will be approximately adiabatic when the water has been transformed into water vapour by the evaporation which must ensue, and has consequently lost the greater part of its capability for absorption. At *d* the mixture enters the second condensation stadium, that of the cirrus clouds. And in this stadium the curve must be truly adiabatic in most cases, as the ice needles do not fall in the form of precipitation, but remain in suspension. The curve of ascent *de* must

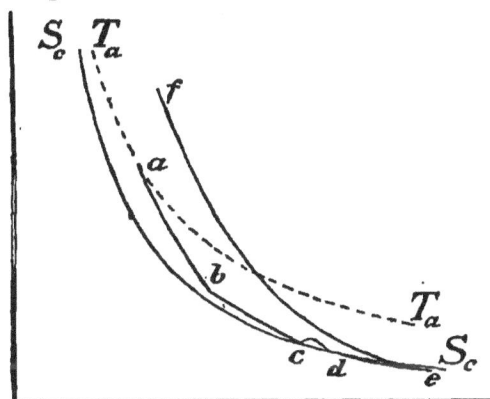

FIG. 77.

very closely approximate the curve of descent; but below this the latter, *ef*, departs more from the former as shown in the diagram.

The curve of descent *ef* is here represented by von Bezold as adiabatic, although this is not strictly true, especially in the lower layer of the atmosphere, in consequence of the warming up of these layers by reason of their nearness to the heated surface of the earth; but, as he points out, this departure from the adiabatic would imply the restoration of the condition of stable equilibrium.

The changes along *ef* take place in the downward current of the anti-cyclone, and the final pressure of the air mass on reaching the ground is greater than the pressure it originally had at the base of the cyclone, and consequently *f* lies farther above the abscissa axis than *a*.   Thus we see that *f* may lie not only above, but to the right of *a*, so that the volume at *f* is greater than the volume at *a*, and consequently the air can be specifically lighter in the anti-cyclone than in the cyclone, notwithstanding the relatively higher pressure of the former.   This is possible because of the compensating influence of the difference in temperature.

We thus see quite clearly that in the passage of air from the cyclone to the anti-cyclone, it is not entirely a question of the specific gravity of the air masses, but the dynamic relations play a very important part as has been insisted on by Hann[1] in his discussion of the observations made at the Alpine Sonnblick Observatory.   Von Bezold considers that the region above the rain clouds, and especially that at their upper limit, should receive the most careful consideration at the hands of investigators, as the conditions existing there have not been properly taken into account.

The air, then, as it approaches the earth in the anti-cyclone is warm and dry until it reaches the lower layer, where it rapidly receives moisture from the evaporation from the earth's surface which is exposed to the unhindered radiation from the sun.

The air may now pass through the conditions *fa* and again become the feeding current of an atmospheric depression—under the same assumed conditions as

---

[1] *Meteorol. Zeits.*, vol. v. p. 15.

the first cyclone—having in its lower course (*fa*) added
to its store of moisture.

In winter (and similarly the analogous night period)
the changes of condition are different from that just
outlined. Since the temperatures are not so high as
those reached in summer, and therefore the isotherms
lying at a considerable distance from the axis are not
reached, at least the beginning and final changes of
conditions occur closer to the co-ordinate axes (see
Fig. 78). At the starting-point *a* in the cyclone,
the condition is that of lower pressure and higher
temperature, while in the
final condition at *d* in
the anti-cyclone, there
is higher pressure and
lower temperature ; so
that in the diagram, *d*
must be above and to
the left of *a*. Since we
have assumed that the
distance of the point
representing the con-

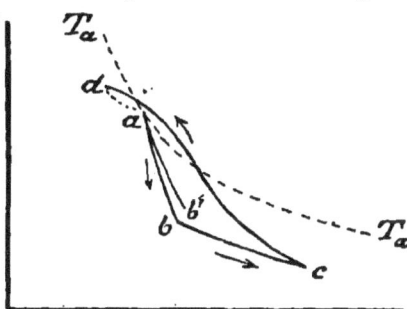

Fig. 78.

dition, from the axis of the plane of projection as
shown in the diagram, depends on the amount of
absolute moisture, then the curves which represent
the changes of condition will not reach such a distance from the axes, since the amount of moisture
remains less in the winter than in the summer.

In the early part of the ascent in the cyclone the
condition is nearly adiabatic, that is, *b* nearly coincides
with *b'* (where *ab'* is the adiabatic course) for the
path up to the upper limit of the cloud layers, as
below this the radiation and irradiation is of secondary
importance ; but if there is a deviation from the

adiabatic, it must be in the opposite direction to that of the summer, and the lines will fall more than in the latter season.   In the accompanying diagram (Fig. 78) this deviation is assumed, and it is supposed that the transformation takes place directly from the dry stadium to the snow stadium.   From this point onward the curve of condition sinks more gradually, but with continually increasing inclination, on account of the greater cooling at great altitudes, up to the turning-point *c*, where expansion ceases and compression commences.   The return path is marked by the line *cd*.   So far as we are capable of judging, the curve of descent is adiabatic at high altitudes, and the adiabatic compression proceeds as for the dry stadium ; but in the lower atmosphere, where the outward radiation is strong, the consequent cooling causes a departure from the adiabatic, inward towards the ordinate axis, as is shown by the lower part of the curve *cd*.   We thus have the schematic graphical presentation of the inversion of temperature which occurs, especially in anti-cyclones, on clear winter days.   In the last part of the descent, the path *cd* approaches and even intersects the dew-point curve near *d*, and in the latter case the condensation assumes the form of ground fog.   But with the beginning of fog-formation the radiation increases, and so the temperature decrease becomes the more marked with the approach of the descending air-current to the ground.   Whether the path *cd* can be throughout an adiabatic, requires further investigation.   If it is so, then there must be a place at some elevation in the anti-cyclone where the same pressure and temperature exists as at a lower elevation over the cyclone ; because in this case the projection of the curve of condition, *i.e.*, the curve drawn in the plane of the diagram, must intersect itself.

§ 4.—*The Hertz Graphical Solution of the Adiabatic Process.*

Although the mass of the mixture of air and vapour decreases steadily with change from one stadium to the next, as has been shown by von Bezold, and although Hertz has not considered the pseudo-adiabatic changes, yet the graphical table of the latter writer can be used for approximate solutions in so many cases and with such ease, that it is given here for the benefit of any who may make use of it, but to whom the original is inaccessible. This will prove of assistance even in the solution of non-adiabatic changes, for any practical investigation of these will probably require the auxiliary use of the adiabatic changes. The original paper by Hertz bears the title, *Graphische Methode zur Bestimmung der adiabatischen Zustandänderungen feuchter Luft*, and was published in the *Meteorologische Zeitschrift*, Nov.–Dec., 1884. While reference has been made to it by various writers on meteorology, it is believed that the very important diagram accompanying that paper has not heretofore been reproduced elsewhere.

The cross-sectional groundwork of the diagram just mentioned, Fig. 79, is so arranged that the horizontal argument (abscissas) represents the atmospheric pressures (in millimetres) to which the air mixture may be subjected ; and the vertical argument or ordinates represent temperatures between —20° C. and +20° C. But the relations are not so simple as may seem evident, for the construction is such that a chosen increase in the length of the co-ordinate corresponds to the similar increase in the *logarithm* of the pressure or absolute temperature ; the numbers themselves not being used ; as, in case they were, curved lines, and not straight or approximately straight lines, would have to be used, which is inconvenient in any graphical work.   The dotted lines passing from right to left across the page indicate the number of grammes of water present in each kilogramme of atmospheric air under the various conditions : each fifth line is made heavier than the others, and marked on the left with its appropriate number (of grammes) 5, 10, 15, 20, 25.

The courses of the adiabatics in the dry stadium are shown by the line marked *a* (see at the top on the right), and the lines parallel to it ; the courses in the rain stadium are shown by the line marked β (at the right near the top), and the lines parallel to it, but it is to be noticed that they stop at the line of 0° C. temperature, or the freezing point ; in the hail stadium the courses coincide with the isotherm of 0° ; in the snow stadium the courses are shown by γ (on the left near the bottom), and the lines parallel to it, and it is seen that they nearly, but not quite, coincide with the lines β of the rain stadium, but they stop at the line of 0° C., for at a higher temperature than this the snow would be converted into rain.

The use of the graphical table is best illustrated by an example, and the one given by Hertz for this purpose is as follows :—At the sea-level a mass of air has barometric pressure of 750 mm., a temperature of 27° C., and a relative humidity of 50 per cent. ; this mass of air is transported upwards under adiabatic conditions (*i.e.*, without addition or subtraction of

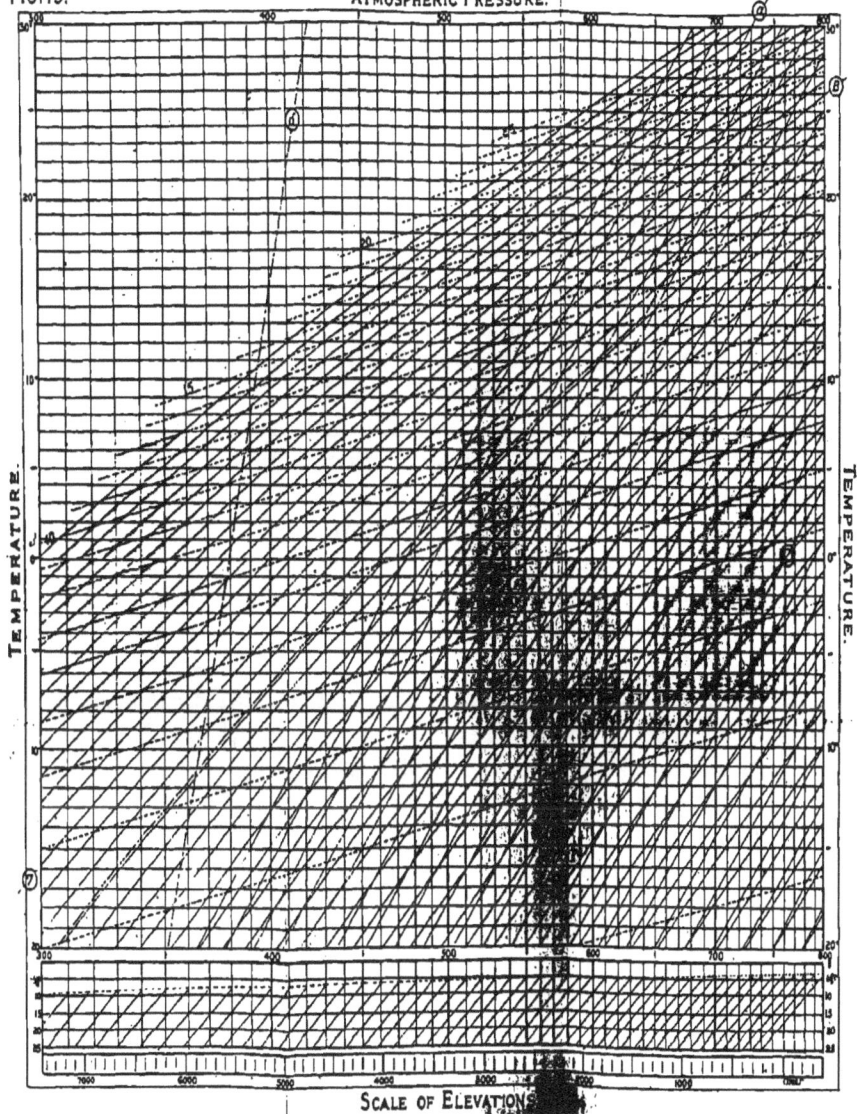

FIG. 79.   ATMOSPHERIC PRESSURE.

SCALE OF ELEVATIONS

heat), and it is desired to know the changes which it undergoes, and where it enters the successive stadia which have been discussed.

The initial or starting condition of the air mass is shown on the graphical table near the upper right-hand corner at the point where the vertical isobar of 750 mm. intersects the horizontal isotherm 27°C. This point is almost exactly on the dotted line (these give the water content at saturation in grammes) 22, as read by the argument on the left hand of the table, so that, if the air mass were *thoroughly* saturated and consequently had a relative humidity of 100 per cent. at the starting-point, it would contain 22 grammes of water per kilogramme of air ; but it has only 50 per cent. of saturation, and consequently contains only half of 22 grammes, or 11 grammes of water per kilogramme. Following the isobar 750 downwards to the bottom of the table, we find that it is about 100 metres above the zero of the scale of elevations given in metres, as this zero of sea-level has been assumed at 760 mm. air-pressure, so that the initial pressure is that at 100 metres elevation on the adopted scale, *and this amount must be subtracted from all altitudes obtained by using the table in solving the present problem.*

Now let the air mass be gradually elevated, and its condition will be represented by the broken line drawn parallel to the lines *a* from the point at 750 mm. and 27° C. on the diagram. If now we wish to find the condition of the air mass at an elevation of 700 metres above the starting-point, the altitude line at the bottom of the diagram is selected at 700 + 100 = 800 metres, and this is followed upwards to the intersection with the broken line which is found to

be at 687 mm. barometric pressure and 19°·3 C. (of
course this broken line is drawn merely to represent
the present problem, and a straight-edged ruler would
in other cases be used in place of drawing a line).
No condensation would take place until the broken
line 11 of the water-content series is reached, and
at this point, which has a barometric pressure of
640 mm. and a temperature of 13°·3, the condensation
stadium would be entered upon. And now the
direction of the broken line changes, and it will run
parallel to the lines (β), which belong to the condensation stadium. But, before following this, it must
be noticed that, if the line of 750 mm. pressure is
followed downwards to the intersection with the line
11 of the water content, the temperature corresponding to this point will be 15°·8 C., and this, then, is the
original dew-point temperature, but it is a somewhat
higher temperature than the 13°·3 C. found as the
actual dew point. This difference is due to the fact
that the air has increased in volume as well as
decreased in temperature in the ascent. The pressure
640 mm. is at an altitude of about 1270 metres, as
found by means of the corresponding altitude scale
at the bottom of the page, and at this altitude the
cloud-forming commences. Following the broken
line showing the changes in condition in the
direction of the (β) lines in the condensation stadium,
it is noticed that its inclination towards the axis
of abscissas is less steep than in the former course,
and consequently the temperature changes more
slowly with the elevation, this being due to the
freeing of the latent heat of the water vapour, and
at an altitude of 1000 metres, *within* the condensation
stadium, the temperature is still 8°·2 C., which is only

0°·51 C. decrease per 100 metres of elevation. At this elevation, too, the water-content line is 8·9, so that 2·1 grammes of water must have fallen out of each kilogramme of air, unless this water has been supported by an upward vertical current of air. Following the broken line to the intersection of the 0° C. line, or line of freezing, it is seen that this point lies at an altitude of 3750 metres, which corresponds to the pressure of 472 mm. For dry air, however, this temperature would have been reached at an altitude of 2600 metres, as is seen by continuing the broken line marking the path of change in the dry stadium down to the 0° C. line. At the 0° C. line, however, it is seen that about 4·9 grammes of water have become condensed. At the 0° C. line the water begins to freeze, and the temperature does not fall farther until all of the condensed water present is frozen, so that now for some distance the path is along the isotherm of 0° C. This distance is determined by aid of the small auxiliary table at the bottom of the diagram. We follow the isobar 472 mm. downwards to the dotted line drawn on the auxiliary table, and through this intersection draw a broken line parallel to the course in the condensation stadium. This line terminates opposite the 11 gramme point of the argument shown on the right of the page, and this termination is at the pressure of 463 mm. This last-found number indicates the pressure (along the 0° C. line of the main table) at which the water-freezing process ends, and in this case it extended through a layer of about 150 metres thickness. The decrease in temperature now continues from the elevation where the pressure is 463 mm. onward. The water now freezes as it is condensed, and as the amount of vapour becomes less, the temperature falls more rapidly as the elevation increases.

The course of the broken line is now parallel to the lines (γ), and by it the condition may be obtained for any desired elevation. For instance, at an altitude of 7200 metres and a barometric pressure of 305 mm., the temperature would be —20° C., and the amount of moisture would be 2 grammes of water vapour per kilogramme of air, the other 9 grammes having been condensed. The relation of the density at any point to that at the place of starting may be determined by drawing lines through the points parallel to (a), and noting the place of intersection with the isotherm 0° C. This occurs at the pressures 330 mm. and 680 mm. in the present case, so that the densities are as 33 : 68. To reduce this to a normal standard of 760 mm. pressure and 0° C., the densities would be as 33 and 68 to 76.

The supposition has been made by Hertz that all of the condensed water remains in the air in the hail stadium, but this would not be a matter of fact in most actual cases. If one can, by judgment or a more accurate means, determine the amount lost from the original amount of moisture by condensation and falling out, then the graphical table of Hertz will be of use in the actual case as well as in the ideal case. Suppose, for instance, it has been determined that half of the *condensed* water has fallen out, then on reaching the 0° C. isotherm there would be only 8·5 grammes of water present in a kilogramme of air, so that the 8·5 gramme line must be used instead of the 11 gramme line, and this would be reached at a pressure of 466 mm. By changing the initial conditions, as, for instance, supposing, in the example which has been used to explain Hertz's table, the relative humidity had been but 10 per cent., many interesting variations which must occur in nature can be traced out. In this case of only 10 per cent. relative humidity, but at 750 mm. barometric pressure and 27° C. temperature, the total amount of water would be but 2·2 grammes per kilogramme of air, and the so-called condensation would take place at a pressure of 455 mm. and a temperature of —13°·6 C., by which neither rain nor hail would have been formed, but there would be a sort of sublimation of the water particles from a vapour to a solid substance. This very interesting topic cannot be followed out any farther in the present connection, but enough has been said to show the usefulness and future possibilities of Hertz's method.

§ 5.—*Potential Temperature.*

In a very important recent memoir[1] von Helmholtz has introduced the term *Wärmegehalt*, which we may translate *heat-content*, and which is defined as the absolute temperature which a mass of air assumes when it is brought to the normal pressure under the adiabatic process. Since, however, it is a temperature and not a certain amount of heat which this term is intended to express, von Bezold proposes to replace the term *Wärmegehalt* by that of *potentielle Temperatur*, and which we may call the potential temperature. Von Bezold has applied this idea to some meteorological conditions, and has extended the limitation of this original definition to include what he calls pseudo-adiabatic changes. And this is a very im-

---

[1] *Ueber Atmosphärische Bewegungen, Sitz. ber. d. Berlin. Akad.,* 1888.

portant proviso in all questions of applying the
principles of thermodynamics to atmospheric changes,
for the truly adiabatic condition of change must be the
exception. As before stated in an ascending current
of air in which condensation has taken place, a truly
adiabatic process would occur only when the rain
does not fall, but is carried along with the current
until changes in pressure cause re-evaporation to take
place, and the moisture again exists as a vapour. In
the case where the raindrops leave the mass of air

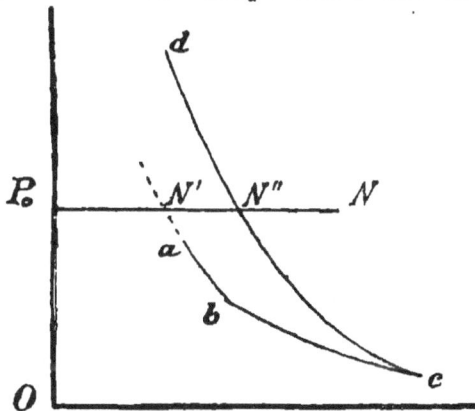

FIG. 80.

by descent, as is usually the case in nature, the process
of change in the condition of the air would then be
what von Bezold calls pseudo-adiabatic.

This potential temperature has also been illustrated
by von Bezold's graphical method. For a constant
pressure, the absolute temperature is proportional to
the volume, or, in von Bezold's diagram, the abscissa
OV.

Von Bezold gives the following rule :—

"For any given condition such as is represented by

a in the diagram (Fig. 80) one can find the corresponding potential temperature by drawing an adiabatic through a and finding its intersecting point N' with the line P_0N drawn parallel to the abscissa at the distance P_0 from it. The distance of this point of intersection from the ordinate axis, that is, the abscissa of the point of intersection furnishes a measure for the potential temperature."

In the dry stadium the potential temperature corresponding to any definite point of departure has a constant value, but in the stadia with precipitation the potential temperature will increase with the amount of precipitation which leaves the air mass.

In the same diagram (Fig. 80) let the adiabatic of the dry stadium which passes through a intersect the dew-point curve (not shown) at b, and let further expansion of the air take place, then the path bc will be over the adiabatic (or pseudo-adiabatic) of the rain or snow stadium. If, now, it is the object to find the potential temperature for the point c (as conveniently placed on the diagram), which must be done by reduction to the normal pressure by adiabatic process, then the path of this adiabatic cd will not be back over the course abc, on account of the water which has fallen out and has thus been lost to this mass of air; and the normal pressure of the dry stadium which is reached over the new path of condition cd will be one with a less amount of water vapour than before. Let the point at which cd reaches the normal pressure be N'', and let the line ab be produced in the reverse direction until it intersects P_0 at N'; then P_0N' is less than P_0N'', or, in other words, the original potential temperature T' is less than the final temperature T″, which belongs to the air mass after the water has fallen from it.   After pointing out this, von Bezold proposes the following theorem :—

"In adiabatic changes of condition of moist air, the potential temperature remains unchanged as long as the dry stadium is not departed from, but it increases when condensation occurs, and the degree of increase is the greater the more water the air loses." Briefly but more generally stated this theorem reads :—

"Adiabatic changes of condition in the free atmosphere—excluding, however, evaporation—leave the potential temperature either unchanged, or else increase it."

The author of this important theorem calls attention to Clausius' somewhat similar but not identical theorem concerning the increase of entropy.   This comparison will be readily understood by a reference to p. 163 of Clerk Maxwell's *Theory of Heat* (6th edition, 1880), where, after explaining the meaning of the term entropy, he says that Clausius remarks that "the entropy of the system always tends towards a maximum value."

§ 6.—*Vertical Temperature Gradient.*

While the horizontal air currents are in general subject to only slight thermodynamic changes, the vertical currents are subject to powerful changes, on account of the very rapid expansion of the air which takes place in ascending currents, and the rapid compression in the descending currents.   It is true that the inclination of the ordinary winds to the horizontal is very slight for a free flowing current of air, and in these the change in altitude is gradual ; but there are numerous causes which give rise to more nearly vertical currents in which the change in altitude is so rapid, or even almost sudden, that very great thermodynamic changes must take place within a few moments.

The changes in the amount of heat received or given off by radiation by the greater portion of the air, are so small as to be negligible, except in certain regions, such as that close to the earth's surface, where the superheating of the ground, and vapour dew, and frost-formation affect the changes in the air temperature ; and also in the fog or clouds where the changes due to absorption or emission of heat and evaporation readily take place, and especially at the upper surface of the clouds. Commonly, then, the changes in the air moving in vertical currents may be considered as adiabatic, with the exception of such regions as have just been noted. And even in these regions, as von Bezold points out, the adiabatic condition is a sort of average between the conditions when the absorption or emission of heat is in excess. Still, for application, the times when there is a preponderance of emission of heat must be examined separately from those when the amount of heat absorbed is in excess. The former, as shown in Fig. 78, can be applied to the winter season, the night-time, or the cold zone ; and the latter to the summer, the day-time, or the warm zone of the earth.

The average or normal condition for the ascending and descending air currents is given in Fig. 80. There, as before stated, *ac* represents the whole condition during the ascent, the portion *ab* is confined to the dry stadium, and *bc* is in the stadium of condensation ; and *cd* represents the condition during the descent. It is seen that, as might be expected, this figure bears a close agreement to that representing the Föhn phenomena as given in Fig. 76. There is some difference however. The branch *cd* of the curve in the normal plan (Fig. 80) is larger than that in the Föhn diagram, because in the interchange between a cyclone and an anti-cyclone the final point *d* at the base of the anti-cyclone has a greater absolute pressure than at the starting-place in the cyclone, which is not the case in the Föhn. So that the process in the latter case really forms only a portion of the process which occurs for the former.

Referring again to the normal plan (Fig. 80), and applying the idea of the potential temperature, von Bezold finds that :—

I. " In the ascending branch (*ac*) the potential temperature increases steadily from the beginning of condensation, and in the descending branch (*cd*) it remains constant at the maximum value reached in the entire process. This maximum value corresponds to the highest point which the air has reached in its ascent."

II. " The potential temperature of the higher layers of the atmosphere is in general higher than that of the lower."

This last theorem is founded on the fact that, in the continued interchange between cyclones and anti-cyclones, the potential temperature in the lower layers must possess some mean value between that of the maximum value T'' and the lesser temperature T' at *a*, the base of the cyclone ; and this mean value must be less than that of T'' at the highest point *c*.

From this it can be shown that the vertical tempera-

ture gradient, or the temperature decrease with elevation (per 100 metres, for example), is on the average
less than that theoretically derived for the dry
stadium.   This last is 0·993° C. per 100 metres.

We have signified by T the potential temperature,
and we will now let the actual temperature be represented by t, and let ta be the temperature at a at the
base of the cyclone, and td that at the point d at the
base of the anti-cyclone. Now while in the dry stadium
the temperature ta will not differ much from Ta, nor
td from Td, yet they will differ slightly and in definite
directions.   At the base of the cyclone the atmospheric pressure pa will be lower than the normal
pressure to which the potential temperature is
referred, and at the base of the anti-cyclone the
pressure pd will be higher than this normal pressure.
Consequently the corresponding temperature ta is
less than T', and td is greater than T'' ; because in
referring them to the normal pressure ta must increase
and td decrease.   This reasoning applies of course to
a well-developed cyclone and anti-cyclone, in which
the normal pressure lies somewhere between that of
the two.   But as T'' is greater than T' on account of
the greater pressure in the anti-cyclone, so td must be
still more in excess of ta.

If the temperature at c (or tc) is taken from that at
a (or ta), and that at c (or tc) is taken from that at d
(or td), we have the total decrease in temperature with
elevation for the rising and falling currents.   This difference must be the less for the rising current since we
have just shown that ta is less than td, at least for a
purely adiabatic process.   The vertical gradient in
the condensation stadium is less than in the dry
stadium.   Von Bezold establishes the following
theorem :—

" For purely adiabatic ascension and descension with entrance into the condensation stadium, the average vertical temperature gradient in the ascending branch is always less than in the descending."

When the ascending and descending current occurs in the same locality (*i.e.*, very close to each other), then the temperature gradient for the air will have a value between the nearly constant gradient in the descending and the variable temperature gradient in the ascending branch of the current. It must be remembered that the temperature in the descending branch is nearly that derived from theory, while that in the ascending one varies with the temperature at the start, and the amount of water in the air. The following theorem is then established :—

" The average vertical temperature gradient under adiabatic change of condition is less for moister air, which reaches the condensation point, than for drier air." And this furnishes at least one explanation of the actual deviations from the computed gradient for dry air, which observations show exist in nature, still considering the change to be truly adiabatic. In considering the variations from this ideal condition of change, the conception of the " potential temperature " renders a very important service. The ideas which have just been elaborated if expressed in the nomenclature of " potential temperature," are as follows :—

" If the potential temperature is constant throughout the air-layers under consideration, then the vertical temperature gradient is that derived by theory."

" If the potential temperature in the upper layers of air is higher than in the lower, which is generally the case, then the temperature gradient is less ; and is less in proportion as the difference in potential temperature is greater for any given difference in altitude."

Now, if the lower layers become markedly cooler the "potential temperature" decreases, and consequently the corresponding temperature gradient decreases, and up to a moderate altitude can even become negative, so that the lower layers are cooler than those above, as in those cases of temperature inversion which occur in nature, and which have caused much discussion as to their true significance.

Such a marked cooling of the lower layers of air occurs at the time when there is greatest loss of heat, which is in the unclouded region of an anticyclone, and in the winter or night time ; so that at these times the vertical temperature gradient is less than in summer or by day. While this lessening, or even reversal, of the temperature gradient can be easily shown to obtain for mountainous regions, by a comparison of the observations made at peak and neighbouring base stations, its complete proof for level regions must depend on observations made in balloons. It is probable, however, that there is a good deal of truth in Woeikoff's [1] assumption of the occurrences of this inversion in the great winter anticyclones of East Siberia ; but he carries it too far when he insists on its application, in the use of thermometric data, as the basis of the isotherms for that region.

The actual influence of the outward radiation of heat, on the diminishing, or even reversal, of the temperature gradient, needs more careful investigation ; the discussion of vertical differences of temperature

[1] *Met. Zeits.*, band i. p. 443.

addition of moisture from the earth's surface. This air, when it comes from the upper part of the anti-cyclone, possesses a higher potential temperature and a higher absolute temperature than the ascending air with which it mixes in the cyclone, so that the rate of cooling of this last mentioned air is retarded, and the same retardation takes place in the commence-ment of condensation, for this air entering from the anti-cyclone has a less water content.

So we see that below the cloud limit, in the region of ascensional cyclonic air-masses, the temperature gradient will not be greater than that derived accord-ing to the law of adiabatic change in the dry stadium, and without the mixing with additional air masses. In the descending current of the anti-cyclone the same is true ; however, in not to so marked a degree, for the greater part of the air in the upper half which comes over into it from the cyclone has not reached the highest layers belonging to the cyclone, and has not been subjected to the maximum effects of condensa-tion, and has not reached the maximum " potential temperature." So that in reality the actual conditions of ascent and descent must deviate somewhat from the schematic presentation already given, and they more nearly approach an average condition compounded by considering the upward and downward motions as a single system. The fact that for not very great alti-tudes the observed temperature gradient, even with cloudiness, is so frequently far below that computed for the dry stadium, is cited by von Bezold as an example of this.

In the actual circulation we also find the original heat, if we may so term the amount of heat with which the air starts upwards in the cyclone, aug-

mented by the heat, set free in the condensation
which takes place, which assists in the work of
expansion, thereby causing the actual total cooling
to be less. When, at the base of the anti-cyclone, the
air is colder than that due to adiabatic change, and
even in the case of inversion of the temperature gradient, then the temperature at this terminal point is
higher than it would have been if the air had simply
glided along horizontally from the base of the cyclone
to this point and not been subjected to the changes
due to the change in altitude.

Those regions at which the descending current
reaches the earth are, consequently, the ones benefited
by the heat of condensation—the freed latent heat.
This principle is strongly brought out by von Bezold,
who compares the whole process with that of a steam-
heating apparatus. We have, then, in the cyclone
the moist air ascending and cooling until the stadium
of condensation is reached, when the cooling does not
proceed so rapidly, on account of the heat of vaporisation being supplied to aid in the work of expansion,
and the heat thus saved is not lost in the descent in
the anti-cyclone, and makes itself felt by an increased
temperature at the base of the latter.

Von Bezold proposes to call " compound convection," that circulation in which, besides the transformation of heated or cooled masses, there is an actual
change in the aggregate condition ; such as when
vapour forms in one place and is precipitated in
another, or when ice is falling in the form of snow or
hail. Making use of this definition, von Bezold
arrives at the following theorem :—

" As a consequence of ' compound convection,' the
temperature in an anti-cyclonal region is always

higher than it would be in the case of a simple conduction."

In making an application of this to the warm zone
of the earth (broadly considered), it must be remembered that the consideration of the general circulation
and pressure-distribution of the atmosphere has shown
a ring of high pressure just outside of the tropics, and
between these is the calm zone. According to the
conditions prevailing at these rings, the above theorem
shows that in them the temperature is much higher
in consequence of " compound convection," than it
would be if only dry air were concerned in the circulation, or if no vertical currents existed.

To this circumstance the warm zone owes a considerable increase in its extent, and a decrease in the
temperature differences. In the calm zone the use of
heat in evaporation, the sheltering from solar radiation, and the descent of water from cooler regions, all
combine to hinder the temperature from becoming
excessive ; while the heat used in evaporation at the
earth's surface shows itself again in the region of the
clouds as work, in the heat of condensation, and so
diminishes the cooling of the ascending currents, and
comes to light again in the two zone girdles having
descending currents.

The practical application of these ideas which von
Bezold has so ably developed will probably cause
considerable change in the now accepted explanation
of the transformations of atmospheric conditions
which are due to thermodynamic causes. Thus we
have been taught that the descending trade wind,
since it becomes cooled, deposits in the higher latitudes the vapour carried from the calm zone ; while
from the foregoing, the water carried upward in the

calm zone must fall there as tropical rains, and the air now made drier must sink again warmer than when it started, unless some outside influence cools it.

§ 7.—*Air Mixture.*

From the time of Hutton's communication of his famous rain-theory [1] down to a quite recent date, the mechanical mixture of masses of air of different temperatures and humidities was regarded as accounting for many of the phenomena which are now explained on an almost purely dynamic basis.   Espy's views, promulgated half a century ago, did not receive the attention which they merited, and we had to wait another quarter of a century before any real value was attached to the importance of purely dynamic changes.   In Wettstein's paper [2] we find the opposite extreme view taken, that the mechanical mixture of the air would never alone cause the condensation of vapour and the formation of rain, but he would explain the whole phenomena as a result of dynamic changes entirely ; which is also untrue, but not so wide of the truth as Hutton's theory.   Hann [3] seems to have been the first person to lay before meteorologists the proper relative importance of these two theories ; he considered that while the mechanical mixture can produce precipitation, yet it must be very slight in comparison with the amounts actually observed, and which the dynamic theory is not unable to explain.   That there is undoubtedly a joint action of the two processes he was willing to admit, but the

[1] *Edinburgh Transactions, &c.,* 1788.
[2] *Vierteljahrsschrift der naturfors. Gesellschaft,* Zurich, 1869.
[3] *Oesterreichischen Zeitschrift für Meteorologie,* 1874.

latter far overshadows the former in importance.
A later mathematical treatment of the problem of
mixtures of air and vapour, by Pernter,[1] has added to
the clearness with which we are now able to see the
conditions and results of the process. Von Bezold
has taken up the subject as a continuation of his own
memoirs on the thermodynamics of the atmosphere,
and has given us a general development [2] of the subject, which is certainly a step in advance of the other
contributions which have been published on this same
general topic. Some of the matter presented in this
memoir will now be considered.

Von Bezola's Explanation of Air Mixture.—As a
result of the motions of the air, some in a vertical and
some in a horizontal direction, there must result a frequent mingling or mixture of air masses which have
very different amounts of moisture, and different temperatures. The barometric pressure will be practically the same for each of the masses of air, as they
must be at the same altitude ; and this equality and
practical constancy of pressure, greatly simplifies the
study of the process and results of mixing of air-
masses.

The relations of the component air-masses and the
final mixture is well shown by von Bezold by means
of a diagram. In Fig. 81 the temperatures are
marked off on the horizontal line of abscissas, and
the vapour quantities are represented by the vertical
ordinates ; the origin O being as usual on the left.
OT_x represents the temperature, and T_xF the quantity

[1] *Oesterreichischen Zeitschrift für Meteorologie*, 1882.

[2] *Zur Thermodynamik der Atmosphäre—Dritte Mittheilung*, Von
W. von Bezold ; *Sitzungsberichte der K. Preussischen Akad d. Wissenschaften*, Berlin, 1890.

of vapour of one component mass of air ; OT_2 the temperature, and T_2F_2 the vapour of the other component ; and OT_3 the temperature, and T_3F_3 the quantity of vapour in the mixture. Then the line $F_1F_3F_2$ is a straight line. T_1F_1', T_2F_2', T_3F_3' are the respective amounts of vapour necessary in the two components and the mixture, in order to make the saturation complete. The line $F_1'F_3'F_2'$ is a curved line, because the amount of vapour necessary for saturation increases with the temperature ; and this line is called the curve of the saturation quantity belonging to the common pressure of the air-masses.

It is readily seen that the straight line, $F_1F_3F_2$, will approach nearer or recede from the curved line, $F_1'F_3'F_2'$, ac-

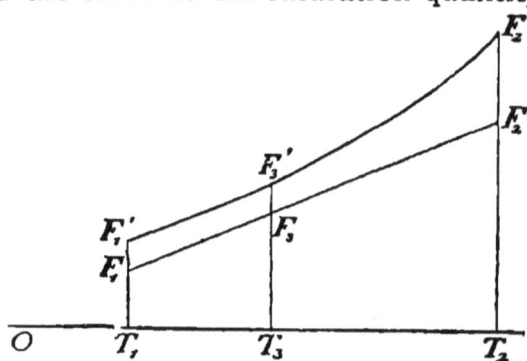

FIG. 81.

cording as the amount of vapour present is a greater or less quantity ; and it will have various inclinations to the horizontal line, T_1T_2, depending on the temperatures of the two component air masses. It is also seen that the line F_1F_2 will not touch or intersect the curve $F_1'F_2'$ as long as the actual amount of moisture present does not too closely approach the amount necessary for saturation ; but when intersection does take place, numerous conditions of condensation and precipitation will arise, depending on the magnitude and form of the intersection.

18

It will naturally be asked what will be the amount
of humidity necessary in one or both of the component air masses at given temperatures in order to
cause condensation.   Von Bezold finds an aid to this
solution in the determination of the minimum amount
of the mixture necessary for condensation when the
components have the same relative humidity.   In this
case, as shown in Fig. 82, the line F_2F_1 extended
to the temperature axis will meet the prolongation of
the line $F_2'F_1'$ at P.   When, for a common humidity
of the components, the proper mixture has been

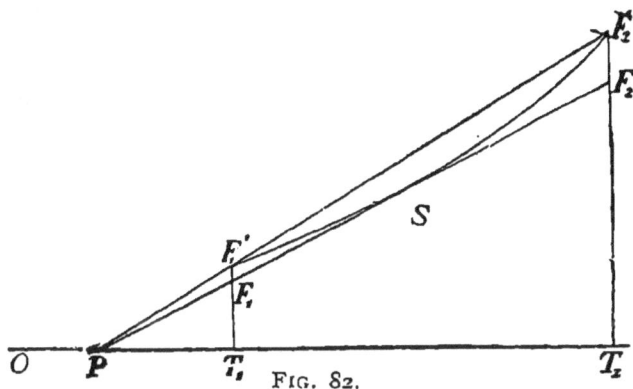

FIG. 82.

reached which will just produce saturation, then the
straight line PF_1F_2 must just touch the curve of the
saturation quantity, $F_1'F_2'$, at some point S ; and the
location of this point shows directly the temperature
of the mixture at which saturation occurs.   As is
easily seen from the diagram, the relative humidity
of the mixture must exceed that of at least one of the
component air-masses, before condensation can occur.

It may also be a question as to the limiting amount
of relative humidity which must be exceeded in one
component air mass, when that of the other com-

ponent is given, in order to produce condensation in the mixture.

Suppose in this case the relative humidity of the cooler component air mass is 100, that is, saturation is complete; then if in Fig. 83, $F_1'F_2$ is drawn tangent to the curve of saturation, $F_1'F_2'$, the point F_2 will give the required minimum moisture necessary for the warmer component to possess in order to produce saturation in the mixture, and condensation will occur in a degree proportional to the elevation of the point F_2 (not shown), indicating the actual moisture in the warmer component, above the point of intersection, F_2, shown in the diagram.

If the warmer component was saturated, then draw the line $F_2'F_1$ tangent at F_2' to the curve of the saturation amount, and the point F_1 will show the minimum amount of moisture necessary for the cooler component, in order to produce saturation in the mixture.

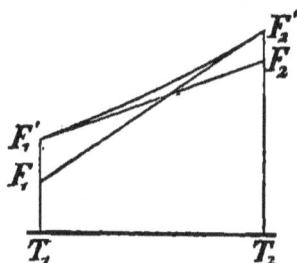

FIG. 83.

By a consideration of this, von Bezold arrives at the following result: when warm saturated air is mixed with cooler, the latter can possess a high degree of dryness and still condensation can take place for a suitable mixing.

It is found that when saturated warm air is mixed with unsaturated cool air, condensation occurs much more easily than when saturated cool air is mixed with unsaturated warm air.

A beam of saturated warm air penetrating into a mass of cool air will be accompanied by a stronger condensation than in the case of a beam of saturated cold air entering into a mass of warm air. This is well illustrated by the examples offered by opening the door into a laundry on a cold day, and then opening the door to an ice-house on a warm day. In the former case fog is instantly formed, but not in the latter.

When the point F_3 is above the line of saturation, then some amount of rain water will be condensed, and will fall out, under ordinary circumstances. It was formerly supposed that all of the water represented by the distance F_3F_3' would fall out as rain ; but comparatively recent investigations, especially those of Wettstein and Hann, have modified this idea, and show that the actual amount of rainfall is less than this amount. The reasoning by which this result is obtained is based on the supposition that just after the mixing the water is present as supersaturated vapour, and that as the actual condensation takes place gradually, heat is liberated, and the temperature is raised, so that the capacity for moisture is raised somewhat.

A graphical representation of this is given in Fig. 84, where the previously used lettering has the same significance. Here T is this increased temperature of the mixture, and F the corresponding point on the curve of the saturation quantity. So that F_3i is really the amount of water lost out.[1]

It is especially interesting to know the conditions under which two air masses of given temperature and

---

[1] The line F_3F is so drawn through F_3, that it makes with OT_2 an angle whose arc tangent is $\frac{1000}{r}c$; where c is the heat capacity of moist air and r is the heat of vaporisation. The cotangent of this angle or $\frac{r}{1000}c$ may be placed roughly at 2·5 when the temperature is above freezing, or 0° Centigrade ; but this value reaches 2·9 for temperatures under 0° Centigrade.

humidity must be mixed, in order to produce the maximum amount of precipitation. This will occur when the line F_3F reaches its maximum value, and consequently happens when the tangent to the curve at F is parallel to F_1F_2. This special F for the maximum precipitation being found, a line is drawn through it parallel to the line FF_3 as shown in the diagram, and the intersection with the line F_1F_2 gives the F_3 for the maximum precipitation; and when this F_3 is found, then T_3, and consequently T_1T_3 and T_3T_2 are at once obtained. (See previous footnote for determining the direction of FF_3.)

Von Bezold has computed a little table (given in his original memoir) showing the conditions resulting from some selected

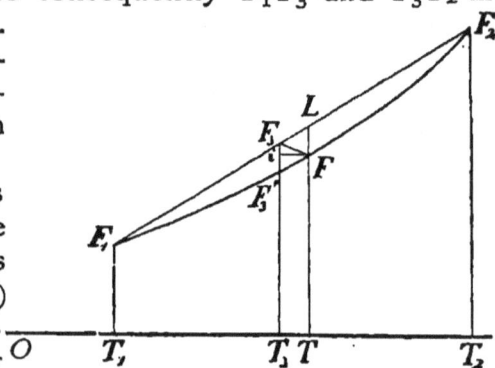

FIG. 84.

cases of air mixture. It is found that there are many cases in which no condensation can occur when the cooler component air mass is dry; and it is plainly shown, in agreement with the results obtained by Hann and Pernter, how small the amounts of precipitation really are which are produced by the process of air mixture alone.

Processes in forming Precipitation. — The causes which give rise to condensation of water vapour in the atmosphere are, direct cooling by contact with colder bodies or through radiation; adiabatic expansion, or expansion with insufficient additions of heat;

and the mixture of moist masses of air having different temperatures.

In causing actual precipitation, these various processes are effective in the order in which they are given, as the following example given by von Bezold will show.

Let it be supposed that at a barometric pressure of 700 mm., two masses of saturated air become mixed together, the one mass having a temperature of 0° C., and the other a temperature of 20° C. The greatest amount of rainfall that is possible from this mixture is 0·75 grammes per kilogramme of mixture of air and moisture, and the warm air will be cooled down to 11° C. In order to obtain this same amount of rain from the warm air alone by adiabatic cooling, it would have been necessary to lower the temperature from 20° C. to only 18°·4 C., which could be accomplished by allowing the air to flow upwards through a vertical distance of 310 metres. Or if the warm air at 20° C. had been cooled down to a temperature of only 19°·2 C., by contact with a cold body or by loss of heat through radiation, this same amount of rain would have been produced.

Of course in actual experience, any or all of these processes may be in operation at any one time; but it is readily seen that the method of mixture is the one which has by far the least efficacy in producing actual rainfall. It does, however, have a very important *rôle* to play in the formation of clouds.

On the other hand, the direct cooling of the air by contact with colder bodies, or by radiation, is the most efficacious, and a very slight cooling by this means is sufficient to produce rainfall. It is probably only in very extreme cases that a difference in temperature can exist in two adjacent masses of air, sufficient to produce rainfall at all, when they are simply mixed together.

In order to build up this theory of rainfall, it was found necessary to assume that a condition of super-saturation exists before the condensation into rain actually commences. This supposition of Hann's is rendered very plausible by the researches of Aitken, Coulier, Mascart, Kiessling, and Robert von Helmholtz on the subject of air moisture.

It has been found by laboratory experiments that when air is perfectly freed from dust particles, it may become super-saturated with moisture, but when fine particles of matter are suddenly introduced precipitation takes place ; and probably also electrical discharges will produce this same result. But there is still a great deal to be proven experimentally, however, before much of a superstructure of theory can be built on this foundation.

That sudden rainfalls, in which a considerable quantity of water is precipitated within a short time, may be the result of this condition of super-saturation is not improbable, but that the vast quantities of water precipitated by cloud bursts can be held up by this alone, is not conceivable. There must exist in these cases a supporting power given by upward air currents.

Aside from the actual precipitation resulting from the causes just mentioned, their influence in cloud and fog formation must be examined more closely. The cloud formation must precede the phenomena of precipitation, and is, then, the result of the same general conditions ; and the dissipation of fog or cloud, which results from these processes of condensation, is accomplished by the reverse conditions : direct heating by contact with warmer bodies or radiation ; adiabatic compression without the necessary abstraction of heat ; and the mixture with other air masses of proper temperature and humidity.

The method of direct cooling and direct warming up is that which causes the formation and disappearance of the fog which extends from the ground upwards. The ground becomes cooled by radiation of heat, and the air in contact with it becomes cooled, causing condensation : the next air layers then become cooled, and .condensation takes place in them. The building process of ground-fog formation proceeds, then, from the ground upwards ; and its dissipation takes place in the reverse manner. The reason that no considerable precipitation occurs as a result of this cooling, is, that the continual formation of a higher fog layer prevents the further excessive cooling of the lower layers by radiation. Condensation by direct radiation can occur in the upper layers of the atmosphere, only when a cloudiness has first been produced by one of the other processes, or by means of a ground work (or frame work) of smoke or other impurity in the air.

Von Bezold also represents the formation and dissipation of clouds on their upper limit, especially in the case of stratus clouds, as being similar to the already cited case of fog near the earth's surface ; the loss of heat by outward radiation being formative, and the gain of heat from the sun being dissipative.

The summer cumulus clouds with horizontal base, the thunder clouds, and the rain clouds are the result of the adiabatic expansion which must take place when ascending air currents are present ; and their dissipation will result from the compression which occurs in descending air currents.

It is seen that in the case where the cooling or warming up is directly accomplished, the formation and dissipation of clouds is comparatively a simple progressive process ; but when this is accomplished according to the method of mixtures, the process becomes more complex. The condensation may not occur until the last stages of the cooling by the mixing process, and likewise its dissipation may be retarded.

In breathing into cold air, the air expelled is saturated when it leaves the nostrils, but the condensation into a fog does not take place until actual contact and mixture with the cold air ; and then by further mixture with the dry (though cold) air, the fog is dissipated. By a careful consideration of the conditions of the mixing of air, von Bezold has shown that when saturated cool air is mixed with larger masses of saturated warm air, then the warming up of the mixture takes place rapidly at first and then more slowly: the reverse is true when the cold air is in excess, the mixture then cooling more rapidly as the process continues. A similar condition prevails when the amounts of condensed moisture are considered, and the maximum amount occurs when the cool mass of air is in excess.

Put into the form of a theorem as expressed by von Bezold, this idea is stated as follows :

" Condensation takes place more rapidly when a beam of cool moist air enters a large mass of warm air, than when a beam of warm moist air is blown into a mass of cool air."

The following classes of fog and cloud are considered by von Bezold as the result of mixture of air masses having different conditions of temperature and humidity :

1. The fog over warm moist surfaces where the cooler air is also present, as in the case of the surface ocean fogs during the cold season.

2. The clouds which are formed at the common boundary of two air currents moving with different velocities, by which a regular succession of clouds is formed as a consequence of a wave motion, as first recognised by von Helmholtz and named by him *Luft Wogen ;*[1] in which, however, the adiabatic condensation arising from the surging upwards must also be considered.

3. The layers of stratus clouds formed at the juncture of two such surfaces of air of different velocities and conditions, and which frequently at first have the form of atmospheric waves (*Luft Wogen*) but afterwards become connected.

4. The banner-like clouds which form and dissolve on mountain peaks, and at mountain passes, when the contour is such that warm or cold masses of air may have currents of air of other temperatures penetrating or flowing into them.

5. The tattered or loose clouds which occur frequently with strong winds, and especially in thunderstorms, and which are constantly undergoing a change of form.

Although the combinations of the three processes of cloud formation, which have been mentioned, may produce cloud formations which it is impossible to

[1] Perhaps this should be translated " Air Billows," on account of the huge dimensions of the air waves.

relegate to any one class, and such in fact is the actual state as shown by observations, yet it does not seem probable that any satisfactory study of the physiology of clouds can be made unless the causes of their formation are studied with these simple processes as a basis. When it is known, from observations outside of the clouds themselves, under just what conditions the clouds are formed, then the most active agency in their formation can probably be pointed out.

In making studies as to the causes of the peculiar shapes of clouds, von Bezold very justly remarks that too much reliance must not be placed on laboratory experiments, which are serviceable for showing the direction of motion of air, but which take no account of the physical changes which have just been shown to be of the utmost importance in cloud-building. For instance, the beautiful experiments of Vettin with clouds of smoke are very useful for following the motions of the air, but of unsaturated air, and yet this is the very condition in which cloud or fog formation does not occur. It is quite possible that out-of-doors experiments on cloud formation will be conducted in the future, in connection with the futile attempts at artificial rain production, which are receiving the support of governments and individuals.

It is impossible to interpret correctly the motions of clouds unless their mode of formation has been studied. Recent studies by Clayton at the Blue Hill Observatory (Massachusetts, U.S.A.) have shown this very plainly; and many old and more widely-known cases can be cited in the phenomena of the Föhn-cloud, the Table-cloth on Table Mountain, a smoke-like cloud of Ceylon, and others.

CHAPTER IV.

THE GENERAL MOTIONS OF THE ATMOSPHERE.

§ 1.—*Development of Modern Theories of the Primary Atmospheric Motions.*

HISTORICAL.—The history of the development of the existing theories of the cause and maintenance of atmospheric circulation is a most interesting one to those who have followed up each step of its progress, and as some of the later investigations are but little known to English readers in general, it may be well to give a somewhat full account of what has been done. The proverbial variability of the wind, especially in the latitudes occupied by civilised man, indicated the apparent hopelessness of finding any laws governing the atmospheric motions. But as reports concerning the wind were accumulated from various parts of the earth, and especially by navigators, it was found that there were regions where the winds were not continually shifting, and sea captains found that if they could once get their ships into the paths of these regular winds they could be depended on with almost as much confidence as we now place in steam. These "Trade Winds," as they came to be called, were found to be so regular that physicists concluded that they at least must be governed by some law. The Englishman, Hadley, whose services are so far forgotten that even the comprehensive *Encyclopædia Britannica* fails to give a sketch of his life and work, published a short paper in the *Philosophical Transactions* for 1735, in which he advances a theory of the "Trade Winds"; and this explanation was so plausible that it was accepted unquestioned for over a century.

Hadley's theory was well worthy of the high esteem in which it was held, for his conception was as nearly correct as the knowledge of his day would allow. He gave the great difference in temperature between the tropical and polar regions as the prime cause of these great atmospheric currents, and explained that their observed direction was due to the action of the rotation of the earth, by which a mass of air moving from the equator poleward retained the velocity of motion of the parallel from which it started out, and as it moved poleward it would have a greater easterly velocity than the parallels which it crossed, and consequently it would pass from the meridian on which it started to those east of it. Conversely the counter-current of air passing from the polar regions towards the equator proceeds from the meridian of its starting-point to those westward of it. But it was assumed that only those currents moving along the meridian were affected in this way, and that a current passing along a parallel of latitude would suffer no deviation. It has since been shown that this view was erroneous. Hadley's error was due to incomplete knowledge of the mechanics of the problem. Mathematicians had not yet solved the problem of determining the effect of the earth's rotation on the motion of a free body moving along its surface. It was necessary to wait a century after Hadley's theory had been published before a complete solution of this effect of the earth's rotation was given to the world. In 1839, Poisson, the eminent French mathematician and physicist, published a lengthy memoir in the *Journal de l'Ecole Polytechnique* bearing the title " Recherches sur le Mouvement des Projectiles," &c.

In this paper, which was read before the Paris Academy in 1837, it was first shown and definitely announced that the effect of the rotation of the earth on its axis on a free-moving body passing along near its surface, is such as to cause a deviation to the right in the northern hemisphere (and to the left in the southern hemisphere), no matter in what direction the body may be moving.[1]

If Poisson had realised the importance of his discovery when applied to the motions of the atmosphere, his genius would probably have enabled him to effect that development of aero-dynamics which was given to the world twenty years later, but which did not secure a general recognition until nearly forty years after Poisson's memoir was published. As it was, Poisson merely deduced the very slight effect the earth's rotation has on projectiles, and the matter was dismissed from his mind, and apparently also from the minds of his readers. They probably all thought that such a small force could not play a very important *rôle* in any practical matter.

In the meantime Dove, the most eminent meteorologist of his day, using Hadley's theory as a basis,

---

[1] The gradual evolving of the principles of dynamics, which at last led to the solution of this problem, has been traced in a semi-popular paper by Günther in the German magazine *Humboldt* for 1882.

had mapped out a theory of the primary circulation of the air, usually referred to now as the Hadley-Dove hypothesis. This emendation of Hadley's theory was generally accepted by the meteorologists of the time, and even now it is clung to by most compilers of works on physics who include meteorology in the circle of sciences treated in their volumes. Maury's study of the winds on the oceans gave a better opportunity than had before existed to realise the actual conditions of the general atmospheric circulation ; and his *Physical Geography of the Sea* was widely read, especially in America. This book came into the hands of a school teacher then residing in Nashville, Tennessee. This teacher, whose name has since become known to every student of meteorology, was William Ferrel. It was probably no mere chance that he read the book, as a few contributions he had made to *Gould's Astronomical Journal* show that he was deeply interested in such questions. In referring to Maury's work in conversation with his friends, Mr. Ferrel expressed the belief that the explanation of the causes of these atmospheric motions was erroneous, and that the action of the forces at work had not been correctly understood. The editor of the *Nashville Journal of Surgery* being one of this circle of friends, asked Mr. Ferrel to write out his objections to Maury's theory, and also to give his own views. The result was the paper published by Mr. Ferrel in that journal in 1856. In this paper, which was written in a popular style, we find outlined his views of the general circulation of the atmosphere, and the causes which produce it. A prominent feature of this paper was a diagram (given later as Fig. 87) showing the motions of the air according to this theory, and the present writer remembers hearing the most eminent of German meteorologists remark, some years ago, that he considered it truly wonderful that Ferrel should have been able at that time to make a sketch which should so nearly agree with that representing our present knowledge of the subject. It must not be thought that it was a lucky accident by which Ferrel stumbled on a correct explanation of the atmospheric circulation. It was the result of patient study and thought, and he had mastered the mathematical principles of forces and motions which alone could have guided him to the solution of this problem. Although at that time but a school teacher, yet the progress he had made in mathematics is shown by the fact that he had purchased a set of La Place's *Mécanique Céleste*, and had read portions of it with such diligence and understanding as enabled him to make actual contributions to certain branches, in his papers published in *Gould's Astronomical Journal*.

If Ferrel had given us only this first paper on meteorology it would probably have remained almost unnoticed, as was the case with a paper by Tracy, which cannot be passed over without mention. In *Silliman's Journal of Science* (New Haven, Conn.) for 1843, Mr. Tracy, a young graduate of Yale College, published a paper in which he advanced some original views as to the causes of the great atmospheric currents. His line of reasoning was somewhat similar to that in portions of Ferrel's first paper, and he certainly was on the right track, but he did not push his investigations farther, and the existence of his single paper was totally forgotten until Davis called the attention of meteorologists to it a few years ago

Mr. Ferrel next published, in several numbers of the Cambridge, U.S., *Mathematical Monthly* (edited by Runkle) during the years 1858 and 1859, a more complete and mathematical development of his views concerning atmospheric motions. This paper may be considered as the pioneer paper of modern dynamical meteorology. In it, Ferrel deduced the equations for the horizontal motions of the atmosphere relative to the earth's surface, which, by varying the altitude and consequently the density of the air, became general in their application. Applying to these motions the condition of continuity, and also the theorem of the preservation of areas, he was able to show that there must exist a current of air towards the pole in the upper atmosphere, and towards the equator near the earth's surface, and with a neutral plane between the two. Neglecting, in general, friction, Ferrel found that there must be a recession of the air from the poles, with a slight depression at the equator, and with a heaping up of the air between these two regions ; the maximum height occurring at the latitude of 35°. This last is also the latitude of no eastward or westward motion : but between latitude 35° and the pole there must exist an easterly motion, while between this latitude and the equator the motion must be westerly. Without modifying influences, great velocities would be attained at the equator, and very great velocities near the pole. In moving toward or from the equator the air does not retain the same lineal velocity in its whole course, as had been generally assumed by former theorisers, but in approaching the polar axis of rotation its easterly velocity is increased.

Perhaps fully as important as this general part of the

19

investigation was Ferrel's study of the motions of the air on a limited portion of the earth's surface. He has shown that in certain cases there must exist inter-changing motions which are in a measure duplications of the general system of air motions. There is the same decrease of pressure towards the centre of rota-tion ; the same ring of high pressure at a certain distance from the centre, and which separates an inner region where there is a gyratory motion from left to right from the outer region where there is a gyratory motion from right to left, there being no gyratory motion at the transition ring of high pressure (see the schematic diagram, Fig. 110, given later). There is also a tendency for the whole mass of air to move towards the north when there is a gyratory motion from right to left, and towards the south when the gyratory motion is in the opposite direction. These directions of gyration apply to the northern hemisphere, and are reversed in the southern hemi-sphere, as are also those for the general atmospheric motions. Incidental to this investigation Ferrel published (in June, 1859) his independent and, as he then supposed, entirely original theorem :—

"In whatever direction a body moves on the surface of the earth, there is a force arising from the earth's rotation which deflects it to the right in the northern hemisphere, and to the left in the southern."

Using these just mentioned results as a basis for more special investigations, Ferrel has worked out the general circulation of the atmosphere and the ac-companying distribution of atmospheric pressure, and sketched the diagram reproduced in Fig. 88. He also gave, and for the first time, an analytical investigation of the air motions in the upper atmosphere ; and he even derived from theory roughly approximate numerical results.

In this same memoir Ferrel gave briefly the theory of cyclones, tornadoes, and water-spouts, which he has since so fully elaborated ; and he proved *why* the winds follow the rule which Buys Ballot had deduced concerning the direction of blowing of local winds.

While Ferrel was engaged in the publication of this paper, there appeared two other short non-mathematical papers in the proceedings of the British Association. One was by Professor James Thomson, and the other by Mr. Vaughn, of Cincinnati, U.S.A. These papers expressed somewhat of the same dissatisfaction with current theories that Ferrel had printed in 1856, and their line of reasoning as regards a new theory was much of the same nature, but not so complete as Ferrel's. These papers are now chiefly valuable historically, and in judging of them it must be remembered that we have only an abstract of Thomson's paper, and there is almost no probability that he had seen a copy of Ferrel's already printed paper ; while, on the other hand, the close proximity of Nashville to Cincinnati, and a certainty that a number of the copies of the *Journal of Surgery* would find their way to the latter place, would indicate abundant opportunity for the earlier paper to have fallen into Vaughn's hands.

About this same time the question of the effects of the earth's rotation on terrestrial motions came up for discussion in the Paris Academy of Science, and the reports printed in several numbers of the *Comptes Rendus* show that the ideas, on this subject, of a number of mathematicians who took part in the discussion were rather hazy, to say the least. During several meetings, various members of the academy gave their views concerning the matter, and they were by no means in agreement. Finally, however, one of the members pointed out that the mathematical expression for this force had been deduced by Poisson twenty years before, and in a manner which admitted of no doubt as to its correctness.

A small edition of Ferrel's mathematical paper of 1858-9 was reprinted separately in 1860, and an account of his main results was given in *Silliman's Journal* in 1861 ; but, with two or three exceptions, no one made any use of this very important contribution to geo-physics for a period of more than fifteen years. Cleveland Abbe, the now well-known American meteorologist, had been attracted by Ferrel's second paper shortly after it was published, and the more he studied it the more he was convinced of the great importance and originality of the views advanced in it. During a two years' residence in Europe, in 1865-67, Abbe lost no opportunity of bringing Ferrel's views to the notice of European meteorologists ; but Dove's ideas had become so deeply rooted in their minds that the new theory gained no foothold, if we may judge from the utter lack of influence to be noticed in meteorological literature of that period. One or two British scientists seem to have early recognised the great value of the paper. Professor Everett, in his translation of Deschanel's *Natural Philosophy*, says that Ferrel's is the best presentation he has seen of the subject treated ; and he has shown by other references and remarks that his knowledge of the matter was by no means a superficial one.

Little was published on the dynamics of the atmosphere until about 1875, when a period of great activity and inquiry commenced. Peslin and Colding had previously investigated the motions of the air and ocean currents, but failed to take into account the full effects of the earth's diurnal rotation.

In 1876 Professors Guldberg and Mohn, of Christiania, published Part I. of a remarkable paper titled *Etudes sur les Mouvements de L'Atmosphère*. Part II. appeared in 1880. In these papers we have a combination of the meteorological knowledge of Professor H. Mohn, and the mathematical skill of Professor C. M. Guldberg. These *Etudes* must have been a considerable time in course of development, for such results are not to be obtained without much consideration. We may perhaps assume that Part II. was four years in preparation, but Part I., being of a less original and simpler nature, might not have required the same expenditure of time.

Almost coincident with the appearance of the works of Guldberg and Mohn, Ferrel published his *Meteorological Researches*, Part I. 1877, Part II. 1880. These memoirs were elaborations of his views advanced in 1858–59, and as they were prepared for and published by the United States Coast Survey, they were widely distributed among the scientific institutions the world over.

The appearance of Part I. of each of these papers, and the publication in the Austrian meteorological journal of an abstract of the Guldberg and Mohn paper, aroused the interest of a number of Continental meteorologists in the question of the theoretical treatment of atmospheric motions, and numerous papers were written on this topic, chiefly in German, however. Perhaps the most interesting and valuable of these papers were those by Dr. Sprung, then of the

Deutsch Seewarte, Hamburg, and Professor Finger of Vienna. Sprung has probably done more than any other single writer towards bringing to the notice of meteorologists the importance of the contributions made by Ferrel and others to atmospheric mechanics. Finger has given us a model theoretical mathematical investigation of the effects of the earth's rotation on the currents of the air and sea; but his application of the results obtained by him to the actual atmospheric conditions is not so successful. An important paper by Hann appeared in the *Meteorologische Zeitschrift* in 1879, and also deserves special attention, as it in a measure connects the new ideas with the older.

After the appearance of the second parts of the memoirs, by Ferrel, and Guldberg and Mohn, in 1880, with the exception of two papers, little of importance, so far as originality is concerned, was contributed to this topic during a period of five years. Professor Oberbeck, in Germany, published in 1882 a paper in which he discusses the general motions of the atmosphere, but more particularly the relations of cyclones to anti-cyclones. His paper may be taken as a continuation of portions of Guldberg and Mohn's memoir, and it is gratifying to see how, by the use of the latest forms of mathematical analysis, he arrives at conclusions which strengthen our acceptance of the results furnished by those writers and Ferrel.

Marchi, in Italy, in 1883, published a paper dealing principally with the cyclonic motions, the chief characteristics of which are his introduction of the ideas of Helmholtz concerning vortex motion, together with the assumption of a change in the density of the air, which others had only partially considered on account of the increased difficulty of treatment of the subject when this supposition was made.

In all of these papers which have been mentioned, individual treatment of the subject was the order of precedure, and in some of them no reference whatever was made to the writings of other workers in the same line of progress.   Such independence of work was confusing to most students of the problem under consideration, and a consolidation of results was much needed.   In 1885 Dr. Sprung published his *Text-book of Meteorology* (*Lehrbuch der Meteorologie*), which gives us as good an idea as could be obtained of the condition of the science of Atmospheric Mechanics at that time.   He has succeeded fairly well in his attempt to systematize the results obtained by others ; but it has required much original work on his part to even partially harmonise the various methods of work and notation to be found in the different memoirs.   In some cases, however, it was found necessary to give these results in disconnected paragraphs.

In 1887 Professor Ferrel's *Recent Advances in Meteorology* was issued by the United States Signal Service.   He has not in any sense duplicated Sprung's work, for the first part of the book is given up to a connected development of his own views on Atmospheric Mechanics, and this portion must be considered as a piece of original work, and a work that was much to be desired, as it brought into a single volume the main results of his contributions to meteorology during a period of thirty years.   In 1889 Ferrel published *A Popular Treatise on the Winds*,[1] in which he has followed out the atmospheric circulation in great detail, commencing at first principles (as considered

[1] Published by Wiley and Sons, New York, 1889.

by him), and going so far as to show the effects of
the circulation on various meteorological phenomena
influenced by it.   The following is an outline of his
general theory, nearly as he has given it in this work
on *Winds*, and it is introduced here in order that
the reader may get some idea of the general air
motions, before reading the necessarily incomplete
account of some of the most recent investigations in
this same field which I shall give later.

§ 2.—*General Circulation of the Atmosphere as outlined by Ferrel.*

The primary cause of the atmospheric motions is
the unequal distribution of the temperature on the
earth's surface produced by the solar heat.   If the
temperature were everywhere the same, then we
should have an atmosphere at rest.   But we know
from observations that there is a large, but not
constant, difference in the temperatures of the air at
the poles and equator, amounting, at an elevation of
but a few feet above the earth's surface, to about
45° C. for the average for the entire year.   The heating up of the air at the equator causes its expansion,
and consequent increase of bulk, but does not increase
its weight or pressure at the earth's surface ; it does,
however, elevate the successive isobaric surfaces, and
this causes differences of level, or gradients, which
cannot exist in fluids without a motion ensuing in
the direction of the lowest level.   Thus it happens
that, in order to restore the equilibrium which would
exist if the temperature were the same at the poles
and equator, a flow of air takes place in the upper
atmosphere from the equator towards the poles.   But

just as soon as this flow has commenced there is a decrease in the actual weight of the atmosphere at the equator, and a counter current of air sets in along the surface of the earth from the pole towards the equator in order to preserve the equilibrium of pressure. In order to satisfy the condition of continuity, which must be satisfied in all such cases of fluid motion, vertical currents must connect these two horizontal currents; the one at the equator being ascending, and that in the region of the pole descending. It is evident that between these counter currents there must be neutral planes in which there is no motion. In the case of the upper northerly current in the northern hemisphere, the greatest velocity would be in the upper portions, and this velocity would decrease to nothing at the neutral plane. For the lower southerly current, the greatest velocity would occur at some unknown distance from the surface of the earth, and would decrease towards the neutral plane above, and also, because of the friction between the air and the earth, towards the earth's surface.

There are two principal modifying conditions which must be taken into consideration before anything like a true conception can be formed of the actual relations of the vertical and horizontal currents, viz.: (1) The fact that as the upper currents go poleward they are directed toward a common centre and the space that can be occupied by any particular portion of this air becomes gradually more limited with the narrowing of the circles of latitude; and *vice versa*, the air going toward the equator can spread out more with the gradually increasing size of the circles of latitude; (2) The rotation of the earth on its axis has such an effect on these currents, that an ideal circulation of the atmosphere without this rotation is entirely different from the actual condition which at present exists ; and it is only by taking into account the effects of the diurnal axial rotation that meteorologists have been able to reach a satisfactory explanation of the phenomena which have been observed. There are still other modifying influences which also enter into the question, but they are secondary to the two just mentioned : the distribution of land and water surfaces, the periodical shifting of the region of highest temperature from one side of the equator to the other, and others of still less importance.

It being known, by the law of the expansion of gases with increase of temperature, that air will increase $\frac{1}{273}$ part of its bulk for each increase of 1° C., it is easily seen that any given isobaric surface will be about $\frac{45}{273}$ or $\frac{1}{6}$ of its altitude higher at the equator than at the poles. This difference in elevation gives the gradient force which causes the interchanging motions between the polar and the equatorial regions. If the earth were at rest there would be no easterly or westerly components of motion, but it would all take place along the meridians. But in the case of the revolving earth, there is the force arising from this rotation which causes a free-moving body to depart to the right of its original direction in the northern hemisphere, and to the left in the southern hemisphere. It is this condition which mainly gives to the existing circulation of the atmosphere its complex character.

Air Circulation when there is no Axial Rotation of the Earth.—In order to get a clear idea of the theory of the atmospheric motions, it is best to consider first the motions which would arise without the rotation of the earth on its axis, and then investigate the modifying effects of this rotation. Under this limitation it was found that a neutral plane must exist between both the horizontal and the vertical currents of the system of interchange of air between the equator and the poles. The elevation of this horizontal neutral plane, between the currents of flow and counter-flow, depends on the condition of continuity and the relative amounts of friction in the two currents. In the present case, this plane would have about half of the mass of the atmosphere above it and half below it. On account of the narrowing of the circles of latitude (consequently the distance between the meridians) toward the poles, the vertical neutral plane would be in the neighbourhood of the parallel of 30°. In following the motion of an elevated mass of air starting out from the equator, we should find that its motion would be accelerated for a time until it has reached some intermediate latitude, when it would be gradually retarded until it has reached its polar limit ; and with this retardation it would be gradually falling downward, until, at its polar limit, it would have passed the horizontal neutral plane and started on its return journey in the lower counter current. Here the motion is at first accelerated until it reaches an intermediate latitude, after which the horizontal motion is retarded and a gradual rise takes place until it reaches its equatorial starting-point. Thus the mass of air has something like an elliptic orbit, of which the horizontal and vertical neutral planes are on the major and minor axes. Some of the air will move in large orbits and some in limited ones, depending on the distance from the neutral planes. As the difference between

the temperatures at the poles and the equator is a
constant (but not uniform) quantity giving rise to a
steady gradient force, if it were not for friction (and
interference) there would result a continually accelerated motion.

*Modifying Effects of the Earth's Rotation on Atmo-
spheric Motions.*—We have next to consider the
theoretical modifying effects of the earth's rotation.
The force arising from this rotation will make itself
felt by causing the free-moving atmosphere to turn to
the right of its course in the northern hemisphere and
to the left in the southern hemisphere ; and if the air
is not free to move in this direction, then it presses
in this direction and causes an atmospheric gradient.
This force, then, causes the upper poleward current
to be deflected towards the east in both hemispheres,
and the lower equatorial current to be deflected
towards the west.

There thus arise contrary motions of the air ;
towards the west near the ground and towards the
east at higher altitudes, thus forming *relative* velocities between the strata.   If the force were impul-
sive, the friction of the layers would destroy their
relative velocities, and a common absolute velocity
below and aloft at any parallel would result.   But as
the deflective force is constant in action the relative
velocities are maintained in spite of friction.

It is assumed, at first, that the surface of the earth
offers no frictional resistance to the motion of the
air ; then the velocities towards the east and west,
arising from the forces already considered, would have
to satisfy this condition : that the gyratory velocity
multiplied into the radius vector, is equal to the
average value which this product had for all the

air of the hemisphere while the air is at rest relative to the earth's surface. For in this last condition, the air is acted upon by a constantly exerted force in the plane of the meridian, and it also has a gyratory velocity around the earth's axis, and when at last motion does set in, the principle of equal areas becomes applicable ; and because, moreover, the action and re-action of the particles of air among themselves must of necessity be equal. At the pole the radius vector, or perpendicular distance to the pole, becomes small, and the gyratory velocity proportionately large.

Still assuming only these same forces as acting, but considering also that there is friction between the earth and the lower layer of the atmosphere, the layers so subjected to friction would have no easterly or westerly motion, because, on account of the condition of continuity, the mass multiplied by the velocity must be the same for the counter equatorial current as for the polar current ; and the same with the deflecting forces ; there would be no force to over-come this friction. And if this layer subject to friction has easterly or westerly relative motion, then this same amount of motion must be applied to the relative motions of the upper layers in order to obtain their absolute velocities. The force which does cause motion in this ground layer, as we may call it, comes from the vertical interchange, which is discussed a little later.

Let us now consider the easterly motions of the upper poleward current.

With any increase of the absolute east velocities, the relative velocities, and consequently the friction, will become greater ; but any such increase reaches its limit when the friction counterbalances the force causing the motion. But we must also take into account the deviatory force due to the earth's rotation acting on these easterly motions, toward the right in the northern hemisphere and toward the left in the southern hemisphere, and consequently toward the equator in *both* hemispheres. This force then acts against the force, arising from the temperature gradient from the equator towards the pole which causes the original interchange of motions between the equator and the poles. Now, as this deviatory force increases with increase of velocity of the air current, it is evident that if the easterly velocity of the upper current is great enough, the deviatory force could counterbalance the gradient force due to polar and equatorial differences of temperature, and in this case there could be no current from the equator towards the pole; and yet without this last motion there would be no force to overcome inertia and friction and keep up the easterly motions. So it is evident that the easterly velocities can never reach the limit just described, for this would entirely stop the upper current from the equator poleward. (Still, if there were no friction to be overcome, this limit could be practically reached, because in this case the interchanging motion would have been established, and there would be no need of any force to maintain the motion by overcoming the friction.)

We thus see that there is a practical limit to the velocities of the polar and equatorial currents; for as these increase in velocity the east component of velocity will be also increased; but we have just seen that if the east component becomes too great it will cause the current from the equator poleward to cease. So that there is, as Professor Ferrel has so aptly put it, a sort of controlling governor to the atmospheric circulation.

In the case of the lower current from the pole towards the equator a reverse of this process occurs, and part of the deflective force to the westward, due to the earth's rotation, is spent in overcoming friction between the easterly current above and the earth's surface below. In the higher latitudes the westerly force is not sufficient to overcome the easterly, but in the lower latitudes a westerly motion actually sets in. An exact quotation from Ferrel's *Winds* seems to be necessary at this point, in order to avoid the possibility of misrepresentation.

" In the higher latitudes, therefore, where there is an east component of motion at the earth's surface, the amount of east momentum lost by any portion of air in its passage above from any given latitude toward the pole until it returns below to the same parallel, is the kinetic energy which it has contributed toward overcoming the frictional resistance of the earth's surface, to the east component of motion at the surface between that parallel and the pole ; and in the lower latitudes, where there is a west component of motion at the earth's surface, the amount of west momentum gained, or considered algebraically, the amount of west momentum lost, by any portion of air from the time it leaves any given parallel of latitude in its passage toward the equator below until it returns to the same parallel above, is the amount of kinetic energy which it has contributed towards overcoming the frictional resistance of the earth's surface to the west components of motion between that parallel and the equator."

Vertical Motions of the Air.—We have now to consider the vertical motions which must form the condition of continuity between the polar and equatorial horizontal currents. The comparatively large easterly motion which the upper poleward current has near the pole, is gradually overcome as the air settles toward the earth, as the momentum is used up by overcoming friction between the successive layers of the air, the final influence of which reaches to the lowest layer of air and overcomes the friction between this and the earth's surface ; but this upper easterly velocity is so great that all of its force is not spent, and there results a small easterly motion of the air near the earth's surface.

We see, therefore, that at the earth's surface in the higher latitudes there exists an eastward motion, and in the lower latitudes a westward motion. There is, then, what Ferrel calls a sort of torsional force, due to the earth's rotation, which acts on the polar and equatorial currents and causes the surface motions just mentioned. There are, however, as in the case before mentioned, the same relative velocities existing between the successive vertical strata, just as if this torsional force did not exist.

As the relative easterly velocities of the upper strata depend on the difference of temperature between the equator and the poles, or the temperature gradient, they must be greater in winter than in summer, and in the middle latitude of the northern hemisphere the January velocities would be more than double those of July.

When the air is descending from above towards the earth's surface, in the higher latitudes, it approaches nearer to the axis of the earth, and consequently, according to the theorem of the preservation of areas, the air will have a greater velocity relative to the earth's surface, and this increase of velocity will be slightly augmented ; and the contrary will take place in the lower latitudes where the ascending current is to be found. Again the exact wording given by Ferrel seems to be necessary.

" The relation between the amount of the east components of velocity of the air at the earth's surface in the higher latitudes, and that of the west components in the lower latitudes, is determined by the condition that the sum of all the forces of the air particles of the higher latitudes acting in an easterly direction upon the earth's surface through friction or resistance of any kind, multiplied into their distances from the axis of rotation, and called moments of couple, must be exactly equal to similar products in the lower latitudes, where there is a west component of motion, and the forces by means of friction and other resistances act in the contrary direction."

This principle was recognised by Hadley, for he remarks that on his theory if these motions did not have their counter motions, the rotation of the earth upon its axis would be changed by the action on the earth's surface.

We must next see how this equality of frictional force exerted by the easterly and westerly surface motions may exist. Since both of these forces are effective in proportion to their distance from the earth's axis of rotation, then either the easterly velocities near the pole must be much greater than the westerly velocities near the equator, or else the amount of air having an easterly motion must be the greater. If this last is the case, then the place of change of direction must be between the equator and the parallel of 30°; because this parallel divides the actual amount of air in either hemisphere into two equal portions.

If it were not for friction, these east-westerly surface velocities would become very great; but as the forces producing them are small, and most of this force is used in overcoming friction, the resultant effect of the motions is small. On account of the excess of land in the northern hemisphere over that in the southern, these velocities are greater in the southern

hemisphere than in the northern, as the friction between air and water is much less than between air
and land.

In the case of no frictional resistance, as we have
seen, the easterly velocities will approach a certain
limit, and after the circulation has once commenced,
it is not necessary that any force be derived from the
meridianal motions between the equator and the poles
in order to overcome the friction of these east components.

" In the case of friction, therefore, the less the
friction, and the greater the deflecting force belonging to a given amount of meridianal motion, the less
of this motion is required to overcome the friction of
the east component of motion." So that as, except-
ing the layers of air directly in contact with the
earth's surface where the friction is great, little
meridianal motion is required to overcome this frictional resistance to the east components, we should
have an almost easterly motion for the air in the
higher latitudes where the meridianal motion is very
small in comparison with the easterly components.
According to the principle of resultant motion from
two components, this motion would be a little north
of east for the upper current, and a little south of east
in the lower strata.

In the latitudes lower than those we have just been
considering, where the deflecting force due to the
earth's rotation has considerably decreased, for it is
greatest at the poles and diminishes to nothing at
the equator, the relative velocities between the strata,
and the absolute east velocities of the upper current
become less than those in higher latitudes, so that
the deviation from a nearly easterly resultant is

greater. Also here the air which occupies the region
between the upper east component and the lower
west component can have very little motion ; but
here the west component of motion of the lower
atmosphere, combined with the motion towards the
equator, gives as a resultant the atmospheric motions
known as the Trade-winds.

Air Motions near the Equator.—In the immediate
vicinity of the equator both the deflecting forces and
the equatorial motions vanish, but " there may be a
westerly motion at a considerable altitude above the
earth's surface, arising from the unchecked momentum
of the west component of motion acquired as the air
moves in from both sides (hemispheres) toward the
equator."

In the equatorial regions where the great ascending current exists, there may be a slight current of air
toward the west at high elevations, due to the effect
of the earth's rotation on the air moving upwards.
This air must describe equal areas in equal times, and
the radius vector (the distance from the axis of the
earth) is increased in the upward motion by the altitude attained above the earth's surface, or starting point
of vertical motion ; so that the east velocity (absolute
motion) is decreased in like ratio, and as the earth
revolves, the air which has ascended is left a little to
the west of the starting-point. This velocity towards
the west is very small however, as theory makes it
only about one and one half miles per hour at an elevation of about eight miles. But there is at times a
cause for a strong wind towards the west at the
equator, in the oscillation of the line of maximum of
temperature from one side of the equator to the other,
following the sun in its annual march. In the

summer time in the northern hemisphere there must
be a flow of air above from the region of maximum
temperature in the northern hemisphere over into the
southern hemisphere, with a contrary lower current.
This upper current would be deflected towards the
right (west) as long as it remained in the northern
hemisphere, and this would give rise to a wind
towards the west, and this direction might be maintained for some distance after crossing the equator.
That motion towards the east does not occur in the
lower layers of the atmosphere is due to the greater
amount of friction to be overcome at the earth's
surface.[1]

During the summer of the southern hemisphere the
reverse of this would take place.

*Effect of the General Motions of the Atmosphere on
the Atmospheric Pressure and Isobaric Surfaces.*—If
the earth's surface had a uniform temperature, then
the centrifugal force due to the rotation of the earth
on its axis would just be sufficient to cause the same
ellipticity in the isobaric surfaces that exists for the
earth's smooth surface, and they would be, therefore,
parallel to the latter. But when, as in the actual
condition of the atmosphere which we have just been
considering, there is an east or west component of
motion at the earth's surface, due to differences of
temperature and the earth's rotation, this uniformity
of isobaric surfaces no longer exists. On the earth's
surface there arise differences of atmospheric pressure,
and the isobaric surfaces are inclined to the surface

[1] This explanation is thought to partially account for the westerly
motion which carried the ashes from Krakatoa so rapidly westward
during the eruption of August, 1883, as this is the period of year when
a maximum effect would be felt.

on which the air rests, and pressure gradients with reference to this surface are produced. These pressure gradients at the earth's surface depend only upon the east and west components of velocity at the earth's surface, and where there is no easterly or westerly motion there exist no surface gradients; but these gradients may still exist at higher altitudes however.

Barometric Gradient.—The general mathematical expression for the barometric gradient for any velocity of motion, and taking into account the dependence upon the deflecting forces corresponding to this velocity, is a function of the velocity, the sine of the latitude, and the ratios of the standard pressures and temperatures to those observed.[1] From this relation Ferrel shows that in the northern hemisphere, in higher latitudes when the motion is easterly (*i.e.*, assumed to be positive in direction), there is an increasing pressure as we proceed from the pole equatorwards; a maximum must be reached at 30° or 35° where, as we have seen, the velocity becomes zero, and gradually becomes a westward one (*i.e.*, negative in direction), and as this latitude is passed and the equator is approached we find the sign of the gradient is changed, and the pressure must decrease from there to the equator. Here the sine of the latitude becomes zero, and there is no gradient force, and consequently the pressure is at a minimum. South of the equator the sine of the latitude changes sign and the gradient is reversed, so that the gradient increases again to about latitude 30°, where it changes direction with the change in direction (and consequently

---

[1] Gradient = constant (0·1571) × (velocity of motion) × (sine of latitude) × (function of the pressure) × (function of the temperature).

the sign) of the motion. At this latitude of disappearance of the motion there must then be maximum of pressure ; and from thence towards the south pole the decrease must take place, and this occurs the most rapidly in the middle latitudes where the velocities are greatest. Since the sine of the latitude is small near the equator the gradient (rate of change) is small there, and the minimum of pressure is consequently only slight. Concerning the cause of this equatorial minimum Ferrel says :—

"It seems difficult for some to conceive how the equatorial depression can be due to the westerly motion of the atmosphere, since this motion prevails at the earth's surface and in the lower strata of the atmosphere only, while above the motion is easterly. But it has been shown that in the case of no east or west component of motion at the earth's surface, the east components are necessary to keep the air above from flowing away from the equatorial region, and thus from diminishing the pressure there. If, then, the lower stratum next the earth's surface has a west component of motion, we have seen that the velocities of all the strata above, to the top of the atmosphere, have their velocities changed by the same amount, and the effect upon the pressure at the earth's surface is precisely the same as if there were no temperature gradient and no east components of velocity above, and all the strata from the bottom to top should receive a west component of motion."

The greatest pressure is to be found at about the parallel of 30°, where the east-westerly motion vanishes at the earth's surface ; but this maximum pressure which occurs at the place of disappearance of the gradient does not occur at this latitude for all altitudes. In the air up above the earth's surface, this gradient force vanishes at some lesser latitude than the one found in dealing with the air at the earth's surface ; because after the velocity becomes negative, or towards the west, a term in the complete mathematical expression for the gradient force, depending upon the altitude, being always positive in sign in the northern hemisphere (and always contrary in sign to the product of the east-westerly velocity × sine of the latitude in these latitudes) only vanishes when the negative value of the velocity × sine of the latitude is equal to the term depending on the altitude. So that when the altitude increases there must be a corresponding numerical increase in the velocity × sine of the latitude. The sine of the latitude decreases very slowly towards the equator, but this decrease is more than counterbalanced by the increase of the westerly velocity as the latitude of 30° is receded from in passing equatorwards, so that the corresponding numerical increase just mentioned is attained by moving equatorwards as the altitude is increased. Therefore the greater the altitude the nearer to the equator is the latitude where the gradient vanishes and the maximum pressure occurs. Where the altitude becomes so great that the velocity term can no longer counterbalance it, then the terms only vanish at the equator ; consequently at and above some given altitude the maximum pressure is only reached at the equator, and in these upper regions there is a continuous gradient of pressure increasing from the poles to the equator. Ferrel finally expresses this idea as follows :—

"With no east or west components of motion at the earth's surface, we have seen there would be no pressure gradient at the earth's surface, so far as it depends upon the velocity and the deflecting forces, but at any altitude above the surface there would be a gradient of increasing pressure from the pole to the equator, and this at great altitudes is so large that the west component of motion of the atmosphere at the earth's surface, and the consequent change of velocities by the same amount at all altitudes, are not sufficient, by means of the deflecting forces, to reverse this gradient except in the lower strata of the atmosphere up to

a certain altitude. Above this altitude, therefore, there is no barometric minimum at the equator, but a maximum, arising from the
gradual nearer approach of the maxima on each side towards the
equator as the altitude increases.

"There is, then, at the earth's surface, an area of low pressure around
each pole, with its minimum at the pole, a zone of high pressure in each
hemisphere, with its maximum about the parallel of 30°, and an equatorial zone of slightly diminished pressure with the minimum at the
equator. And this same arrangement exists in the lower strata of the
atmosphere up to a considerable altitude, except that the polar depressions are greater and the parallel of maximum pressure comes nearer
the equator as the altitude is increased, until at a certain altitude the
two maxima combine to form a maximum pressure at the equator.

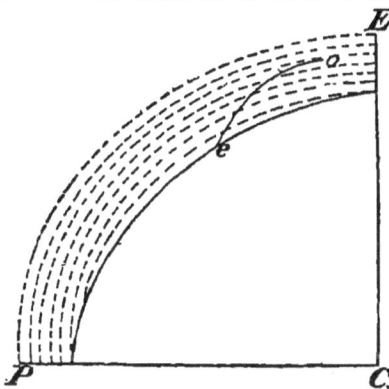

Instead, then, of the isobaric
surfaces being elliptical, and
having an increasing ellipticity
in proportion to the altitude,
they have, in consequence of the
east and west components of
motion of the atmosphere at the
earth's surface, steeper gradients
in the higher and middle latitudes, a bulging up nearer the
equator, and in the lower strata
a minimum altitude at the equator, as represented in Fig. 85,
in which the intersections of the
line, eo, with the isobaric surfaces
indicate the latitudes, at the

FIG. 85.

several altitudes, where the pressures are the greatest and where the
east components of velocity vanish, change sign, and become west
components of velocity. The actual form and position of the line eo,
since they depend upon the west components of motion at the earth's
surface, and the temperature gradient, are quite uncertain towards the
equator."

Concerning the interchange of air between the two
hemispheres, Ferrel says :—

"Since the whole vertical circulation and the east and west components of motion, and the deflecting forces arising from them, from
which result the polar and equatorial depressions and the zones of high
pressure near the tropics, depend upon the temperature gradients

between the equator and the poles, there must be greater polar depressions of the isobaric surfaces and a greater bulging up in the middle and lower latitudes in winter than in· summer, since the temperature gradients are much greater in the former season than in the latter, especially in the northern hemisphere, where they are more than twice as great in January as in July. There must, then, be an annual inequality of atmospheric pressure from this cause, such that the pressure is greater in the winter than in the summer of each hemisphere in the middle and lower latitudes, and the reverse in the polar regions.

" But for another reason, also, there is an annual oscillation of pressure. During the winter of each hemisphere the atmosphere becomes much colder than it is in the other hemisphere, and consequently its volume considerably less, so that the isobaric surfaces lie lower in the colder hemisphere than in the other. There is, consequently, a pressure gradient above by which the air of the higher strata flows from the warmer hemisphere to the colder one and increases the mass and pressure of the atmosphere there a little, and diminishes them by the same amount in the warmer hemisphere, until there is a reverse gradient formed in the lower strata which gives rise to a counter flow of air below from the colder to the warmer hemisphere. There is, in this way, an interchanging motion originated and maintained between the two hemispheres, just as there is between the warm equatorial and cold polar region of each hemisphere. · But the interchange between the two hemispheres is kept up with greater facility, since there are no east components of motion at and near the equator, where the temperature and pressure gradients in this case and the velocity of flow are the greatest, to give rise to counter deflecting forces, as there are in the middle latitudes in the other case, but the whole temperature and pressure gradients are brought to bear in maintaining the flow of air above from the warmer to the colder hemisphere, and the contrary in the lower strata.

" As there is an annual inversion of the temperature conditions which give rise to and maintain these interchanging motions between the two hemispheres, so there is also one in the pressure gradients at the earth's surface and in the lower strata, arising from the flow of air above from the warmer to the colder hemisphere, and in the slight increase of pressure there and corresponding diminution in the other hemisphere, to cause this gradient. There is therefore a little more air, and consequently a little greater pressure at the earth's surface in each hemisphere in winter than in summer, and this causes a slight annual inequality of pressure, the pressure being the greatest in each hemisphere in the mid-winter of that hemisphere. The greatest effect of this latter cause of the annual inequality is at the poles, whereas in the other case the pressure is increased in winter in the middle latitudes, but diminished in the polar regions. It is the resultant of these two effects which is observed."

	AIR PRESSURE.			TEMPERATURE.					AIR PRESSURE.	
	Annual Mean.	Jan.	July.	Annual Mean.		Observed in		Temp. decrease per 100 metres in altitude.	At an altitude of 2,000 metres.	At an altitude of 4,000 metres.
	Gradient.			Observed	Computed	Jan.	July			
1	2	3	4	5	6	7	8	9	10	11	12
90°	...	...	...	...	...	−17,0°	...	...	...	...	...
80	760,5	...	760,4	760,6	−15,5°	−15,8	−31,9°	1,0°	0,19°	582,0	445,2
75	60,0	−0,19	60,2	58,8							
70	58,6	−0,14	59,0	58,2	− 9,8	−10,2	−26,5	6,9	0,25	583,6	446,6
65	58,2*	+0,01	58,8*	57,6*							
60	58,7	+0,15	59,7	57,7	− 1,6	− 2,2	−16,9	13,8	0,40	587,6	451,9
55	59,7	+0,20	61,0	58,4							
50	60,7	+0,18	62,1	59,3	6,3	6,5	− 6,0	18,6	0,57	593,0	457,0
45	61,5	+0,15	63,0	60,0							
40	62,0	+0,07	63,6	60,4	13,6	14,4	4,5	22,8	0,65	598,0	463,6
35	62,4	−0,03	64,1	60,7							
30	61,7	−0,18	63,4	60,0	19,8	20,4	12,9	26,6	0,67	600,9	468,3
25	60,4	−0,25	62,0	58,8							
20	59,2	−0,21	60,6	57,8	25,3	24,3	21,7	29,0	0,68	600,9	469,9
15	58,3*	−0,13	59,3	57,3*							
10	57,9*	−0,03	58,4	57,4	27,2	26,4	25,9	28,4	0,69	600,9	470,7
5	58,0	+0,01	58,0	57,9							
0	58,0	+0,04	57,4	58,6	26,7	26,8	27,3	26,1	0,70	601,1	471,0
5	58,3	+0,11	57,1*	59,5							
10	59,1	+0,20	57,4	60,8	25,9	26,0	27,9	24,0	0,69	601,6	471,1
15	60,2	+0,26	58,2	62,2							
20	61,7	+0,29	59,5	63,9	23,7	23,8	26,6	20,8	0,68	602,7	471,1
25	63,2	+0,18	60,8	65,6							
30	63,5	−0,08	61,3	65,7	19,3	20,2	23,0	15,6	0,67	602,2	469,3
35	62,4	−0,30	60,6	64,2							
40	60,5	−0,51	59,1	61,9	14,4	14,9	17,6	11,1	0,65	597,1	463,1
45	57,3	−0,73	56,3	58,3							
50	53,2	−0,91	52,7	53,7	8,8	8,2	11,1	6,4	0,59	588,0	453,7
55	48,2	−0,97	48,2	48,2							
60	43,4	−0,83	...	..	1,8	0,9	3,6	0,0	0,46	577,0	443,9
65	39,7	−0,56	...	...							
70	38,0	...	...	...	...	− 5,8	...	...	0,32	569,9	437,2
80	...	...	...	...	..	−10,6	...	...	...	...	...
90	...	...	...	...	...	−12,4	...	...	...	...	...

(Column 1: Northern Hemisphere 90°–0; Southern Hemisphere 5–90.)

Sprung's Table of Meridional Distribution of Air Pressure and Temperature. — The above table, from Sprung's *Meteorologie*, gives the average barometric pressures and the temperatures for the average on various latitudes of the northern and southern hemispheres — in other words, for the average meridian. It is introduced at this point in order that the explanations offered by Ferrel, of the general configuration of the atmospheric conditions, may be compared with the results obtained by observation.

Col. 1 gives the latitude ; col. 2, the average barometric pressure at sea-level ; col. 3, the gradient or rate of change of pressure for each degree of latitude ; col. 4, the average pressure at sea-level for January ; col. 5, the average pressure at sea-level for July ; col. 6, the average temperature at sea-level for the year ; col. 7, the values of these average yearly temperatures when computed by means of an empirical formula in which the temperature is considered a function of the latitude ; cols. 8 and 9, the average temperatures at sea-level for January and July ; col. 10, the decrease in temperature with each hundred metres of altitude above the earth's surface, on the average for the year ; cols. 11 and 12, the computed air-pressure at elevations above the earth's surface of 2,000 and 4,000 metres, the decrease of temperature with elevation being used as given in col. 10.

We see in this table of pressures at the earth's surface the maximum pressure at the latitude of 30°, and the slight minimum at the equator with another more pronounced minimum at the pole, which Ferrel has deduced from theory. We also see that at 2,000 metres' elevation the maximum pressure is somewhat nearer the equator ; while at 4,000 metres' elevation the maximum is about at the equator, and this elevation is therefore within the upper region shown by the outer ring given below in the diagram No. 89. The southern hemisphere, on account of its greater uni-

formity of surface (being mostly water) agrees best
with the theory.

In order to bring out more clearly the general pressure distribution on the earth's surface as given in the
table, a graphical presentation of these pressures, as
given by Sprung, is reproduced here in Fig. 86. The

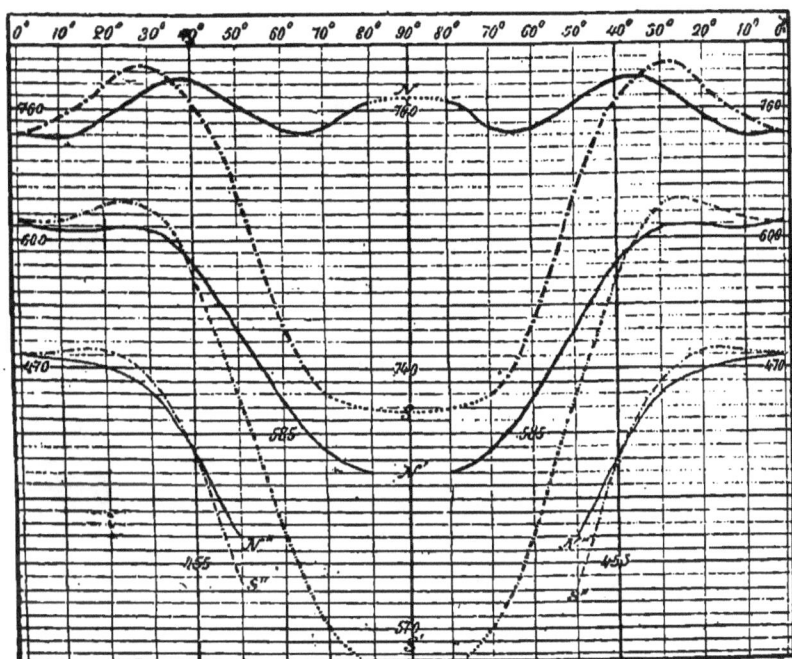

FIG. 86.

curved N and S show the atmospheric pressures at the
earth's surface for the northern and southern hemispheres respectively. Similarly the curves N' and N"
show the pressures in the northern hemisphere at
altitudes of 2,000 metres and 4,000 metres, and the
curves S' and S" show the pressures in the southern

hemisphere at these altitudes. A comparison of the curves N and S show that they differ very much ; and of the two, the curve S must be taken as more nearly representing the ideal pressure which would exist on the earth if the surface were all water. That this difference between N and S is due to this in-

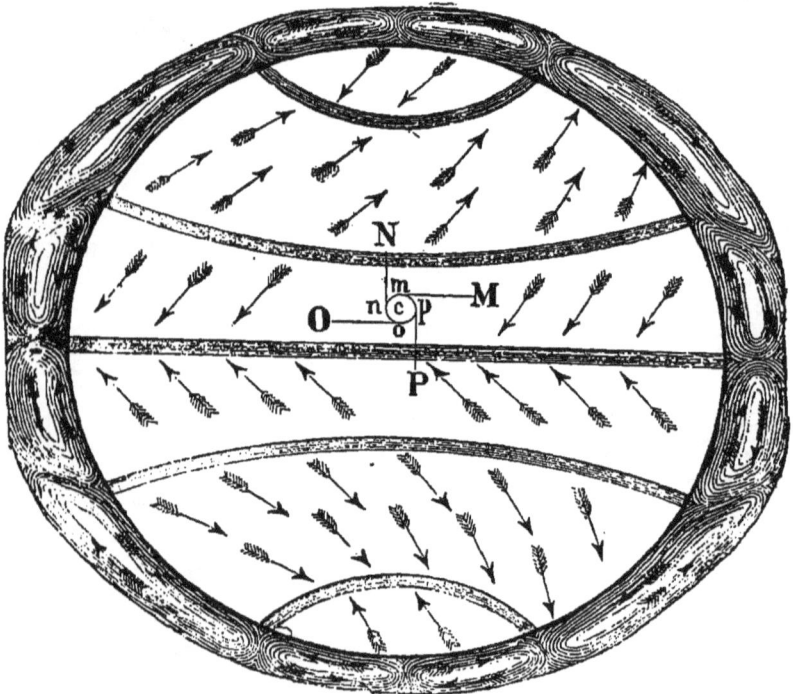

FIG. 87.

equality of surface is clearly shown by a comparison of the curves N′ and S′ and N″ and S″. With each increase in altitude, and consequently with the distance of removal from the unequal surfaces of land and water, there is a greater similarity in the pressure curves for the northern and southern hemispheres.

It is to be noticed, also, that the apparent increase of pressure near the north pole, as shown by observations at the earth's surface, has entirely vanished before an altitude as slight as 2,000 metres is reached.

Graphical Representation of the Air Circulation as outlined by Ferrel.—As before mentioned, in 1856

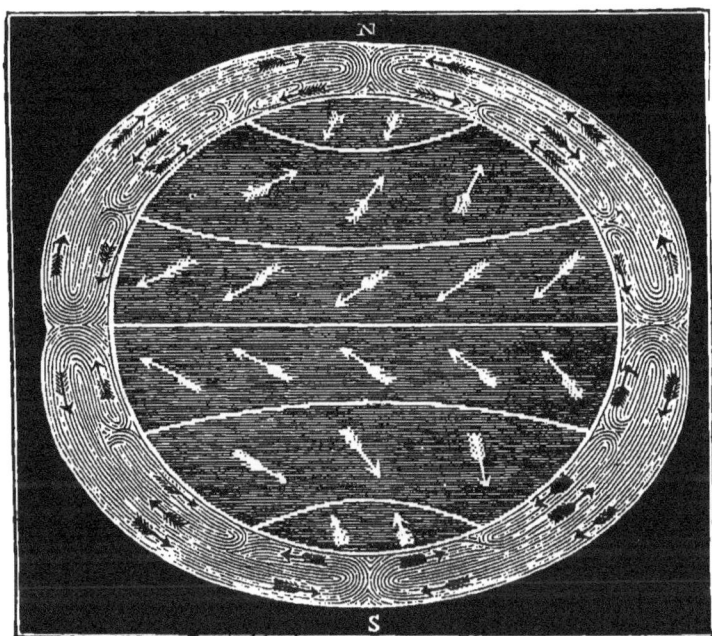

FIG. 88.

Ferrel published a diagram (Fig. 87) of the general atmospheric motions ; and again, in 1858–59 he published another (Fig. 88), which contained the third current at middle latitudes. In some subsequent publications he has also given such schematic diagrams. The accompanying diagram (Fig. 89) shows Ferrel's latest [1] ideas concerning the general circulation of the atmosphere. In the central ball we find the horizontal projection ; the surface winds are indicated by the unbroken arrows, and the upper currents by the broken arrows. A vertical section of the circulation is also shown on the outer circumference of the ball, and the position of an isobaric

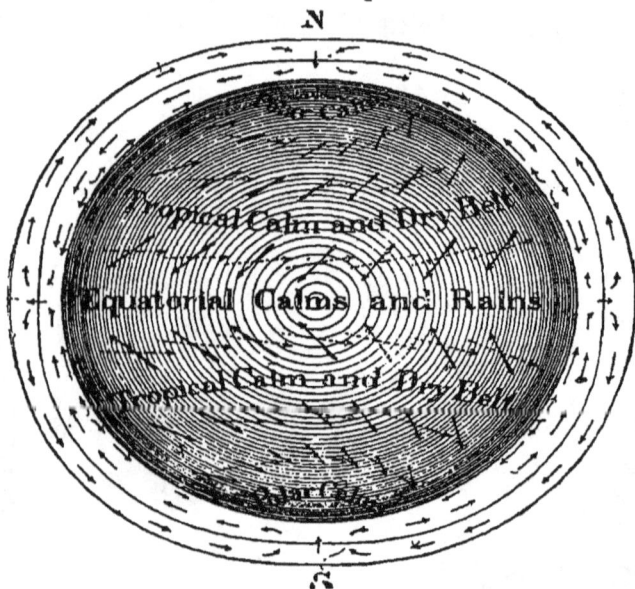

Fig. 89.

surface at low latitudes is shown by the encircling line nearest the ball, while the outer line represents an isobaric surface at high altitudes. The greater velocities are roughly indicated by a lengthening of the arrows. This ideal figure is based on the assump-

[1] Figs. 87 and 88 are of historic interest, and are given close to Fig. 89 to show, by comparison with this, how Ferrel changed his ideas in the course of thirty years' study.

tion of a uniform earth's surface, and is for the average of the whole year.

Computation of East-westerly Velocities of the Air from Pressure Gradient.—The difficulties of obtaining data concerning the wind velocities at any distance above the earth's surface are such that it has been found expedient, and in fact necessary, to investigate them by an indirect process. Ferrel has shown by a process of mathematical reasoning that the easterly or westerly motions may be deduced for all altitudes, except near the equator and the poles, by a consideration of the pressure distribution as determined by observations. He has shown that there is a direct connection between the east and west velocities and the pressure gradient; and from the latter he has computed the table on the following page. (For a full account see p. 146 of *Winds*.)

In this table + motion refers to east velocities, and − to west velocities. The numbers express kilometres per hour; the factor h is the altitude above sea-level expressed in kilometres.

The first column of figures in each set gives the average east and west velocities at the earth's surface, and to these must be added algebraically the corresponding numbers in the second column, multiplied by the number of kilometres above the earth's surface, in order to get the velocities at that elevation. Ferrel and others have found this table to be of great use in studying the mechanics of the upper atmosphere. As Sprung has said, it is most worthy of notice that Ferrel had successfully attempted the investigation of the upper air currents by theoretical deductions from the pressure distribution at the earth's surface, fully twenty years before the European

LATI- TUDE.	EAST AND WEST VELOCITIES OF AIR MOTION.		
	Mean of the Year.	January.	July.
	km.	km.	km.
N+75°	—4·4+4·8 h.	—1·9+4·4 h.	—5·7+5·1 h.
70	—3·3 6·6 ,,	—1·1 7·5 ,,	4·8 5·8 ,,
65	+0·2 7·7 ,,	+1·6 9·6 ,,	—1·2 5·9 ,,
60	3·9 8·3 ,,	5·5 10·9 ,,	+2·3 5·6 ,,
55	5·5 8·5 ,,	7·0 11·6 ,,	4·0 5·4 ,,
50	5·4 8·6 ,,	6·4 12·1 ,,	4·4 5·1 ,,
45	4·9 8·8 ,,	5·5 12·6 ,,	4·1 5·0 ,,
40	+2·6 9·0 ,,	+2·8 12·9 ,,	+2·4 4·8 ,,
35	—1·2 9·3 ,,	—1·0 13·8 ,,	—1·4 4·6 ,,
30	8·6 9·5 ,,	9·1 14·7 ,,	8·1 4·3 ,,
25	14·4 9·4 ,,	16·0 15·2 ,,	12·2 3·6 ,,
20	15·1 9·0 ,,	20·2 15·6 ,,	11·7 2·4 ,,
+15	12·5 5·6 ,,	21·8 10·5 ,,	3·8 0·6 ,,
...	... ...	... ...	... ...
S—15	25·0 8·2 ,,	18·7 4·9 ,,	31·0 11·7 ,,
20	20·9 7·8 ,,	18·8 5·8 ,,	23·0 10·0 ,,
25	—10·3 7·6 ,,	—10·4 6·6 ,,	—16·1 8·7 ,,
30	+ 3·8 7·5 ,,	+ 2·3 7·3 ,,	+ 5·2 7·8 ,,
35	12·4 7·4 ,,	10·0 7·8 ,,	14·4 7·1 ,,
40	18·7 7·4 ,,	16·0 8·2 ,,	21·4 6·7 ,,
45	24·0 7·1 ,,	21·0 8·6 ,,	27·0 6·4 ,,
50	27·5 7·5 ,,	24·5 8·9 ,,	30·5 6·1 ,,
55	27·3 7·5 ,,	+24.6+9·1 h.	30·0+5·9 h.
60	+21·9+7·5 h.		

meteorologists had recognised the great importance of the formula for decrease of the atmospheric pressure with elevation, for the indirect determination of the pressure distribution and air motions in the upper atmosphere.

Ferrel's Conception of the Atmospheric Circulation viewed as a whole.—In summing up the results of Ferrel's reasoning concerning this general circulation, they may be stated briefly as follows :—

1. The atmosphere at the earth's surface has an

easterly motion in the middle and higher latitudes, and the velocities increase with the altitude.

2. The atmosphere at the earth's surface has a westerly motion in the lower latitudes, and the velocities decrease up to a certain altitude where the motion is changed to an easterly direction, and the velocities increase again with the altitude.

3. Deflecting forces arising from this easterly motion in the middle and higher latitudes (see 1) force the air towards the equator. A contrary but weaker effect of the westerly motions in the lower latitudes at the earth's surface forces the air pole-ward, but at high altitudes where the easterly motion exists the air is still forced towards the equator.

4. From (3), then, the isobaric surfaces are depressed at all altitudes in the higher latitudes, and more slightly in the equatorial regions. Consequently there is a heaping up of the air, and a bulging up of the isobaric surfaces at about the latitude 30° in the lower part of the atmosphere ; but in the upper atmosphere there is a maximum pressure at the equator and a minimum at the poles.

Viewed as a whole, the general circulation of the air represents two huge atmospheric whirls, one in each hemisphere, with the poles as centres. The direction of the motion of the air around this is determined by the rotation of the earth on its axis, and in the northern hemisphere is opposite to that of the hands of a clock, and in the reverse direction in the southern hemisphere. But each of these whirls has on its outer circumference an atmospheric ring or belt in which the motion of the rotation of the air is in an opposite direction. In what may be termed the transition zone, which separates the inner whirl from its encircling ring with opposite rotation, there is a heaping up of air, and consequently an increased atmospheric pressure, due to centrifugal force.

§ 3.—*Recent Theoretical Investigations of Atmospheric Motions.*

For a number of years about the time of the appearance of Sprung's *Meteorologie* there appeared to be a great lack of interest in this subject of dynamics of the atmosphere; and even in the Austrian and German meteorological journals there was but little said concerning it, except in the way of reviews. This emphasises the importance of the labours of Dr. Sprung in expounding the older and recent theories, for his critical studies are somewhat isolated by the general paucity of original memoirs contributed during the years 1880–86.

This apparently smouldering blaze of interest was deceptive in its indications, for the study of atmospheric mechanics soon received a remarkable awakening by the appearance of a number of German memoirs written by able thinkers.[1] In 1886 Werner Siemens presented a paper to the Berlin Academy of Sciences, in which he outlines a theory of atmospheric motions, and in which the principle of living force is applied to the explanation of these phenomena. In this paper the author goes back to first principles, and does not make use of the results previously obtained by Ferrel

[1] Only a few of these papers can be mentioned here, but a notice of them is given by Abbe in a Smithsonian Institution Publication in 1891, and by the present writer in the *American Journal of Science*, 1890; and somewhat critical reviews by Pernter and Sprung elsewhere.

and others, so that some of the ideas which the author advances as novel had already been treated in a more or less satisfactory manner. Still, such questions as the origin and maintenance of barometric maxima and minima and the mechanics of spout phenomena need further treatment, and every new view, if based on acceptable principles, is a help towards the final discovery of the actual truth, which cannot be said to be reached at the present time. In Siemen's theory, as in that of Ferrel, Hadley's view of the initial cause of air currents is adopted, but the subsequent reasoning agrees more closely with that of Dove, and in fact it is claimed that, if Dove had known the now accepted theorems concerning *living force*, he would have satisfactorily solved the problem of general atmospheric motions. Siemen's theory may be outlined as follows: We must first conceive the air to be everywhere at *relative* rest; the atmosphere will then possess, by means of its absolute motion of rotation, a certain amount of living force. Now suppose the whole atmosphere to be suddenly thoroughly mixed. Then there will be produced an everywhere uniform velocity of rotation, and its amount is such that the total living force is just the same as before. This uniform velocity of rotation he finds to be 379 metres per second. Combining this with the *absolute* velocity of the air due to the rotation of the earth, the *relative* motion of the air is obtained. Computation then shows that at the equator there would be an east wind of 85 metres per second, which would decrease in velocity until at latitude 35° 16' there would be no wind ; and then a west wind arises, which increases in velocity until at the pole a west wind of 379 metres per second is to be found.

Of course these values are in a great measure schematic, and do not represent the wind velocities actually encountered.

The later views of von Siemens and some other writers are presented in outline in the following pages.

§ 4.—*General Circulation of the Air, according to von Siemens.*

The primary meridional circulation depends on the equilibrium between the acceleration of the air in the equatorial regions caused by the super-heating of the lower air layers in the torrid zone, and the loss of force which the moving air experiences in its course. The atmosphere has acquired the rotational velocity of the earth beneath it without the aid of these meridional currents, in the course of thousands of years ; and when the meridional currents do arise von Siemens thinks that the motions are not, as Ferrel supposes, subject to the law of " conservation of areas," or, as it would be in this application, the " preservation of the moment of rotation," but he thinks that the currents continue with unchanged absolute velocity, and consequently retain their living force. We cannot, however, follow out von Siemens' argument as given in his paper, " Ueber das allgemeine Windsystem der Erde : " [1] but the main results of his reasoning are of great interest as they express in a general way, and one easily comprehended, not only his ideas as determined by an independent method, but much that is accepted by those meteorologists who have made the most complete study of dynamical meteorology.

[1] *Sitz. ber. Berlin Akad.*, June, 1890; reprinted in the *Meteorologische Zeitschrift*, Sept., 1890.

Von Siemens finally sums up his ideas as follows:—

1. All motions of the air are caused by disturbances of the condition of indifferent equilibrium of the atmosphere, and their mission is to restore this condition.

2. These disturbances are brought about by the super-heating, by solar radiation, of the air layers lying next to the ground ; by the unsymmetrical cooling of the upper air layers through radiation, and through the damming up of moving air masses, caused by their meeting with obstacles to their progress.

3. The disturbances are eliminated by ascending air currents, in which there is caused an acceleration, such that the increase of the velocity of the air is proportional to the decrease of the air pressure.

4. To correspond to these upward currents there are downward currents in which there is found a retardation of the velocity of the air, corresponding to the accelerating of the velocity of the upward current.

5. If the region of super-heating in the lower air layers is a limited one, then a local upward current will arise which will extend to the highest layers of the atmosphere, and there will be formed the spout phenomena with vortial motion: an inner spiral ascending current, and an outer similar spiral descending current. As a result of these vortial currents there is a spreading of the excess of heat in the lower air layers, so that the condition of adiabatic equilibrium is disturbed throughout the entire height of the air mass which takes part in the vortex motion.

6. In case the region is very extended in which the indifferent (or adiabatic) equilibrium is disturbed, as in the case when it embraces the whole tropical

zone, then the equalising of the temperature cannot be accomplished by these local vortex currents. Vortex currents must arise which embrace the whole atmosphere. We find in these the conditions corresponding to the accelerated ascent and retarded descent of air in local vortices, represented by the velocity of air motion, caused by the work of heat, in the different elevations, and approximately inversely proportional to the air pressure at these elevations.

7. Since, in consequence of the meridional currents caused and maintained by the continuous work of heat, the whole sea of air must rotate in all latitudes with approximately the same absolute velocity the meridional currents produced by super-heating, combine with the terrestial to such a degree that they embrace the system of currents of the whole earth ; and by this means the whole atmosphere is affected by the great addition of heat in the tropical zone ; and the equatorial heat and moisture are transmitted to the middle and higher latitudes, there to cause the occurrence of local air currents.

8. This last occurs through the production of varying local increase or diminution of air pressure through the disturbance of the indifferent equilibrium in the higher layers of the atmosphere.

9. Maxima and minima of the air pressure are the result of the temperature and velocity of the air currents in the upper layers of the atmosphere.

§ 5.—*General Circulation of the Atmosphere, according to Möller.*

One of the most interesting of the memoirs in which an attempt has been made to deduce actual numerical (although confessedly only approximate) results concerning the general atmospheric motions, is that by a German engineer, Max Möller.[1] While the most important matter treated in this paper is the circulation of the air in lower latitudes, yet a few additional items which he has given with great clearness are also mentioned here.

The first of these is the deviatory effect of the earth's rotation which is best shown by a numerical example.

The well-known expression for this force is 2 v ω sine φ: where v is the velocity of the wind ; ω the angular velocity of the earth's rotation on its axis ; and φ the latitude of the place at which the air is in motion. This formula has been deduced by many different investigators since Poisson first gave it. Suppose a case in which the air is moving with a velocity of 20 metres per second, and the latitude is 30° N. ; then in 100 seconds the amount of the deviation to the right of a straight course would be 7·3 metres. It is readily seen that in moving over a long distance of perhaps hundreds of miles, the result of this deflection reaches a magnitude that cannot be neglected, as many have supposed.

Where temperature differences exist such as are found at the equator and poles, the surfaces of equal barometric pressure are inclined (converge) towards the poles, where the colder air exists. Suppose the difference of temperature between the latitude of 30° and the pole to be 50° C. ; and that the mean temperature of a layer of air of 10,000 metres' thickness

[1] *Der Kreislauf der atmosphärischen Luft zwischen hohen und niederen Breiten, die Druckvertheilung und mittlere Windrichtung.* Von Max Möller. Aus dem Archiv der Deutschen Seewarte. X. Jahrgang, Hamburg, 1887.

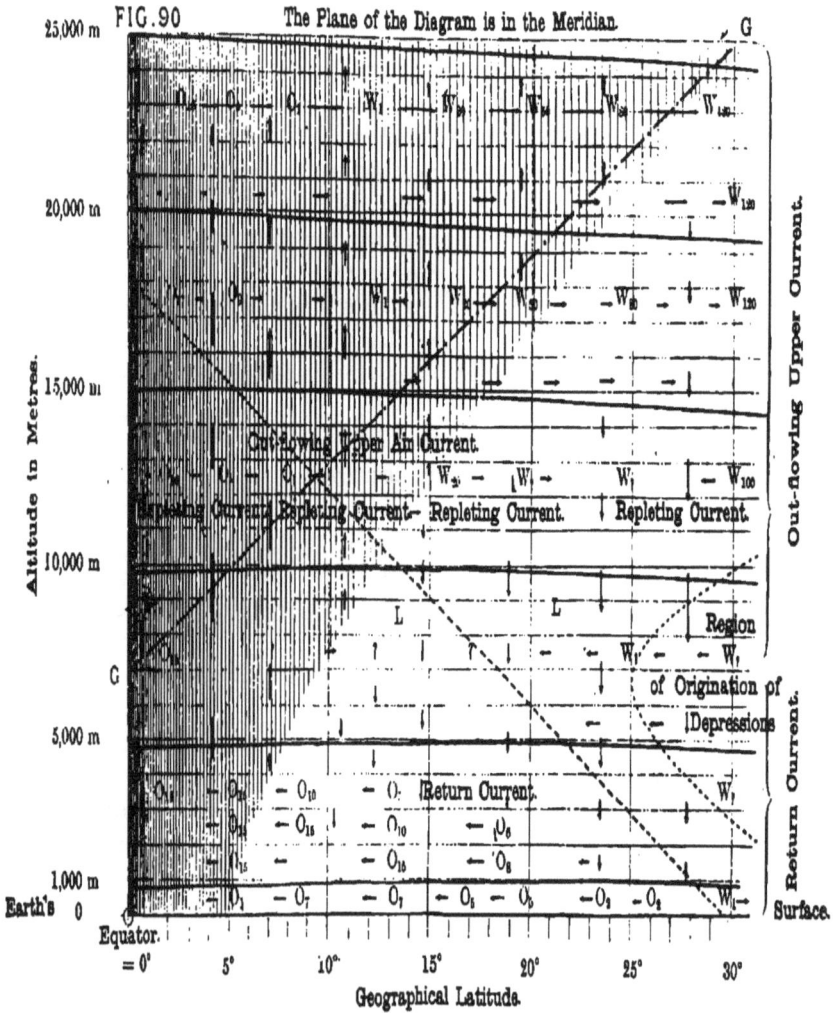

Representation

of the Depleting and Repleting Currents, and their limits; also the Rising and Falling Currents,
and the surfaces of equal pressure of the air circulating in lower latitudes.

FIG.90   The Plane of the Diagram is in the Meridian.

ji
nu
re
is
th
th
fe
cle

rot

sin
an
anc
mo
diff
Su
vel
30°
tion
met
dist
this
neg

four
bar
the
diff
and
per

*L
niedo
Von
Jahrg

is 5° C. at latitude 30°; then the amount of convergence of this upper layer would be 1,800 metres between lat. 30° and the pole. A mass gliding, frictionless and undisturbed, down this incline would have a velocity of 188 metres per second on reaching the pole.

Observations show that the greatest pressure at the earth's surface is at about 30° latitude, and at the equator the surface of equal pressure is depressed about 80 metres; in the northern hemisphere the maximum sinking is about 100 metres at lat. 60° or 65°, and at the poles there is a slight rise; in the southern hemisphere the sinking is at least 200 metres. At an altitude of 10,000 metres the sinking is about 860 metres in the northern hemisphere, and 960 metres in the southern hemisphere.

Aside from pure theory, the results arrived at by Möller as to the general circulation, and his picturing of the air movements, deserve special mention here; although, like all similar attempts, the results are largely schematic.

In the accompanying Fig. 90, Möller has sketched the circulation as he deduces it for the equatorial region up to the latitude of 30°, and for an altitude of 25,000 metres. In this diagram the wind components *in the plane of the meridian* are represented by arrows, the lengths of which are drawn in proportion to the wind velocity, with this exception, that the vertical and horizontal scales are not the same; all of the vertical velocities being grossly exaggerated in comparison with the horizontal currents. The arrows fly *with* the wind. Where there is an east component of velocity (wind towards the east) an O (*ost*) is given, and for a west component of velocity a W is given ; the numerals attached to these letters signify metres per second.

It is to be noticed that in the shaded portion there is an upward movement of the air, and in the unshaded portion a downward movement.

Möller distinguishes between the accelerated current which carries away more air than is received, and the retarded current which carries into any region more air than is carried away. The former he calls a " Saugestrom," and the latter a " Staustrom " ; in English we may say " depleting current " and " repleting current." The boundary between these two currents is indicated by the broken line G. The surfaces of equal barometric pressure are indicated by the heavily drawn nearly horizontal lines. The points of intersection of these surfaces with the horizontal plane are connected by the dotted line extending downwards from left to right.

It is seen that the upper current flowing away from the equator, or the outflowing current, is first noticeable at the equator at an altitude of from 5,000 metres to 7,000 metres, but with increasing latitude the altitude increases, and at lat. 30° the altitude of the lower limit is 15,000 metres. This outflowing current is divided into a "depleting" and a "repleting" current ; the limits of which are defined by the line G. While this dividing line is at an altitude of 7,000 metres at the equator, and 25,000 metres at lat. 30, under ordinary circumstances, yet when local ascending currents are developed at some distance from the equator, then this limit is to be found there at much lower altitudes.

Energy developed by the Air at various Altitudes.— The amount of energy which the air is capable of exerting is approximately proportional to the inclination of the surfaces of equal pressure and the density of the air, and Möller has made a very interesting computation of the relative amounts of this energy at different altitudes. Assuming that half of the mass of the air is to be found below the altitude of 5,000 metres, he presents the accompanying Fig. 91, as showing the relative amounts of energy in the different layers. In this figure the vertical scale shows the altitudes and the horizontal scale (abscissa) shows the energy per unit of space of air, that at 5,000 metres being taken as unity. The values for the upper half of the atmosphere are computed on the supposition of no gradient at the surface of the earth, and Möller finds per unit of space 4·00 kilogram-metres for the whole atmosphere, and but 0·63 for the lower half (*i.e.*, below 5,000 metres). This gives an amount of energy for the upper half of nearly six times that for the lower half; assuming that the horizontal differences of temperature are the same above as below, and no gradient at the earth's surface. It is to be seen from this that the upper atmosphere can by no means be omitted in accounting for meteorological phenomena.

It is seen that half of the energy is above, and half below the altitude of 13,000 metres, which is about the region of the cirrus clouds.

§ 6.—*General Atmospheric Motions, according to Oberbeck.*

There have been advanced objections to the methods followed by nearly all of the various investigators in the study of atmospheric motions. Ferrel has applied the older methods of analysis by

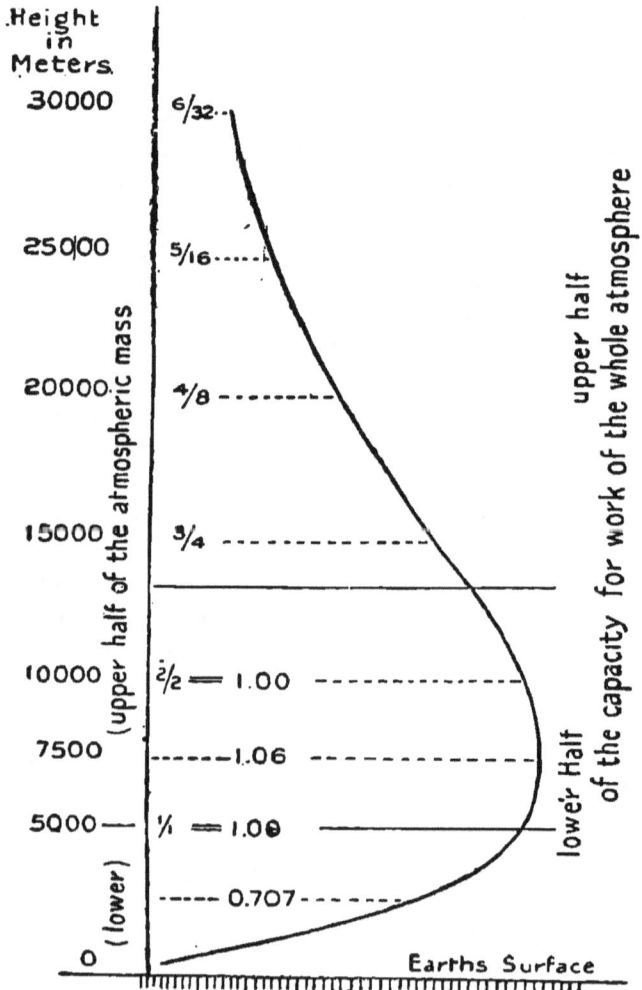

Capacity for work of the
Atmospheric Air. per unit of space

FIG. 91.

modifying the reasoning applied to the motions of a particle so as to take account of the changes pertaining to a fluid, and he has applied the condition of continuity and the effects of friction in such a general way as to leave still to be effected a more exact treatment. Still, he mapped out the path for others to follow up with this greater accuracy. Guldberg and Mohn have started out with the correct hydrodynamical equations, and have deduced the motions resulting from the differences of pressure, the effects of friction being partially considered. The velocities are assumed as proportional to the resistance ; which is somewhere near the truth for the motions of the air near the earth's surface, but which leads to very erroneous results when applied to the upper atmosphere.

Oberbeck [1] has attempted to outline the plan of the general circulation of the atmosphere by avoiding some of the unsuitable assumptions of previous investigators. He starts out with the hydrodynamical equations such as have been developed by Helmholtz, Thomson, Kirchhoff, and others, and against the use of which no objections are to be raised. Proceeding, then, on the assumption that the motions are due to differences in density of the air, Oberbeck applied the deviatory force due to the earth's axial rotation, and the modifying effects of friction. This last he has particularly inquired into. He states that the resistances which masses of moving air meet with

[1] *Ueber die Bewegungserscheinungen der Atmosphaere.* Von. A. Oberbeck (in Griefswald). Sitzungsberichte d. Kön. Preussischen Akad. d. Wissenschaften zu Berlin, 1888 ; pp. 383–395, 1129–1138. Also a popular paper with same title in the *Wissenschaft. Rundschau,* June 9, 1888, and reprinted in the *Meteorolog. Zeitschrift,* Aug., 1888.

vary greatly. If a swiftly moving air mass moves into a quiet mass of air, then the resistance is considerable, being somewhat like that encountered by a solid body moving in an atmosphere.

If the moving mass of air is subject to a continuous driving force, it will separate into beams, or, as Helmholtz calls them, discontinuous currents, which pass through the quiet air with a relatively smaller frictional resistance. Two adjacent parallel moving air currents will exert a slight mutual action, the faster moving one being retarded, and the other accelerated ; this action is very much increased when the air of the two currents becomes mixed by means of any interchanging currents due to local causes. In general it would seem that the law of resistance formulated by Newton can be applied. This is, that the resistance is proportional to the difference in the velocities of the two fluid masses. But this simple law cannot be applied directly to the actual determination of the friction in the atmospheric motions.

Oberbeck now, by the aid of the known laws of the variation of density of air, as dependent on the pressure and temperature, deduces the theorem :—

"If we conceive of a definite temperature distribution prevailing over the whole earth's surface, or a portion of it for a considerable length of time, then there arise continuous air currents, which may be determined by computation when the temperature distribution is known."

This is, indeed, the same idea which has been the foundation of most of the researches covering the question of general atmospheric motions, and the main differences between the various theories have been in the method of its application. In representing the temperature distribution, Oberbeck has advocated the use of spherical functions, which will undoubtedly cause meteorologists to regard, with something more than mere curiosity, the paper written by Dr. W. Schoch in 1856,[1] in which the subject of representing the temperature at the earth's surface by means of spherical functions is beautifully worked out.

As in Ferrel's theory, Oberbeck has found it convenient to find, first, what motions will exist in the case of the non-axial rotation of the earth, and then find the modifying effects of this motion of rotation.

1. Without the rotation he finds :

a. The meridional current on the northern hemisphere is towards the south below and towards the north above. It reaches its greatest velocity at the latitudes 45°, but vanishes at the equator and the poles.

b. The vertical current disappears at the earth's surface and at the upper limit of the atmosphere.

From the equator to latitude 35° 16′ north and south, the air current is an ascending one, but in higher latitudes a descending. Its velocity at the pole is double that at the equator. The vertical current has the same ratio to the horizontal current as the height of the atmosphere to the radius of the earth, and the importance of its action is due to its great lateral extent.

2. Currents with axial rotation of the earth are twofold, and as follows :

a. The motion of one current sets in towards the west at the equator, but changes its direction into an easterly motion at the latitude of 35° 16′.

[1] Translation published in 1891 by the Register Publishing Company at Ann Arbor, Michigan, U.S.A.

b. The motion of the second current is zero at the equator, reaches its maximum at the latitude of 54° 44′, and is easterly throughout. The motion is also zero at the earth's surface, but at high altitudes it is of considerable amount. Both of these motions disappear at the poles.

In Fig. 92 Oberbeck has shown, on a chart drawn

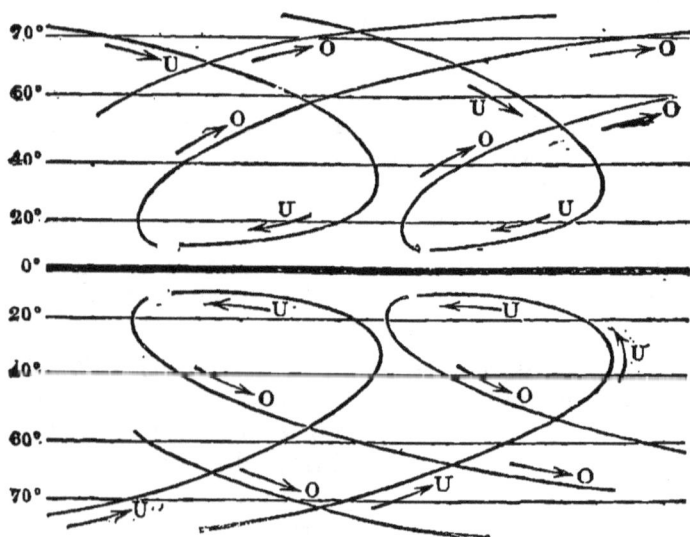

FIG. 92.

in Mercator's projection, the general circulation of the atmosphere resulting from the foregoing deductions.

The poleward currents, marked O, are the upper currents, and those towards the equator, marked U, are the lower ones near the surface of the earth. The horizontal currents are connected by downward currents at the poles and upward currents at the equator. The curves U taken within the tropics, represent with considerable fidelity the trade-winds as found on the great oceans. Oberbeck finds that the agreement of this schematic circulation with that actually observed shows that in general the current mentioned under the heading 2.*a* is comparable in magnitude with the lower current U which flows equatorwards ; and that the current mentioned under the heading 2.*b*, which is the easterly current of the higher altitudes, is of a greater velocity than the lower current. The upper currents O are the anti-trade winds of the tropics. In the higher latitudes an easterly wind (*i.e.*, from the west) prevails in the higher altitudes. At great altitudes the velocities decrease on account of the very small density of the air in those upper regions.

Theoretical Distribution of the Atmospheric Pressure.—Having derived the just explained results concerning the general circulation of the atmosphere, Oberbeck proceeded to examine into the conditions of atmospheric pressure distribution. The extremes of temperature on the earth's surface, amounting to 70° C., would cause, for uniform pressure, differences in density of the air of more than 20 per cent. But since the density and pressure counterbalance one another, it is found that there is a maximum pressure where the temperature is low, and a minimum pressure where the temperature is high. The fact that the actual mean differences of pressure on the earth's surface amount to only 7 per cent., is explained by the fact that air currents are produced which flow in general in the direction of the increase in temperature below, and in the opposite direction above. Concerning the fact that in taking the atmosphere as a whole, the pressure is slightly lower at the equator, increases to a maximum between 20° and 40° latitude,

22

and decreases again at the poles, Oberbeck agrees
with Ferrel in saying it is due to the effects of the
earth's rotation on the air currents. He comes to
the final result, that the existing pressure distribution
is fully explained by the currents of the atmosphere,
and that when the air pressure is known conclusions
can be drawn as to the intensity of the currents.

If there were no axial rotation of the earth, and
the temperature varied only with the latitude and
not with the longitude, then at the surface of the
earth the barometric pressure would increase from
the equator towards the poles ; at middle altitudes
these differences of pressure would vanish, and in
the upper atmosphere the pressure would be greatest
at the equator and least at the poles. The actual
pressure does not, however, follow this configuration.
It is the view of meteorologists in general that the
rotational currents of the air possess a greater intensity than that of the meridional currents, so that
the irregularities produced by these become of prime
importance in outlining the system of atmospheric
pressure and motions ; and Oberbeck found, as
Ferrel had shown many years before, that the
rotational motion of the air in higher latitudes is in
the same direction as the axial rotation of the earth,
but at a certain latitude it reduces to zero, and at the
equator has an opposite direction. Having found the
magnitude of these rotational currents as a function
of the distance from the pole, and knowing also the
amount of the vertical and meridional components,
Oberbeck is able to produce formula for the computation of the pressure distribution.

The angular velocity of the rotational motion of
the air is small in comparison to the axial rotation of

the earth, and has a maximal value of 4·59 metres (per second) at the latitude of 56° 27'.

From the pole to latitude 16° 49' the rotational motion is towards the east, and from thence to the equator towards the west. It will be remembered that Oberbeck has previously considered the total motion of rotation as made up of two members as mentioned under the heading 2, and that each was most effective in specified regions. In the region between 35° latitude and the equatorial region, there would be two currents flowing, the one over the other, but in opposite directions, and this would bring the region of no rotational motion nearer to the equator than 35° latitude.

Oberbeck considers that his results are almost entirely compatible with those derived by Werner Siemens, the only modification being that the easterly motion (*i.e.*, towards the east) in the higher altitudes and higher latitudes has the advantage over the easterly motion of the lower altitudes and lower latitudes, that the former loses through friction a much less amount of living force than the latter.

A few words must still be added concerning the vertical and meridional components of motion.

The vertical component gives rise to an ascending motion at the equator and poles, and a descending motion at the middle latitudes; consequently the actually existing upward current at the equator is increased, and the existing descending current at the poles is retarded.

The meridional component is zero at the earth's surface at the equator; from thence to about the latitude of 24° it is directed towards the south, and from this point polewards it is towards the north, but vanishes at the poles. The equatorial current is consequently strengthened in the tropics and weakened in the higher latitudes.

§ 7.—*General Remarks on Friction of Horizontal Currents.*

What may be termed the sliding friction of the air on other air is so very slight that for all practical purposes it may be neglected when dealing with air currents existing for a short time. A horizontal current of air with a certain velocity may lie adjacent to another parallel current of different velocity, and the frictional action between the two will not, in a long time, make itself felt beyond a very narrow layer at the surface of meeting. A good illustration of the smallness of this frictional action is shown by the following recently published computation of its amount. A current flowing steadily for twenty-six years over a quiet mass of air will succeed in imparting only half of its velocity to the originally quiet air, at a distance of one hundred metres below the surface of contact, if only the friction is taken into account.

The determination of the actual effects of friction on the change of velocity of air movements with change in altitude above the land and above the sea, is a question which will require an immense amount of investigation in order to reach any numerical results which are of real value.

A few experiments in Great Britain by Stevenson and Archibald show that there the increase of wind velocity for a short distance (several hundred feet) above the land, *at certain times*, is shown by the following formula (Abbe) :

Velocity in miles per hour $= 18\cdot7 + 0\cdot13 \sqrt{\text{altitude in feet.}}$

The little practical value of such a formula is shown by the following data obtained from a number of years' observations at neighbouring stations having different altitudes. It was found that between Mount Washington (United States) and a not very distant sea-coast station the increase per one hundred feet was 0·33 miles per hour for the yearly average, while in May it was 0·28 miles per hour, and in December 0·42 miles per hour. The observations for the same region show a greater range for inland stations. In Eastern Massachusetts, on the coast, the increase up to six hundred feet altitude was 37 per cent. of the surface wind for the year, 22 per cent. in January, and 62 per cent. in August.

Comparing Pike's Peak (altitude 4308 metres) with Dodge City, Kansas, it was found that the increase was 1 per cent. in July, 129 per cent. in January, and 63 per cent. for the year. There seems in all cases to be a well-defined yearly period with great range, so that a few observations will not give an average value.

As regards the differences existing for land and water surfaces I think that if an instrument on a fairly good land station gives a wind of 1, a very free high exposure will be 2, a good *sea coast* exposure will be 3, and a mid-ocean exposure will be 4, for the same general conditions of atmospheric disturbances, and near the surface. Of course these numbers are but very rough expressions of these relations.

§ 8.—*General Atmospheric Motions, as Outlined by von Helmholtz.*

Friction of the Atmosphere.—While other investigators have contented themselves mainly with general remarks concerning the effects of friction on the atmospheric motions, von Helmholtz [1] has given us the results of a more particular inquiry made by him into the question. On account of the very minute scale on which experiments can be conducted in the laboratory, it is impossible to consider the matter by this means, as is shown by the fact that if an air layer 1 metre in thickness is experimented on, then the influence of friction for a motion with velocities corresponding to those in the atmosphere must be 8,026 times as great as in the atmosphere, the latter being assumed to be of constant density at 0° C., and extending to an altitude of 8,026 metres.

However, some special cases of the action of friction can be considered by using the already determined friction constant for air. If the air, of uniform density, has a certain specified velocity of motion, it would require 42,747 years to reduce this velocity one half by the action of friction with the ground ; this retardation being communicated to the successive layers of air above. In the case of the actual air the diminution of pressure and temperature would make this computed time still longer. The loss of living force by friction must consequently occur near the earth's surface, and at the limiting surface of secondary irregular or vortex motions.

A somewhat similar computation for the conduction of heat from one layer of the atmosphere shows that it would take 36,164 years to reduce, by one half, by the process of conduction alone, specified differences of temperature existing at the upper and lower limits of an atmosphere of constant density.

[1] *Ueber Atmosphaerische Bewegungen.* H. von Helmholtz, Sitzungs berichte d. Kön. Preus. Akad. d. Wissen. zu Berlin, Part I. 1888. Part II. 1889.

Von Helmholtz shows that an unhindered circulation of the air in the trade zone could not hold out to the latitude of 30°. He finds that if the air of any rotating air girdle, whose axis is the pole, is pushed unhindered, sometimes northward and sometimes southward, the moment of rotation must remain constant. By applying this principle he finds that air which was calm at the equator would have :

A velocity of 14·18 metres per second in latitude 10°
    ,,   ,,   ,, 57·63   ,,   ,,   ,, ,,   ,,   20°
    ,,   ,,   133·65   ,,   ,,   ,, ,,   ,,   30°

So it is seen that the air cannot even reach the latitude of 20° unhindered, without producing almost unheard-of wind velocities relative to the earth's surface. Now, a circulation of air certainly does exist in the trade zone, and it is of interest to see by what means the wind is prevented from reaching these enormous velocities. Von Helmholtz proves that in such a rotating girdle of air there must be a ring where there is a calm, and that the pressure decreases from this ring towards the pole and the equator. His reasoning gives but another independent proof of the cause of the ring of high pressure around the earth in middle latitudes, and of the accompanying calm belt.

Von Helmholtz has also given a general statement of the conditions of atmospheric equilibrium as determined by the heat capacity of the different layers of air. He has shown the form of these layers in various assumed cases, and has given their position in the complex case of continual change of velocity of rotation with the heat capacity ; this last being of special importance in explaining natural phenomena.

" When the layer of air with the greatest heat content lies uppermost in the direction of the pole of the heavens (*i.e.*, parallel to the axis of revolution of the earth), then it must be in stable equilibrium."

This theorem is very important, and its derivation and application by von Helmholtz to the actual atmospheric conditions is very valuable to meteorologists, but it is too complex to present without the use of mathematical notation. It may be noticed, however, that when the air layer lies in the direction of the pole of the heavens the angle of inclination forms a very small angle with the horizon at the equator, but this angle constantly increases towards the pole.

In the case of rings encircling the pole, and where the temperature and latitude vary, and the temperature change is continuous, then in order to have a stable equilibrium, the heat content must increase in the direction of the celestial pole. In other words, the warm ring must be on the polar side of the cold ring. If the rings are superimposed then the warmer must be above.

The gradual change of equilibrium through friction and warming up of air is further elaborated by von Helmholtz as follows :—

If the under side of an air layer is warmed up, as by contact with the warmer ground, then it begins to ascend ; at first in isolated streamlets according to the condition of the heated contact surface, and finally these gradually unite into larger currents. This distribution of heat takes place rapidly through the air layer, and when the latter is in adiabatic equilibrium throughout, or, as Helmholtz would say, of equal heat content, these currents endeavour to distribute the heat to the top of the layer. An analogous process also occurs when the upper side of the air layer is cooled by the abstraction of heat.

When, however, the upper side is warmed or the lower side is cooled, then these convective motions do not arise.

Since conduction of heat in a great atmosphere acts so slowly, and since even radiation may have very little effect on the pure atmosphere, there may exist for a considerable length of time cold air layers at the ground, and warm air layers high up, the equilibrium of temperature being but slowly restored.

Similar differences exist for changes of velocity through friction. In the normal inclination of the air layers the higher end is nearer the earth's axis than the lower. If a layer reaches to the earth's surface and is there a west wind, then the moment of rotation of the lower portion is retarded, its centrifugal force is diminished, and on the poleward side of the layer there is an upward motion with the approach to the axis, in order that the position, which restores stable equilibrium, may be reached at the upper end of the layer.

Such effects must also occur in the cases of vertical currents like those of ascending warm air, but in such cases as the latter the difference of the moment of rotation below and above is slight; and since the action is distributed over the whole air mass, it is less noticeable on the lower side of the layer than if it were restricted to the under portion.

For the east wind the opposite is true. Its moment of rotation is increased by friction with the earth's surface. The accelerated air mass being in the position of equilibrium pertaining to the inside of the air layer, will be, along the earth's surface, pressed into the next layer on the equatorial side. If at the same time it becomes warmed up, then the ascent resulting from this heating will proceed slower than if the air layers were at rest. As a consequence of this, the change of the east wind through friction is confined to the lower air layers, but there it is relatively greater than for the same velocity of west wind. The retarded air layer will generally be pressed equatorwards, and the east wind thus becomes a north-east wind in the northern hemisphere : however, the tendency to remain an east wind is strongly retained since the air in moving equatorward enters zones in which the earth revolves more rapidly. The air above, at the place where there is freedom below, as for instance on the outer border of the trade zone, follows after and appears below as an east wind with an unchanged moment of rotation ; and moving gradually towards the equator, is affected by the influence of friction as described. At the same time it is to be pointed out that the moisture-laden air from the tropical zone enters into the trade wind with a greater rotational velocity, and consequently must lessen this lagging behind relative to the earth's surface.

The lower layers of the Trade wind are forced into the calm zone as soon as differences of rotation, as regards the earth's surface, cease. Consequently they then become part of the calm zone, and increase it in amount so that it is thus allowed to overrun at its limits, by an inclination of the boundary surfaces, the layers below, where there is a disappearance of the east wind. Accordingly it is to be seen, that while at low altitudes there is in general a continuity in the temperature and moment of rotation of the layers there, yet finally at higher altitudes the edges of the spread out calm zone (where the layers have the great moment of rotation of the equatorial air, which would cause in the latitude of 10° a strong west wind, and at the latitude of 20° a violent west wind), come into immediate contact with lower-lying layers having a smaller velocity of rotation and lower temperature. The uppermost side of these last layers will scarcely suffer any change of heat-content or moment of rotation, because they shift towards the equator after the loss of their lower layers.

As von Helmholtz has shown in the memoir on discontinuous fluid motion, which he communicated to the Berlin Academy of Science in 1868, such a discontinuous motion may for a time exist, but the equilibrium at the boundary surface is unstable, and sooner or later there must be a breaking up into vortices, which leads to an extended mixing or mingling of both layers of air. In this case the lower layers are the heavier, and the disturbances proceed similarly to water waves produced by the wind ; as von Helmholtz has been able to show by experimental means. This process is shown in nature by the streaks of cirrus clouds which occur when fog can be precipitated at the boundary of the two layers. In the case of water waves the specific gravity of the fluid is so different from that of the air, that the degree of motion is not the same, but it is of the same nature. Violent storm winds cause the water waves to break and become " white-capped," and the waterdrops are thrown up into the air from the upper edge. For less differences of specific gravity the result of the mixture of the two layers would be the formation of a vortex, accompanied in some cases by copious rainfall.

The mixture resulting from the mingling of the two layers will have a temperature and moment of rotation whose value lies between that of the two components, and its position of equilibrium will be found nearer the equator than that of the colder component. It will, then, fall towards this locality and force back the layers lying to the poleward. At the places which are vacated by this means, the layers from which these downward moving air masses are taken, must extend themselves upwards, and thereby lessen their cross sectional area. In the places where the descending air masses push into and force aside those lying beneath them, there will be formed the so-called anti-cyclones ; and where partial vacancies are formed by the ascension of air masses the so-called cyclones will be formed. The meteorological charts show anti-cyclones, and corresponding barometric maxima, on the irregularly changing border of the north-east trade winds on the Atlantic Ocean. These are found to occur with great regularity in winter under lat. 30°, and in summer under lat. 40°.

Owing to the inclined position of the layers, the rain, which is produced by the mixing of the air (Dove's subtropical rains), falls a little further to the north. There, also, is to be found the zone of cyclones, which however increase in frequency towards the north. It may certainly be assumed that a complete mixing of the air does not take place at the boundary of the trade zone, but portions of the warm upper layers possessing a powerful rotational motion are here in either their original or only a partially mixed state, and they consequently become more thoroughly mixed with other layers farther to the north.

On the whole, in this zone of mixture the west wind must prevail at the earth's surface, as the increase of the total moment of rotation which the air mass is subjected to by friction in the east wind of the trade zone must finally become so great, that somewhere the west wind reaches the ground and sufficient friction is encountered to entirely counteract that increase.   The air masses resting in equilibrium in the air layers, will not, after a time, have a moment of rotation which differs materially from that of the portion of the earth's surface beneath them.   Consequently when air with a strong west wind comes down from above and mixes with them, there will be imparted to them a momentum towards the east. Moreover, the rain which mostly comes from the higher regions of west wind, will carry this motion with it, and impart it to the air through which it passes.   All air zones which are forced polewards by the masses of air falling out from them equatorwards will finally have west winds.

Polar Cold as a Source of Air Motion.—The cooling of the ground at the poles is also a continual source of winds.   The cold layers will endeavour to flow apart at the ground, and form east winds ; which in this case would be anti-cyclonal.   Above, the warmer upper layers would flow in to fill the vacancies created by this, and west winds would result ; which in this case would be cyclonal.   In such a circulation there can be an equilibrium only when the cold lower layers do not gain through friction a faster rotational motion, and by this means be enabled to make a farther advancement.   They must, then, remain close to the ground ; and that this is their actual condition is shown by the frequent occurrence of the winter north-east winds of Germany, whose extreme cold frequently does not reach even to the tops of the low mountains of North Germany.   Moreover, on the preceding edge of these east winds which advance into the warmer zones, the same circumstances exist which will cause discontinuities between the motion of the upper and lower currents as in the case of the advancing trade winds, and so here there will also be a new cause for the formation of vortex motion, *i.e.*, atmospheric whirls.

The spreading out of the polar east winds, even though recognisable in the principal paths, proceeds relatively very irregularly, as the cold pole does not coincide with the pole of the axis of rotation of the earth ; and since the wind does not extend to a great altitude, even the low mountain ranges have a great influence on it.   Thus it happens that fog in the cold zone allows a moderate cooling of thicker air layers, while in the clear air there is a very great cooling of the lower layers.   It is as a result of such irregularities that the anti-cyclonal motion of the lower layers, and the great and gradually increasing cyclonal motion of the upper layers which were to be looked for at the pole, are found to be broken up into a great number of moving cyclones and anti-cyclones, the former being the more numerous.

A most important conclusion to be drawn from these results is that the main hindrance to the motions of the atmosphere becoming extremely rapid, is not so much the friction with the ground as the continual mixing of air layers having different velocities, by means of vortical motion, which results from the rolling up of discontinuous surfaces.   Within such a vortex, or whirl, the air layers which were originally separated become increased in number, and with a corresponding thinning of the strata become twisted spirally around one another, which causes such an extended surface of contact that there occurs a quite rapid exchange of temperature and equalising of their motion through friction.

Wave Motion in the Atmosphere.—When a light fluid, like the air, lies over a heavier, such as water, with a sharply defined line of contact, the conditions are favourable for the formation and propagation of waves. Somewhat similar conditions exist between different layers of the atmosphere, but in this last case the differences of specific gravity are far less. The causes and action of these wave motions occurring in the air have been investigated by von Helmholtz, who has thus fitted another piece to the mosaic of the theory of atmospheric motions.

That such waves do really exist, can be plainly observed when the under layer is so far saturated with vapour of water, that where the pressure is less, in the waves, a formation of fog takes place. When this occurs there will be observed in somewhat regular repetition, parallel lying strips of cloud, of different widths ; these sometimes cover a considerable portion of the sky. It is only under peculiar circumstances of moisture that these clouds are formed, but the wave motion in the invisible air is probably a most common phenomena.

Helmholtz has found that these atmospheric waves or undulations may exist on a most gigantic scale, in which the wave length is several kilometres, and so great is the amplitude that when they occur at elevations of one or even more kilometres above the earth's surface their action is felt at the ground. This explanation accounts for the intermittent wind gusts which are so often noticed in squally weather, and which are frequently accompanied by showers of rain. These wave motions are a most important factor in the mixture of different layers of atmosphere, and, as has been noticed previously, certain conditions may arise in which cloud is formed in the ascending air masses, and a condition of almost unstable equilibrium is formed.

Helmholtz has also given a mathematical study of the wave motion on the common boundary surface of two fluids ; such as the water surface in contact with the air. In this he has taken into account the action of the air on the waves, a point which has been hitherto neglected, probably on account of its difficult treatment. It simplifies the question greatly if we consider such straight-lined wave movements as progressing with a uniform velocity without change of form. The boundary surface then may be considered a fixed (in space) surface ; the upper fluid flows over it in one direction, and the under fluid in the opposite direction. At a considerable distance from this boundary surface, the motion becomes a straight-lined motion of constant velocity, but in the region of the wave-formed boundary surface the direction of the current follows this in direction.

We have not space to follow von Helmholtz in his deductions concerning water waves, but he has shown that if, for a certain wind velocity, water waves of a certain wave length are formed, the wave length of the geometrically similar air waves must be 2630·3 times as great in order to correspond to the same differences of velocity in the currents for the air alone. A wind of 10 metres per second would cause a water wave of about 0·21 metres wave length, or an air wave of 549·6 metres wave length. This is in the extreme case where the store of energy of the wave is equal to that of the straight-lined motion along a plane surface. This relative value is considerably diminished in the cases where the computation is carried out for the lowest waves, when the ratio would be 2039·6 under certain assumed conditions.

Water waves of 1 metre wave length are frequent in a moderate wind, and the corresponding wind would produce air waves of from 2 to 5 kilometres wave length in air layers differing in temperature by 10° C. Great ocean waves of 5 to 10 metres would find their counterparts in air waves of 15 to 30 kilometres wave length, and these last would stir the air at the ground in a similar manner to that which the water undergoes in shoal places when the depth of the water is less than a wave length.

Von Helmholtz states that the principle of "mechanical similarity," by which he is enabled to pass from the condition of water waves to that of air waves, is applicable for all waves which have a forward motion with constant form and constant velocity.

The atmospheric counterpart of the wave motion in shallow water is of great significance in meteorological theories which attempt to explain intermittent phenomena of local winds. The velocity of propagation of small water waves (without wind) in shoal water depends greatly on the depth of the water. When the ratio of the depth of the water to the wave length equals 0·5, then the velocity is diminished in the proportion of 1 : 0·96 ; when this ratio is 0·25 the proportion is 1 : 0·81 ; and for a ratio of 0·1 the proportion is 1 : 0·39.

When there is a calm below, the wind beneath the
wave troughs is contrary to the direction of propagation, but on the wave crests is in the same direction
with it.   According to the law of the decrease of
amplitude at the bottom as compared with that at
the surface, these oscillations can only be felt below
(on the ground), when the depth is much less than
the wave length.   So that changes in the barometer
reading would not be expected unless there are
strong changes of wind noticeable in the passage of
the waves.

Theoretical investigations show that when waves
rise too high, they lose their continuity of surface; but
the "breaking" of the wave surface cannot occur
unless the medium into which they rise is in motion.
When this medium must flow in around them, then
there must arise (theoretically) infinite velocities and
infinite negative pressure at these points, and this
causes a powerful sucking up of the wave fluid, such
as is made manifest by the high-rising and white-
capped waves which one observes.   In the case of
waves moving forwards with equal velocity to the
wind, the crest can even reach an angle of 120° before
it breaks.   The same wind can produce waves of this
kind with different wave lengths, although there are
limits to the differences : the longer have relatively
greater heights.   This depends on the store of energy
in the wave.

When the energy of the waves produced by wind
blowing across a water surface is compared with that
of waves produced at the plane boundary surface of
two fluids which move with equal velocity, it is found
that *a great number* of the possible stationary wave
motions require a less store of energy in the first-

mentioned form than in the case of the corresponding current for a plane boundary surface ; so that the current with a plane boundary surface is in a kind of unstable condition of equilibrium as compared with these wave motions. In some conditions of wave motion the energy is the same in the two cases ; and in still others the energy is greatest in the waves on the water surface.

Two kinds of energy come into play ; the *potential energy* due to the position of the water at the summit of the wave, and which increases with the height of the wave ; and the *living force* which is common to the two systems, and is dependent on the distance from the boundary line. It is to these forces that the above differences are due, but we will not follow out von Helmholtz's discussion of their action for water waves, for he makes but a partial application to air waves. He has, however, indicated by his methods, a most promising field for future research in the extension of the ideas to purely air motions.

The consideration of the conditions of unstable, indifferent, and stable equilibrium, are of the utmost importance in accounting for the origin of these waves. It has been proven by them, that the slightest wind over a water surface must produce small waves which, for specified heights, assume regular forms and velocity. An increase of the wind's velocity causes an increase in the height of the waves ; the shorter ones, breaking and spraying, are able to form new longer waves of a less height.

The counterparts in the air of these breaking spraying water waves, would cause a mixture of air layers, and since the upheaval of air in the truly mountainous air waves amounts to hundreds of metres, precipitation can easily occur ; and this would accelerate and extend higher upwards the movement already begun. While short air waves can occur, yet the not very sharp boundary limits between the two moving air layers would usually cause a very great wave length. The circumstance that the same wind can cause waves of different lengths and velocities, gives rise to wave interferences by which some of the waves observed will be higher and some lower, and this difference may be so marked that the "breaking" phenomena will attend only some of these long waves.

I have now given some idea of the most important methods which have been employed by recent investigators, in their attempts at founding a system of atmospheric mechanics which would satisfactorily explain the general atmospheric circulation. The next step should be, to give as a connected whole such a combination of the results obtained by these investigators as can be united in one general system. A work of this nature is, however, not to be undertaken in the present connection, for it would be no mere compilation, but would require great originality of treatment, and should first be presented for the consideration of professional meteorologists, by publication in a proper scientific channel. Such a task is now more difficult than it would have been a few years ago, because the recent contributions to this subject have thrown considerable doubt on what seemed to be fairly well-established results.

CHAPTER V.

Secondary Motions of the Atmosphere.

§ 1.—*General Ideas Concerning Secondary Atmospheric Motions.*

IT has been shown that in each hemisphere there is a general atmospheric circulation, in which, in the northern hemisphere, for most of the lower atmosphere, the general direction of motion is north-easterly in the middle and higher latitudes, and south-westerly in the lower latitudes. Within these great primary currents there exist local disturbances, usually of great extent, which act with so great an influence that, when single localities are considered, the secondary phenomena of the regions thus affected far overshadow the primary. The study of simultaneous observations extending over large areas, such as continents or oceans, or even a whole hemisphere, is necessary to show the true relation of these disturbances, and the observations must be sufficiently numerous to give a relatively microscopic view of the distribution of the meteorological elements. So· far, it has not been found practicable to bring together *complete* data of this kind. The nearest approach to it have been the synoptic weather charts of Hoffmeyer and Neumayer of the Deutche Seewarte; the charts of

341

the United States Signal Service International Weather Review ; the less extensive charts of storm tracts issued by the various Government meteorological services ; and the daily weather maps now to be found in most of the great European countries and the United States.

These extensive local atmospheric disturbances are characterised by either a deficiency or excess of atmospheric pressure as measured by the barometer. In the former case the motion of the air is, generally speaking, inwards towards the lowest pressure, and the term cyclone is applied to it ; while in the latter case the motion is outwards, and the term anti-cyclone is applied to it. Viewed as a whole, we find these disturbances following each other in a motion of translation in a direction from west to east around the globe in the middle latitudes.

We must limit ourselves mainly to the examination of these huge moving atmospheric disturbances, in which the air pressure is least at the centre—that is, to the cyclones ; although a certain amount of consideration must be bestowed upon the companion but less definitely characterised anti-cyclones. While to the imagination they can be best pictured as counterparts of the whirls and eddies seen in flowing water, yet a close study of them requires us to place on charts of some kind their properties and peculiarities as denoted by meteorological observations.

As a plane of representation the earth's surface is chosen, as most of the observations are made there. The amount and distribution of the air is indicated by its weight as measured by the barometer, and it is customary to reduce these results to what they would be at sea-level in order that the projection of the

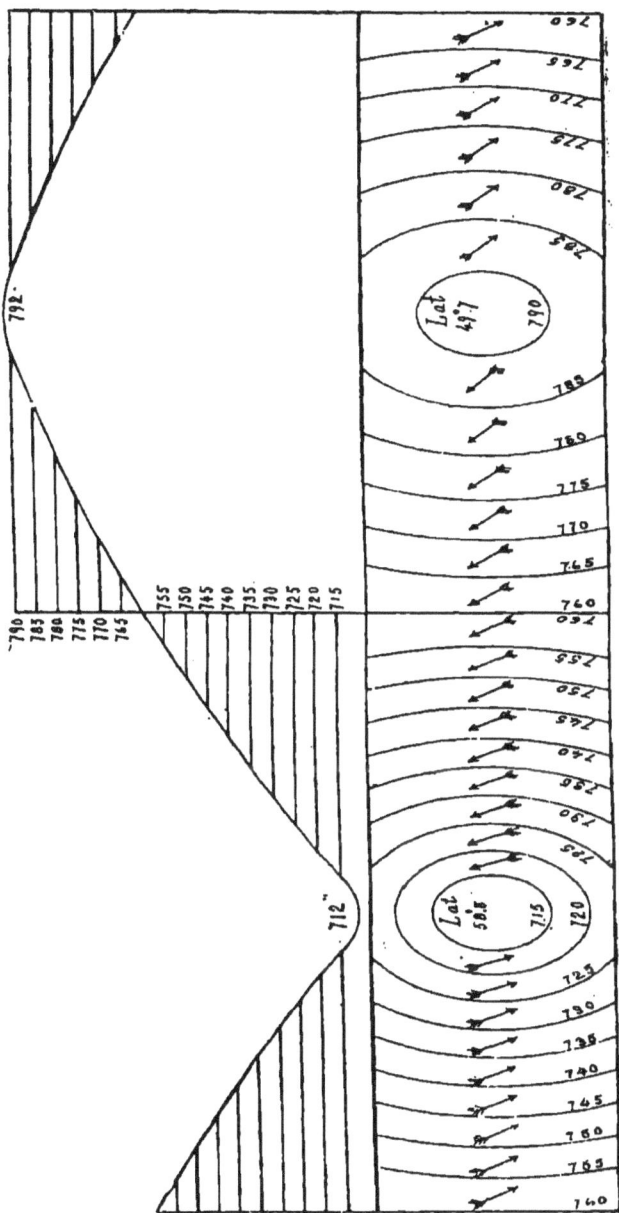

RELATION OF WINDS TO BAROMETRIC GRADIENTS. ATLANTIC OCEAN

FIG. 93.

weights may be symmetrical. The temperatures, however, are usually given as they are observed, but
observations at any great altitude above the surrounding country (as mountain peaks, &c.) are not
used in determining the horizontal distribution of the
temperatures. The other meteorological elements are
presented, as observed, in the usual manner of chart-
ing distributed phenomena.

By far the most important element to be noted is
the barometric pressure ; and the manner of representation is shown by Fig. 93, in which the method of
projecting the isobars, or lines of equal pressure, is
given for both an area of high pressure and low
pressure, *i.e.*, anti-cyclone and cyclone. The lower
part of the diagram shows schematically the (incom-
pleted) concentric isobaric rings, one for each five millimeters pressure, as they are usually drawn on the horizontal plane representing the earth's surface; while the
upper part shows the distribution of these pressures
in a vertical section, taken from left to right through
the centre of the horizontal projection. On the
diagram are also noted, by means of arrows, an
average direction of the wind, at the successive
isobars, on both the preceding and following sides of
the centres of high and low atmospheric pressure.

The relations of the cyclones and anti-cyclones for
a whole hemisphere (northern) are shown by Fig. 95,
in which the isobars are given in the horizontal plan
just mentioned. While this figure shows a simultaneous distribution of these phenomena, the Fig. 94
shows the paths pursued by the centres of a great
number of individual cyclones, in the movements of
translation to which they are subjected, and which is
spoken of in detail in another place. I shall now

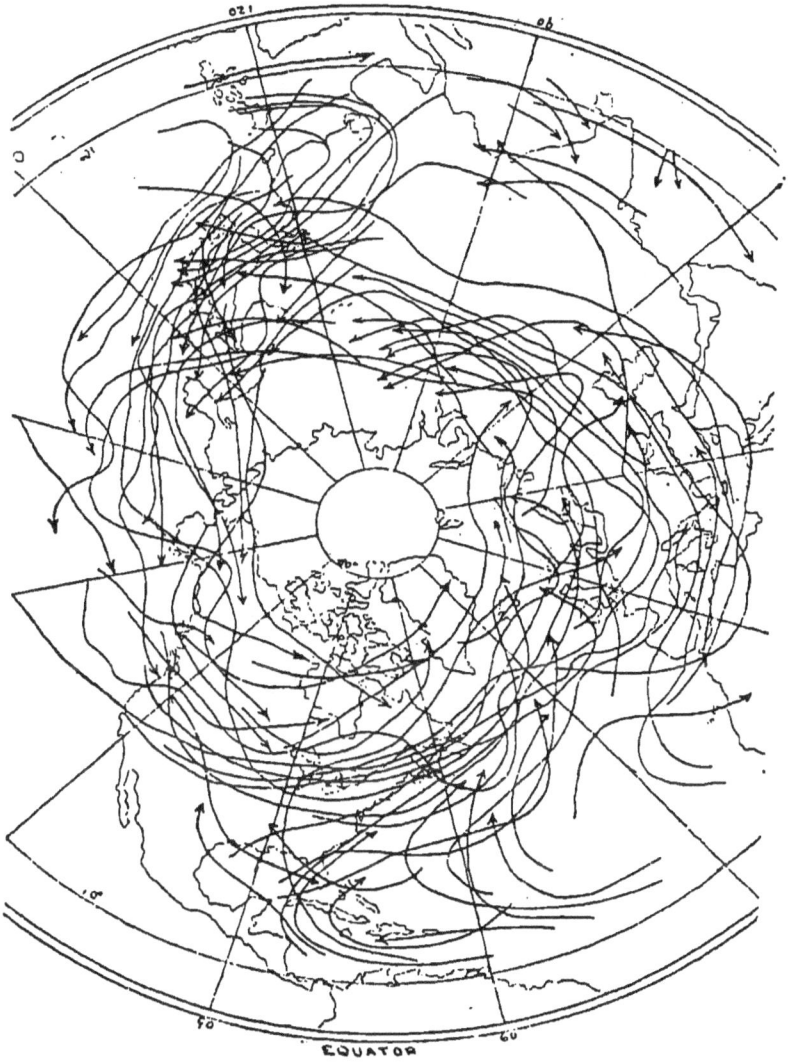

FIG. 94.

narrate some of the facts which have been accumulated concerning cyclones ; after which we will consider some of the theories which have been promulgated by investigators who have attempted to account for these facts.

§ 2.—*Investigators of Facts concerning Cyclones.*

The earlier investigators in the theory were also necessarily collectors of the facts on which to base their theories. They mainly investigated individual cyclones, in which case the data were very incomplete ; or they generalised from the few cases, descriptions of which might be found collected together.

These two general divisions may be made of the recent work in this field of inquiry, but the conditions under which they are carried out are very much changed. Studies of individual cyclones, and of specially great storms, have been made with the greatest care by many meteorologists ; counting among them such investigators as Köppen and van Bebber in Germany, Hann in Austria, Loomis and Upton in America, Hill in India, and others.

The most marked improvement in the methods of investigation has been due, however, to the introduction of the daily weather charts, and the publication by various governments of the courses pursued by cyclones on the continents of Euro-Asia and North America, and the adjacent oceans. These have been carefully studied, mostly by Government meteorologists, but others as individuals have also contributed to our knowledge by their laborious devotion to this work ; and of these outside workers the names of Elias Loomis in America, Vettin of Germany, and

Clement Ley of England, deserve to be mentioned most prominently in this connection ; the work of the two latter, being mainly devoted to the upper air currents, is of a more personal character than that of the former, who, in later years, mainly discussed the data given in the Government weather reports.

The most important relatively recent investigations of the courses of cyclones have been made by Köppen and van Bebber of Germany ; du Bort of France ; Loomis, Abbe, and Finley of the United States ; and a number of Russian meteorologists, and some others; in the tropics, Eliot, Hill, Meldrum, Poey, and Vines have done very important work. The study of upper air currents has been most assiduously followed up by Ley, Vettin, and Hildebrandson ; although many others of different nationalities have also done excellent work in this field.

Only a brief mention will be made of the generally important facts concerning cyclones, as other works on meteorology treat of this subject very fully, and it is mainly the theoretical deductions concerning atmospheric motions that we will concern ourselves with in the present instance.

§ 3.—*Some Facts concerning Cyclones.*

Air Pressure and Isobars in Cyclones.—In treating of cyclones in a general way, it is usually assumed that the isobars are circular in form, but in reality this seldom occurs ; Loomis says, not oftener than once a year for the United States weather maps. The amount of ellipticity as observed during several years is shown by the following ratios of the longest and shortest diameters :

UNITED STATES OF AMERICA.	ATLANTIC OCEAN.
59% 33% 11% 3% were greater than $\begin{cases} 1.5 : 1 \\ 2.0 : 1 \\ 3.0 : 1 \\ 4.0 : 1 \end{cases}$	54% 17% 1% were greater than $\begin{cases} 1.5 : 1 \\ 2.0 : 1 \\ 3.0 : 1 \end{cases}$
Average ratio of diameters, 1·94 : 1	Average ratio of diameters, 1·70 : 1

The longest axis is most frequently pointed towards N. 36° E. for United States of America, and N. 35° E. for the North Atlantic Ocean ; but while it may point to any azimuth, yet 70 per cent. of the cases for the United States, and 64 per cent. of those for the North Atlantic point between North and East.

The weather charts show that the area of low pressure is nearly always preceded and followed by areas of high pressure, and the proximity and relative barometric pressures in these determine not only the relative lengths of the two axes of the cyclone, but also their direction. Fig. 95 shows the great cyclones and anti-cyclones of the northern hemisphere existing at the same time, on a specified day. The atmospheric pressures are given in inches and tenths as shown by barometer readings. We see here a good representation of the disintegration spoken of by von Helmholtz in connection with the general circulation of the air in the middle and higher latitudes.

The amplitude of oscillation of the barometric pressure with the passage of anti-cyclones and cyclones is variable within certain limits, and depends on the intensity of the development of these pheno-mena. Assuming the atmospheric pressure to be about 760 mm. at sea level on the average, the fluctuations are between 700 and 800 mm. In certain extreme cases the pressure within an anti-cyclone has exceeded 800 mm., and in a cyclone has been below

FIG. 95.

ISOBARS NOVEMBER 4, 1881.

34

U
59°
33°
11°
3°

N.
for
to
Ur
At

sur
of
me
rel
als
an
at
spl
sho
rep
Ho
of

pro
cyc
on
me
abc
flu
ext
exc

700 mm., with a range of over 110 mm., or nearly the one-sixth of the minimum air pressure.

On the ocean where the conditions are more uniform than over the continent, it is found that the isobars are more nearly circular and are closer together, and the area of low pressure is greater, and the pressure at the centre is lower.

Sometimes when the isobars are considerably elongated there may be two or more centres of low barometer within the cyclone, and while these centres are most frequently of unequal depth, yet they do not usually differ by more than 0·1 mm. or 0·2 mm. ; and they are sometimes of the same depth.

The decrease in pressure from the outer boundary to the centre of a cyclone is not a wholly uninterrupted descent. Redfield first called attention to the rise of the air pressure on the approach of a hurricane, and Espy considered that unless such a rise of barometer first heralded it, no great storm could approach. It remained for Ferrel to show that each cyclone is encircled by a more or less distinct but narrow ring of high atmospheric pressure, and that this ring is the dividing line of the outer and inner circulation of the air. This ring of higher pressure corresponds to the belt of increased pressure at about lat. 30º in each hemisphere (see the inner ring of Fig. 110).

Temperature in Cyclones.—The surface temperature distribution in areas of low barometric pressure for the middle latitudes of the northern hemisphere is represented by an individual example given on the following chart—Fig. 96—on which the dotted lines denote temperature and the full lines air pressures (the former in degrees Fahrenheit and the latter in inches).

The western and northern sides of a cyclone are the coldest, and the eastern and southern sides the

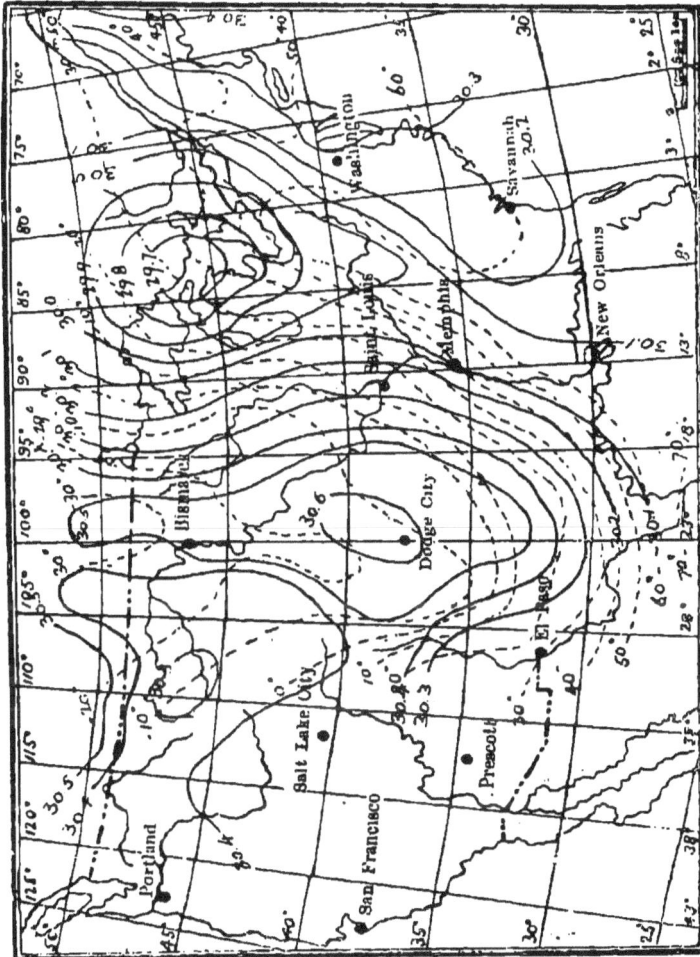

FIG. 96.

warmest, but the location of the area of low pressure with reference to land and water distribution has much influence in modifying these conditions. The effect of water is well shown by the investigations of Hildebrandson for Upsala, and Krankenhagen for Schwinemünde on the opposite German coast, and a full summary of these is given in van Bebber's *Meteorologie*, p. 294.

In the central and eastern United States (Fig. 96) we have about the best examples of the greatest contrasts of temperature in the various quadrants of a cyclone, which attain a maximum in the winter season when these differences on two sides, for large cyclones, may amount to more than 40° C. ; and the course of the isotherms, which in normal conditions should lie east and west, become so bent in a northerly direction that the thermal gradient is towards the south-east, or even nearly the east. The preceding diagram is well calculated to show the cause of the sudden fall of temperature at any place lying in the path of the easterly course of a cyclonic area.

The observations made at a few mountain stations have usually been considered as showing that the mass of air at the interior of a cyclone is relatively warm, and that in an anti-cyclone is cold. There are many circumstances which tend to render the question one of great complexity, and the recent investigations of Hann in Central Europe show us that much additional work needs to be done by meteorologists before the question will be definitely settled. On account of their great importance in the establishment of a theory of cyclone formation and maintenance, the facts determined by Hann are given in that connection farther along.

Cloudiness in Cyclones.—The distribution of cloudi-

ness in a cyclone depends much upon the position of
the supply of moisture, and consequently no law is
found to hold good for all cases. Van Bebber gives
a comparison of observations at Upsala and Schwinemünde, which shows that the relation of the position
of the centre of a cyclone to water distribution has
much to do with the amount of cloudiness. The
various kinds of clouds having different causes and
lying at different altitudes, also makes it difficult to
make generalisations concerning them. Taking the

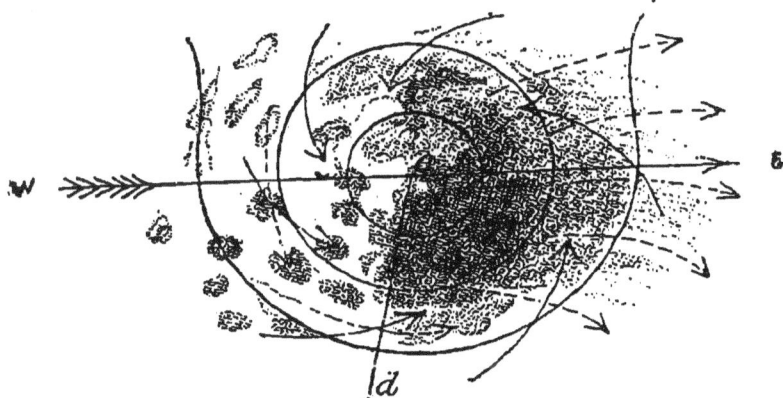

FIG. 97.

clouds as a whole they are most dense at the point
of great rainfall, but they cover a much greater area
than the rain area, being especially abundant on the
preceding or eastern side and lacking on the following
or western side of the cyclone. The octants on the
south-west, west, and north-west of a cyclone are
usually the freest from clouds. Fig. 97 shows the
cloudy areas for a cyclone in Northern Europe : cd
separates the cloudy from the clearing sky ; and the
arrow shows the direction of translation of the
cyclone. In Fig. 98 the kinds of clouds to be met

with in the various regions of a cyclone are specified by name.

The amount of cloud in its relation to cyclones has not been so carefully studied as the direction of motion ; and this last of course is merely significant of the winds direction.

Rainfall in Cyclones.—In dealing with the relation of rainfall to barometric minima, Loomis has investi-

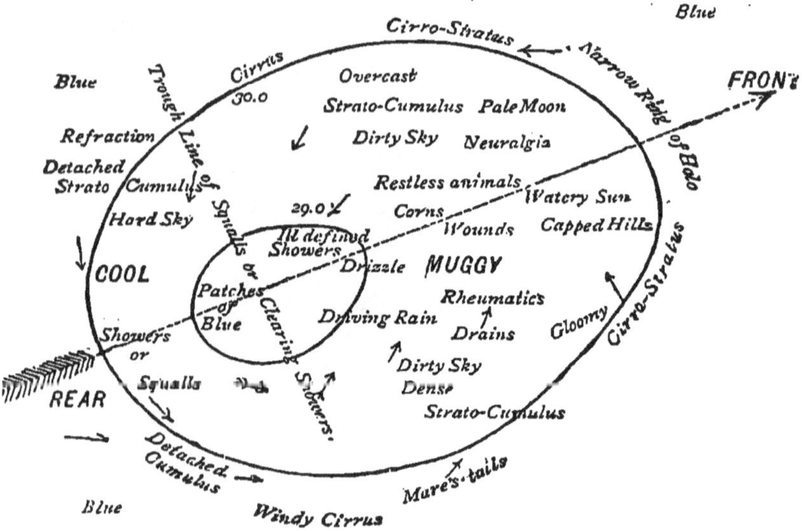

FIG. 98.

gated a great many cases for the United States, and some of his results are given below.

Before mentioning these, however, a glance at Fig. 98 will show the rain regions usually to be found in cyclones passing Great Britain. The inner closed curve shows the centre of the cyclone, and the outer one the outside limit of marked disturbance. This diagram shows not only the physical phenomena attending the passage of a cyclone, but also the

24

physiological effects which are noted in this same connection.

The great downpours of rain which draw our attention the most markedly to the presence of cyclonic phenomena will be noticed first.

The relation of excessive rainfall to barometric minima is, according to Loomis, characterized as follows :—

FOR THE UNITED STATES EAST OF ROCKY MOUNTAINS.

2¼ inches of rainfall in 8 hours occur in frequency on the east and west sides in the ratio of 2·7 : 1.

FOR EUROPE.

2½ inches of rainfall in 24 hours occur on the east and west sides in the ratio of 2·0 : 1.

FOR THE NORTH ATLANTIC OCEAN.

Rainfalls occur on the east and west sides in the ratio 2·6 : 1

These great rainfalls are usually associated with a barometric pressure somewhat below the average.

Loomis found that the following relations exist between rainfall and the rise and fall of the barometer :—

Indianapolis, U.S.	Frequency of rainfall for falling barometer : rising barometer : :		1·32 : 1
Philadelphia, U.S.	Ditto	ditto	2·88 : 1
Seven British Stations	Ditto	ditto	2·08 : 1
Paris and Brussels	Ditto	ditto	1·19 : 1
Pawlowsk (near St. Petersburg)	Ditto	ditto	1·06 : 1
Prague and Vienna .	Ditto	ditto	0·80 : 1

It is seen from this that with an advance westward (in the United States) from the Atlantic coast, the relative frequency of the precipitation when the barometer is rising decreases somewhat rapidly ; and with an advance eastward in Europe the same decrease is to be met with. This seems to show conclusively that the excess of rainfall usually noted on the eastern side of an area of low pressure is mainly dependent upon the distance and direction of the principal supply of aqueous vapour.

It is so usual to associate rainfalls with areas of low barometer that a special mention must be made of the occasional lack of this accompanying feature. Loomis finds that in the United States such cases usually occur in the months from October to April ; very few of them in midsummer. These barometric minima were in most cases thoroughly well marked, and many of them of large extent, but at the centres there was on an average a barometer reading of only about 29·7 inches, and with the temperature usually from 10° F. to 20° F. (average 16° F.) above the normal.

The barometric minima accompanied by heavy rains have strongly marked characteristics, such as steep barometric gradients, violent winds, rapid fluctuations in pressure and rapid progressive movement ; while those having light rains, or no rainfall at all, have these characteristics but feebly marked.

Loomis finds that while the forms of the rain areas are not always regular, yet they are generally elliptical, with the longer axis, usually about double the length of the short one, extending in the direction of progress ; the area may have a length of 2,500 miles. In looking at the rain area as a whole, it is found that there is a centre of maximum rainfall surrounded by zones of diminishing amount. For the United States north of latitude 36° the average distance of the rain-centre from the centre of the area of low pressure was about 300 miles, but it extended to three times that distance in individual cases.

The relative humidity increases with the approach towards the centre of the cyclone ; and the probability of rainfall increases in the same direction, being usually least in the south-west, west, and north-west octants, but this varies with the water distribution at the surface of the earth.

Direction and Inclination of Winds to Isobars in Cyclones.—Inclinations of surface winds to the isobars were noticed by Redfield, but were not measured.

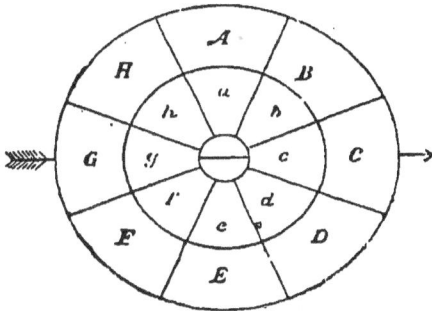

FIG. 99.

Buys Ballot a few years afterwards announced his law that for the northern hemisphere, when the observer faces the direction in which the wind blows, he will find a high barometric pressure on the right, and a low barometric pressure on the left. Rev. Clement Ley made a great number of measures of the angle of inclination of the winds to the isobars in cyclonic systems as shown by weather charts for North-west Europe. His first results showed the average inclination of surface winds to be 20° 31′ for fifteen mixed stations, but when the sea-coast stations were separated from the inland stations the angle for the former amounted to 12° 49′, and for the latter 28° 53′. A subsequent (1877) and much more complete investigation for various parts of Europe gave the following inclinations to the radius drawn from the centre of the cyclone :—

OCTANTS.		SURFACE WINDS.	
		No. of Observations.	Mean Angle with Radius.
N.	A	198	62°
N.E.	B	407	52°
E.	C	511	48°
S.E.	D	675	54°
S.	E	803	66°
S.W.	F	378	76°
W.	G	277	79°
N.W.	H	196	80°
N.	a	195	65°
N.E.	b	391	53°
E.	c	426	58°
S.E.	d	454	55°
S.	e	629	64°
S.W.	f	402	74°
W.	g	250	77°
N.W.	h	204	81°

The various octants mentioned in this table are shown in Fig. 99, in which those of the outer region are designated by capital letters, and the inner region by small letters. The average inclination to the radii was 64°·6 for the outer region, and 65°·9 for the inner region, or an average of about 65°; the inclination to the isobars being the difference between 90° and this, or 25°. In these measures the land stations largely predominated. Loomis first found for the United States an inclination of 47°, but subsequent inves-

tigations gave him 44° at a distance from the centre
of 1,200 kilometres, and 37° at 250 kilometres.
Loomis found that on the Atlantic Ocean the incli-

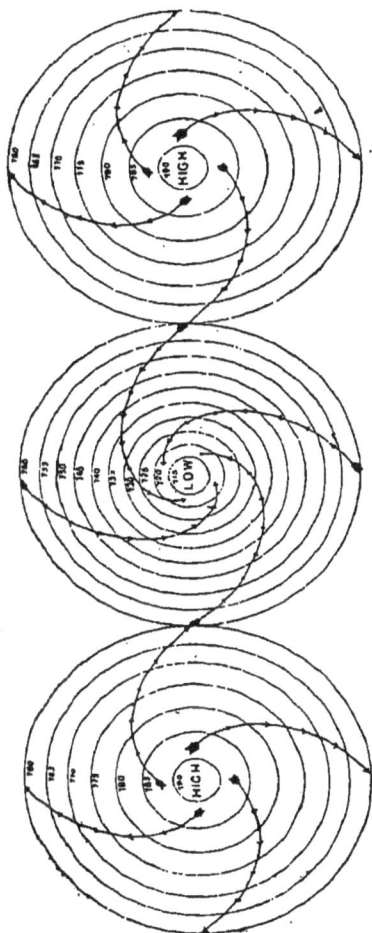

nation was 25°·4 for a
radial distance of 278
kilometres, and 34°·8 for
a radial distance of
1,535 kilometres. Fig.
100 shows the spiral
movement of the surface
winds which circulate
inward around a cyclone
(or Low), and outward
around an anti-cyclone
(or High), as pictured
by Loomis.

At various elevations
above the earth's sur-
face different relations
exist.   Loomis finds
that for the elevation of
Mount Washington
(6,300 feet), high winds
circulated around the
cyclones as at the earth's
surface, but the direction
of the wind is at right
angles to the radius of
the cyclone.

For the higher currents
of air as shown by cloud
movements, Ley found the following angles for the
different octants of cyclones (refer again to Fig.
99) :—

OCTANTS.		UPPER CURRENTS.	
		No. of Observations.	Mean Angle with Radius.
N.	A	51	—5°
N.E.	B	173	163°
E.	C	226	152°
S.E.	D	290	146°
S.	E	328	124°
S.W.	F	199	101°
W.	G	81	96°
N.W.	H	42	99°
N.	a	58	172°
N.E.	b	104	130°
E.	c	94	135°
S.E.	d	141	102°
S.	e	135	73°
S.W.	f	142	51°
W.	g	83	90°
N.W.	h	46	106°

In Fig. 101 we have Köppen's schematic presentation of the isobars at different altitudes (a) ; and the direction of the winds circulating around a cyclone at different altitudes (b). In the former figure the isobars at sea level are shown by full lines, and those in the region of the cirrus clouds by the fainter dotted lines ; while the isotherms are shown by the heavy broken lines : in the latter figure the arrows fly with the wind, and the full heavy arrows show the direction of the wind near the ground, the full light arrows that at the height of ordinary clouds, and the broken arrows that at the altitude of the cirrus clouds. This brings out clearly the result announced by Ley twenty years ago : that with an increase of altitude the wind direction is more and more towards the right.

Van Bebber has given a short summary of the wind

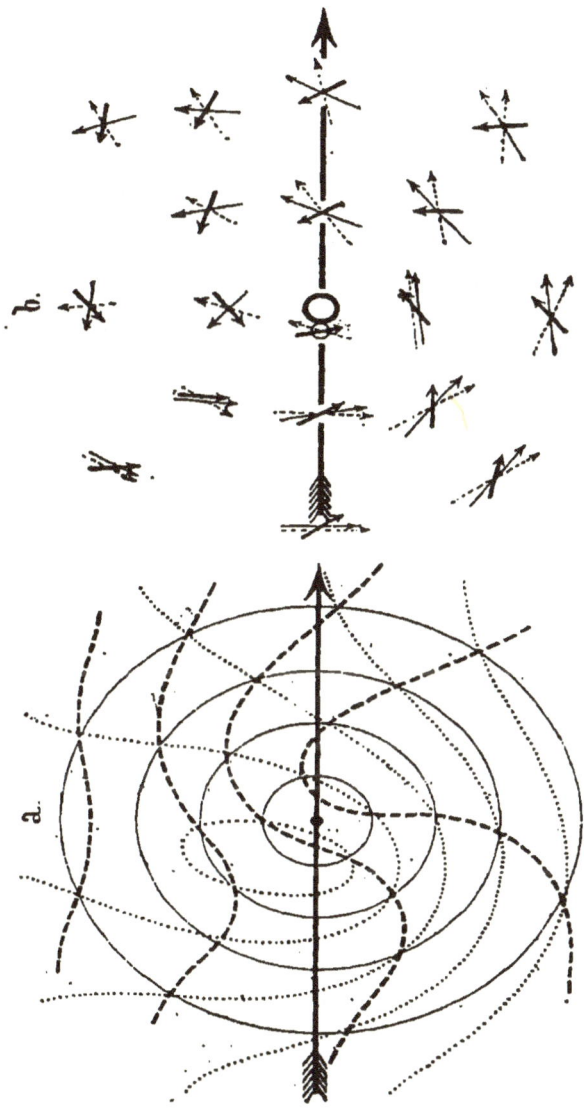

FIG. 101.

circulation in cyclones in his *Meteorologie* (pp. 291, 292). This is about as follows :—Near the surface of the earth, in West and North-West Europe, the steepest gradients are on the south (S.W. and S.E.), and the least gradients on the north side ; with the south-west winds the strongest and the east winds the weakest ; but for *any given gradient* the wind velocity is greater in winter than in summer, and the winds from the north and east are the strongest. The wind direction is inward towards the centre of the cyclone, and its path is approximately that of a logarithmic spiral. The angle which the direction of the wind makes with the direction of the gradient, is greater on the sea than on the land, greater in the warm than in the cold season, greater in a cyclone than in an anti-cyclone, greater on the succeeding than on the preceding side of a cyclone (but a reversal of this last seems to be true for the Eastern United States).

At low altitudes above the earth's surface, that is, at the elevations of the lower clouds, the wind direction is to the right of the surface wind, and is nearly perpendicular to the direction of the gradient, and is consequently nearly tangent to the isobars, except on the preceding side where the angle with the gradient is a little greater than 90°.

At high altitudes, in the region of the cirrus clouds, the wind direction is to the right of that of the lower winds, and consequently carries air away from the cyclone, and with an intensity of motion increasing from the centre towards the outer edge. The out-stream of air above forms about the same angle with the isobar that obtains for the in-stream below referred to the same surface isobar. At the rear of a cyclone (in Europe) the upper winds have nearly the same direction as the lower winds. On the preceding side of a cyclone there exists the maximum surface inflow, and the maximum outflow above. It is supposed that the S.E. surface wind on the preceding side of a cyclone does not reach half the altitude of the N.W. wind on the following side. On the approach of a cyclone the wind vane is influenced by the under air-current earlier than by the upper. The inclination of the vertical axis of the cyclone depends on the temperature distribution, and in Central Europe is inclined forwards, but towards the left in winter, but in summer rearwards towards the left.

Amount of Air Feeding Barometric Maxima and Minima (Anti-Cyclones and Cyclones).—Many investigators have busied themselves with the question of the circulation of air around barometric maxima and minima, but we owe to Dr. Vettin, of Berlin,[1] the most successful attempt at measuring the amount of air drawn into the minima below and expelled above; and the reverse process for the maxima. He has been able to obtain relative values only, but these give us an approximate idea, at least, of this interchanging air motion. Vettin has been observing the direction and relative velocity of the various cloud layers for a number of years, and he has applied some of these data to the question now being considered. He has divided the motions of the layers of the atmosphere into strata of different altitudes as follows :—

The wind as observed by means of exposed instruments, and which he designates by P ; the motion of

[1] Vettin, *Ueber die Volumina der in die Barometrischen Minima und Maxima hinein und aus denselben heraus strömenden Luft. Aus d. Archiv d. Deutschen Seewarte,* Band XI. 1888.

the loose lower clouds ⌒⌒ ⌒⌒ (elevation 1,600 feet) ;
the well-rounded compact lower clouds ⌒ ⌒ ⌒
(elevation 3,800 feet) ; the higher clouds usually presenting a broken appearance, but without much space

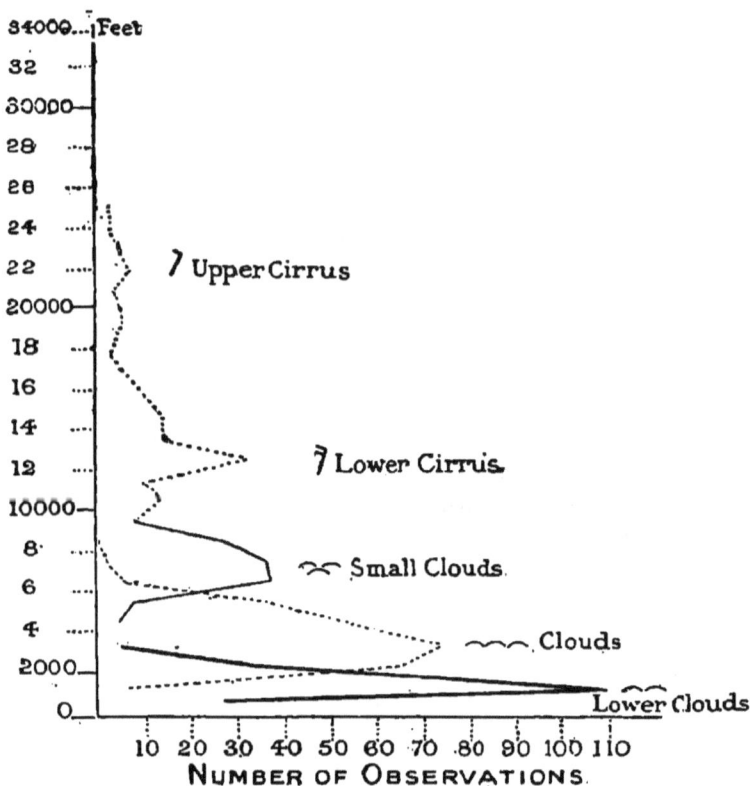

FIG. 102.

between the fragments ⌒⌒ (elevation 7,200 feet) ;
the lower cirrus ⟩ (elevation 12,800 feet) ; and the
upper cirrus ⟩ (elevation 23,000 feet). The frequency of cases for each class is shown by Fig. 102.

The apparent relative velocities (Fig. 103) of these various layers are observed with reference to the location of the centres of barometric maxima and minima, the positions of which are determined by the weather charts, or the charts of storm tracks. By adopting a certain unit of cross section for a given direction, and knowing the angle which the observed

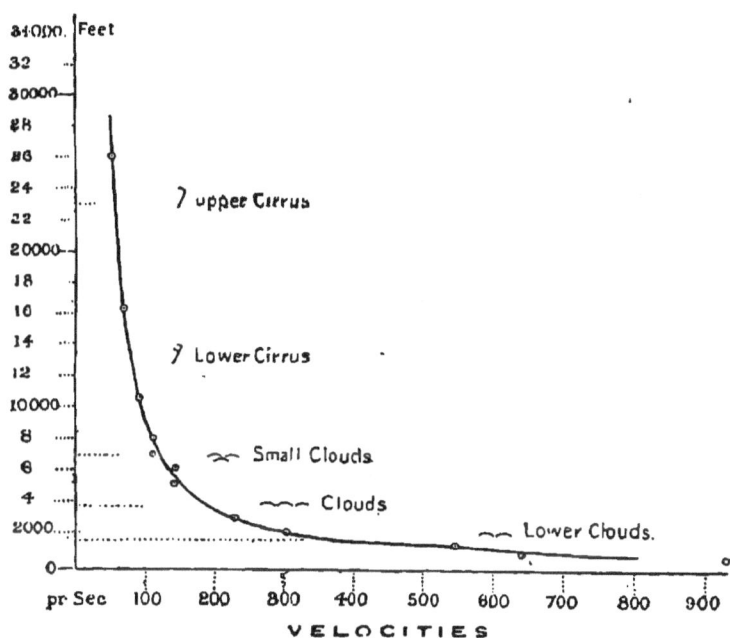

FIG. 103.

air current makes with respect to this, the amount of air, counting by volumes, which enters or leaves the region can be computed when the angular velocity and position (distance) of the current are known.

The details of this computation cannot be explained here, but the following table shows some of Vettin's results; + signifies an inflowing current, and — an outflow.

		⌐	⌐	~~	~~~	~	Observed Surface Wind. P
Summer	Bar. minima	+ 2·5	+4·7	— 2·3	+9·7	— 2·3	—14
	Bar. maxima	—24	—6	+13	—1·1	+14	+11·5
	Average	—10·7	—0·6	+ 5·3	+4·3	+ 5·8	— 1·2
Winter	Bar. minima	—4·4	+45	+52	+31	— 7·5	—15
	Bar. maxima	—8·5	—17	—24·5	+12	+38	+21
	Average	—6·4	+14	+13·7	+21·5	+15·2	+ 3
Year	Bar. minima	— 6·8	+19	+33	+27	— 8·3	—15
	Bar. maxima	—15	— 9·9	+ 1·1	+ 0·4	+23	+16
	Average	—10·9	+ 4	+17	+13·7	+ 7·3	+ 0·5

The meaning of the numbers in this table is best shown by the accompanying diagrams (Figs. 104, 105, 106). The dotted lines refer to the minima, and the unbroken lines to the maxima ; the negative sign to the left of the zero indicates, as given in the table, the centripetal motion, and the positive sign the centrifugal motion. The line at the base indicates the ground layers of the atmosphere where the barometric pressure is 760 mm., and the successive horizontal lines above indicate the regions where the pressure has decreased by a fraction of 760 mm., as shown by the left-hand argument.

We see, then, from the table, and more plainly from the diagrams, that for the barometric minima (cyclones) and maxima (anti-cyclones) the vertical distribution of the air circulation is as follows :—

Barometric Minima (Fig. 104).—The layers of air in which the inflowing current preponderates extend from the earth's surface up to about 640 met. (2,100 ft.) altitude, and includes, therefore, the lower clouds. Above this elevation the reverse condition begins to prevail, and more air leaves the minimum than enters it, and the outflow increases up to an elevation of 2,300 metres, where the maximum outflow is reached ; but from this point upward it decreases, and at the height of the upper cirrus clouds it ceases. The difference in the outflow in the upper regions for summer and winter is very marked. In summer it is at all elevations very weak, and there are traces of a reverse motion

BAR. MIN. (CYCLONE).   BAR. MAX. (ANTI-CYCLONE).

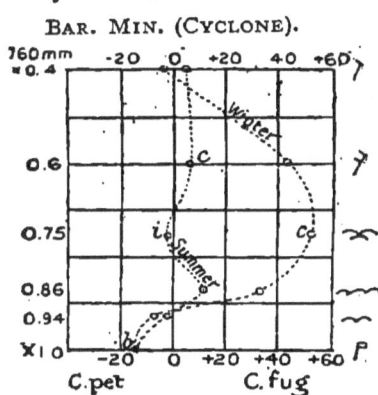

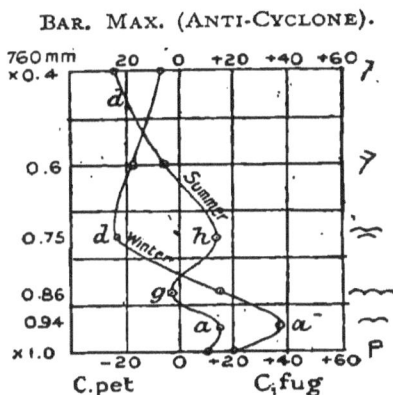

FIG. 104.   FIG. 105.

at about the elevation where the outflow is strongest in winter.

Barometric Maxima (Fig. 105).—Here the outflowing currents preponderate at the earth's surface, and become strongest at an elevation of 1,600 feet, in both winter and summer ; but in the winter they are far greater than in summer. This outward motion diminishes from this altitude upward to about 4,000 feet elevation, where there is neither inflow nor outflow. Then a little higher up, at about 7,000 feet altitude, we find in summer a strong outflow, and in winter a strong inflow.   Above the altitude of about 10,000 feet there is an inflow both winter and summer.

It is to be noticed that in the lower atmosphere, for both barometric maxima and minima, in the winter there is a single maximum of outflow and inflow, with almost no motion at the elevation of the highest cirrus clouds, while in the summer there is a double maximum.

Fig. 106 shows the curves for the average for the

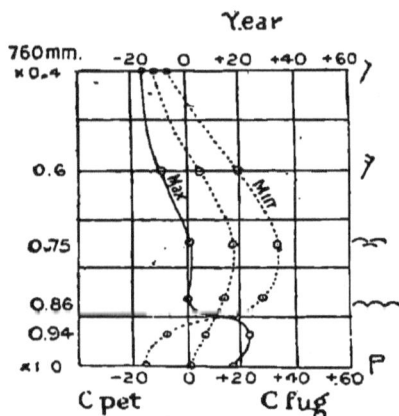

FIG. 106.

year.   It is seen that for the barometric minima there is an inflow at the earth's surface, decreasing up to a height of perhaps 3,000 feet, where it changes to an outflow, which reaches its greatest force at a height of about 7,000 feet, from which elevation it decreases to a neutral or slightly inflowing motion at the height of the upper cirrus clouds.   For the barometric maxima the lower outflow is reduced to nothing at about 3,800 feet altitude, and from thence upward to an elevation of about 7,500 feet there is little motion either way, but above this there is an inflow. The broken medial line, which represents the average for the barometric maxima and minima, shows a neutralisation of the outward and inward motion at the earth's surface; but from thence up to an elevation of perhaps 14,000 feet there is a preponderance of outflow, and above this point the inflow is the greatest.

A glance at Figs. 107 and 108 will show. the necessarily compensatory nature of the interchanging motion between barometric minima and maxima. At those elevations where there is marked inflow or outflow in the one system of atmospheric circulation the reverse motion takes place for the other system.

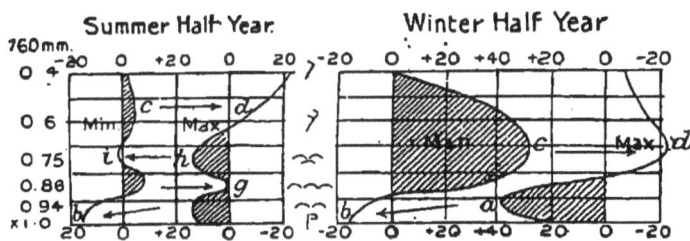

FIG. 107.   FIG. 108.

It must be remembered that these interesting results were obtained for the locality of Berlin, and also that the densities of various layers of the atmosphere are not considered, although to a certain extent the progressive increase in the thickness of the layers with increase of altitude counterbalances the decrease in density. It seems quite probable that the absolute elevations of the clouds as obtained by Vettin are too low (at least for the lower clouds), as the observations of others show that the inflow for barometric minima and the outflow for barometric maxima near the earth's surface extend to higher regions than those indicated by his results as given here.

Direction and Velocity of Motion of Cyclones.—The chart by Loomis (Fig. 94) is well adapted to show the paths of the centres of cyclones in the northern hemisphere. The body of each arrow shows the actual course passed over by some individual cyclone, and it is seen that while for the middle latitudes the direction is, in the main, easterly, for the lower latitudes it is westerly; the latter courses being much shorter than the former. The tracks do not approach the equator within 6°, but it is difficult to fix upon the northern limit.

It is seen that there are but two main regions of tropical cyclones in the northern hemisphere, viz., in the Atlantic Ocean near the West Indian Islands, and to the south and south-east and east of Asia. Some of these tropical cyclones form and disappear within the tropics, while others moving westerly for a time gradually attain a northerly direction, and, after passing into the middle latitudes, assume the easterly motion prevailing in those regions.

For the first mentioned class, the average direction was found to be W. 26½° N., velocity 12 miles per hour, for the American cyclones; and direction W. 27½° N., velocity 9 miles per hour, for the Asiatic cyclones. For the second class, for the West Indies the westerly direction was about the same as for the other class, but the velocity was 17·4 miles per hour, while after the course had become eastward (on the average E. 38½° N.) the velocity had increased to 20·5 miles per hour: but for the East Indian cyclones, in their westerly course the average direction was W. 38° N., velocity 8 miles per hour; for the northerly

25

direction the velocity was 9·3 miles per hour, but after assuming an easterly course (on the average N. 35° E.) the progress was but 9·8 miles per hour.

Only a small number of these tropical cyclones are used by Loomis in obtaining these average results, but the values given are probably not far removed from those which would be obtained by using a large number of cases.

It is, however, with the motions of cyclones of the temperate zone of the northern hemisphere that modern meteorologists are best acquainted. Loomis found the average direction of the cyclones in the United States to be about N. 81° E.; a little more easterly than this in summer and a little more northerly in winter. In the central part of the United States the direction is slightly south of east, and in the eastern part is slightly north of east. On the North Atlantic Ocean the direction is about N. 67° E. in the western part, and N. 80° E. in the eastern part. In Europe there is a greater irregularity in direction than in America.

There is, undoubtedly, a tendency for cyclones which come from the same regions and have somewhat similar conditions of environment to move in a common path; and the combination of the tracks of many of the cyclones which pass over a continent will allow the determination of the position of a number of these most frequently pursued courses.

Fig. 109 shows the paths most frequented by cyclones in Europe, as determined by van Bebber. The width of the path is made proportional to the frequency with which the path is visited.

Path I., which is the most frequented, is that mostly taken by cyclones in autumn and winter, but seldom in the spring. At the north of Norway it breaks up into branches Ib, Ic, Id, and the positions of neighbouring cyclones and anti-cyclones probably determine which of these sub-paths the cyclone will take.

The Paths II., III., IV., and V., break up into branches in like manner. These various paths and

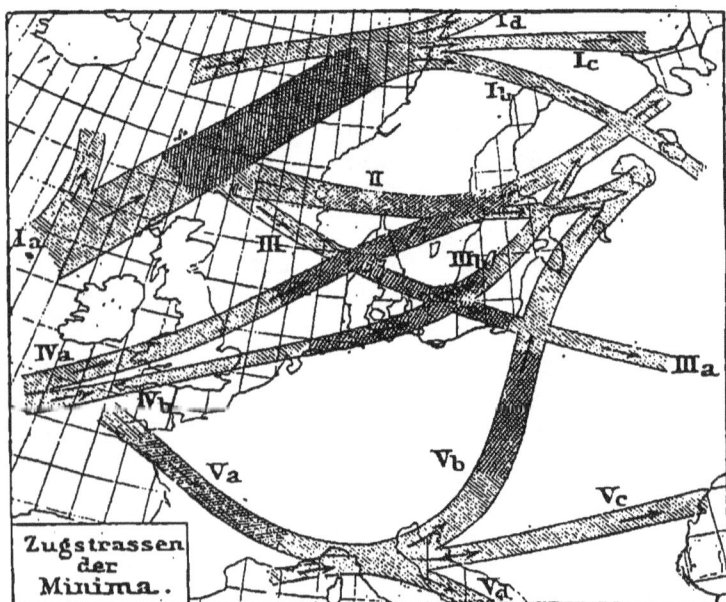

FIG. 109.

branches sometimes intersect, sometimes unite, and sometimes pursue independent separate courses In the case of Path V. there is a union of two totally distinct paths, which is followed very quickly by a sub-division.

Köppen and Finley have also published charts showing the frequency of cyclones in the various regions from the western part of the United States to Eastern Europe.

The velocity of the cyclones is not the same for the different regions. The following table by Loomis shows the average velocity in miles per hour, obtained by combining several years' observations.

MONTH.	UNITED STATES.	ATLANTIC OCEAN. MIDDLE LATITUDES.	EUROPE.
	Miles per hour.	Miles per hour.	Miles per hour.
January	33·8	17·4	17·4
February	34·2	19·5	18·0
March	31·5	19·7	17·5
April	27·5	19·4	16·2
May	25·5	16·6	14·7
June	24·4	17·5	15·8
July	24·6	15·8	14·2
August	22·6	16·3	14·0
September	24·7	17·2	17·3
October	27·6	18·7	19·0
November	29·9	20·0	18·6
December	33·4	18·3	17·9
Year	28·4	18·0	16·7

For the entire year the velocities of cyclones in the five regions are:

United States	28·4 miles per hour	
North Middle Atlantic Ocean	18·0 ,,	,,
Europe	16·7 ,,	,,
West Indies	14·7 ,,	,,
South and south-east of Asia	8·5 ,,	,,

§ 4.—*Investigators of the Theory of Cyclones.*

The observations of cyclonic phenomena and theories explaining them have progressed side by side since the first quarter of this century. The early labours of Dove, Redfield, Piddington, Espy, Loomis, Poey, and others, laid the foundation for the later work of Buys Ballot, Ferrel, Köppen, Hann, Guldberg and Mohn, Clement Ley, Abbe, Eliot, Hill, Meldrum, Blanford, Hildebrandson, van Bebber, Reye, and others.

Dove early tried to refer the action of cyclonic phenomena to mechanical action and the supremacy of the great currents and counter-currents, while Espy took the opposite view of a purely physical cause, and applied the laws of dynamic heating and cooling, and suggested the great source of energy to be the freeing of latent heat through condensation. It remained for Ferrel to point out the true method of analytical investigation of the problem, and picture the phenomenon of atmospheric whirls as a result of the combination of forces which can be considered as governing their formation, maintenance, and movements. While Ferrel's methods were open to some criticism on account of their crudeness, he undoubtedly grasped the problem as no one else had done, and opened the way for the more elegant mathematical analysis of Guldberg and Mohn, Oberbeck and Marchi. These last two have abstained from discussing the physical side of the question, and have treated of the dynamics mainly in continuation of the mathematical work of Guldberg and Mohn, who have unfortunately introduced the influence of condensation as being of the utmost importance to their theory; a position which we understand they now no longer hold.

The quite recent studies of several well-known continental European meteorologists and physicists have greatly tended to strengthen the hypothesis that the dynamic action is of primary importance in causing cyclonic phenomena, while the studies of the English meteorologists in India point to a physical cause (the effects of condensation) as predominant. The American meteorologists, Loomis and Abbe, seem to have also retained the local condensation theory as of the greatest importance ; and Ferrel's ideas, too, appear to be so interpreted, although I think his view of the *rôle* played by condensation makes it subservient to the influence of the great primary air motions. Ferrel's continued comparison of the local cyclonic action with that of the great hemispherical cyclone (in which condensation has been so little considered) seems to show conclusively his idea of the relatively small influence of condensation.

It seems quite possible that, as Hann considers probable, the great cyclonic disturbances of the middle latitudes are best explained by the dynamic theory, and are distinct from the more limited cyclones of lower latitudes to which the condensation theory is applicable. If this is true, then the continental European theorists are right in their view of the formation and action of the cyclones of middle latitudes, while the American theorists must, with the Indian theorists, limit their views as applicable to the cyclones of the lower latitudes, although they have applied them likewise to those of the higher latitudes. These two classes of cyclones, then, need separate discussion, and space compels me to limit the present presentation to those of the middle latitudes.

Concerning the condensation theory of storm (cyclone) movements, the following summation by Abbe shows briefly, perhaps the most advanced ideas on this subject :

"The statistics of American storms, as given by Loomis, and those of Indian storms, as given by Eliot, unite in a confirmation of the views that the causes that contribute to the motion of the storm, are :

"1. The unbalanced northward pressure attending a low [cyclone], as deduced by Ferrel.[1]

"2. The drift of the general current of atmosphere that carries the air and the storm along together.

"3. The insolation that stimulates uprising currents on the sunny side.

"4. The orography that promotes cloud growth and rain on the windward side of mountains and coasts.

"5. Oceans and lakes that promote evaporation and moisture.

"6. The geographical distribution of the areas of high pressure.

"7. The precipitation of rain that leaves heat free in the cloud.

"Of all these, the last (7), when it occurs, becomes at once the leading factor in determining the progress of the whole disturbance."

Since, however, this theory is not considered by most modern students of this branch of meteorology as completely applicable to the cyclones of the middle latitudes, its further elaboration here is omitted.

§ 5.—*Ferrel's Theory of Cyclones.*

During a period of more than thirty years Professor Ferrel made additions to and elaborated the theory of cyclones which he first published in 1858–60. Most of the memoirs published by him were unsuited for the general reader, as they were written for, and published in, science journals or public documents. In his *Treatise on the Winds*,[2] however, he has devoted about two hundred and fifty pages to discussing

[1] Since the deflecting force due to the earth's rotation increases with the latitude, it must be greater on the polar side than on the equatorial side of a cyclone ; and since this force would act towards the right in the northern hemisphere, on the westwardly moving air current on the north side of the centre of the cyclone, it would cause a poleward displacement of the air mass. The deflecting force in so far as it depends on the latitude would be the same on the east and west sides of the cyclone, and these would therefore balance each other.

[2] John Wiley and Sons, New York, 1889.

cyclones and their attendant phenomena, and has given there, for the first time, an account of his theory which was available to the public. This account is of great scientific value as well as of popular interest; and as it may be taken as expressing Ferrel's latest ideas, I extract from it important parts of a general outline of his theory of cyclones, giving in his own words the portions of it which require special precision.

When, at any locality, the air up to a considerable altitude becomes warmer than the surrounding air, there will arise an upward current in this warm mass of air, and an outflow, or perhaps a better expression would be an overflow, of air takes place aloft. This causes a decrease of air pressure in the warm area and an increase in the adjacent region which receives the overflow. This difference of pressure gives rise to a gradient of pressure at the earth's surface, in which the region of least pressure is within the warm area. There thus arises an inflowing current at or near the earth which acts as a feeder for the ascending current in the warm area. This circulation is continued as long as the condition which gives rise to it continues to exist.

Usually the inner upward current is over a much smaller area, but is more intense than the outer downward current, which is generally a mere settling down of the air.

When the air at any place is in an unstable condition as compared with the surrounding air, and a local upward motion takes place, then this air by dynamic cooling becomes warmer than the air which surrounds it above; the unstable condition is thus extended, for a time, and a circulation of air is maintained until the restoration of stable or indifferent equilibrium. In case the air is moist, there is an additional source of energy in the freeing of latent heat from the aqueous vapour, by any condensation which may arise in the ascending current. The continued feeding of the upward current by the outer drier cooler air will, however, soon cause the disappearance of any vertical temperature gradient, and consequently of the vertical circulation maintained by it.

The general circulation between the inner, warm, and the outer, cold, region, is in many respects comparable with that existing between the warm equatorial region and the cold polar region ; they are, however, reversed, as in the more local case the central region is the warmer. The same ideas of condition of continuity, planes of neutral motion, and horizontal and vertical motions, which have been mentioned in the general circulation of the atmosphere, hold good also in this more limited interchange of air.

If the earth had no rotation upon its axis, the air moving in towards the warmer centre where the air pressure is less, would move directly towards the centre ; but the deviatory force due to the earth's axial rotation will cause it to move a little to the right of this centre in the northern hemisphere (and to the left in the southern), and there results a gyratory motion around the centre. The direction of this atmospheric circulation is then opposite to that of the hands of a clock ; and this is to be seen on almost any chart of the surface winds in the region of a barometric depression or cyclone in the northern hemisphere.

The velocity of gyration increases towards and becomes very great near the centre. The horizontal motion, however, ceases at some distance from the centre.

The air aloft, in flowing away from the warm centre, is also deflected towards the right in the northern hemisphere, but at first this deviatory force is used in overcoming the gyratory velocity which the air attained below. Ferrel says :—

"Where the system of vertical circulation and gyratory motion becomes fully established, and the air which flows out above is drawn in again toward the centre below, this gyratory motion from left to

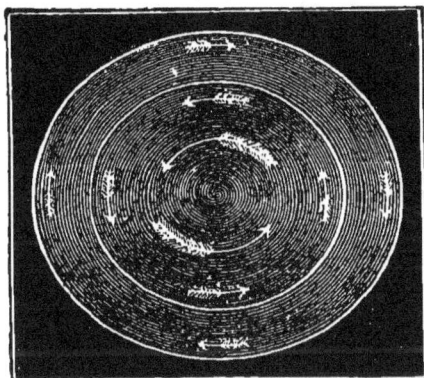

FIG. 110.

right has first to be overcome by the deflecting force before a gyratory motion from right to left begins to be generated. In connection, therefore, with every vertical system of circulation there is produced by the deflecting force of the earth's rotation two kinds of gyrations—the one, mostly in the interior part, from right to left in the northern hemisphere, and the other, mostly in the exterior part, in the contrary direction. The interior, and by far the most violent part, is called a cyclone, and therefore the exterior and comparatively gentle part is properly called an anticyclone; and the two always go together [Fig. 110 shows Ferrel's first diagram of this circulation]. The gyrations of the former are properly called cyclonic gyrations, and those of the other, anti-cyclonic gyrations; it being understood that each of them is exactly reversed in the other [southern] hemisphere.

" The deflecting forces upon which the gyrations depend being greatest at the poles and diminishing as the sine of the latitude toward, and vanishing at, the equator, of course the gyrations arising from any given temperature disturbance or vertical circulation, all other circumstances being the same, are most violent in the polar regions, less so with decrease of latitude, and vanish at the equator, the motions there (at the equator) being directly toward and from the centre, the same as would be the case everywhere on the earth's surface if the earth had no rotation on its axis."

In case of no frictional resistance the air would have this cyclonic motion at the interior and anticylonic motion at the exterior at all altitudes, and the zone of change from one motion to the other would be the same distance from the centre at all altitudes. But the effect of friction is such as to make the gyratory velocity below less, and that above relatively greater at the same distance from the centre, so that in reality the greater the altitude of the disturbed stratum of air, the nearer the zone of change from cyclonal to anti-cyclonal gyrations approaches the centre. Comparing these results with the great hemispherical cyclone having a cold centre at the pole, it is seen

"that they are very similar, except that in the general circulation of a hemisphere the gyratory or east components of velocity are greatest above, while in the cyclonic they are least above ; and in the general circulation the distance from the pole at which the easterly velocities vanish and change to westerly ones increases with increase of altitude, while in cyclones the distance from the centre, at which the gyrations vanish and change from the cyclonic to the anti-cyclonic, decreases with increase of altitude. This arises from the difference in the distribution of temperature and of the horizontal temperature gradients ; in the case of the hemisphere in the general circulation, the polar or central region is the colder, and in the cyclone the central part is the warmer, and consequently the directions of the vertical circulations are reversed in the two cases."

For a complete account of the mechanics of the

cyclonal and anti-cyclonal motions, the reader must
be referred to Ferrel's *Winds*, and I shall only add
that the centripetal motion of the lower air gives rise
to the force which causes the cyclonic gyration, and
the centrifugal motion in the upper air gives rise to
the force which counteracts the cyclonic motion, and
by reversing this produces the upper and outer anti-
cyclonic gyration.

Air Pressure in Cyclones.—The observed barometric
pressure in cyclones shows a steep gradient at the
centre but decreasing towards the outer edge, and
this, as explained by Ferrel, is due to the fact that the
gradient is proportional to the force which causes it;
so that it depends almost entirely on the gyratory
velocities which are small at the outside but comparatively large at the centre.

The ring of high pressure which is found in the
hemispherical cyclone near the latitude of 30°, has
its counterpart in the smaller cyclone. Concerning
this Ferrel says :—

"Since the deflecting forces of the cyclonic and the anti-cyclonic
gyrations are *from* the centre in the former, and *toward* it in the latter,
the greatest pressure is at the distance from the centre where they
vanish and change from the one to the other, at least so far as the
pressure and pressure gradients depend upon these forces. The difference of pressure between the lowest at the centre and the highest where
the gyrations vanish, depends of course upon the summation of the
pressure-gradients, and as these are comparatively steep in the cyclonic
part, the difference between the highest pressure and that at the centre
is generally very much greater than that in the anti-cyclonic part
between the highest pressure and that of the general undisturbed surrounding pressure.

"What has just been stated with regard to pressure at the earth's
surface is true of those at any altitude above the surface. For the
atmosphere above any given level can be regarded as a separate atmosphere, and the gyratory velocities of the general atmosphere at that
level as those of the base of the atmosphere above that level. At any

given altitude, therefore, the highest pressure in the plane of that altitude is also, or very nearly, where the gyrations vanish and the cyclonic change into the anti-cyclonic. But the greater the altitude, the nearer the centre, as we have seen, does this change take place, and the conditions may be such that at considerable altitudes the gyrations may be anti-cyclonic at all distances from the centre, and in this case there is no central area of low pressure (at this altitude) but the greatest pressure is in the centre, and the gyrations are all anti-cyclonic; and considering the part of the atmosphere above this level, we have here an anti-cyclone alone, with no exterior cyclone, while at lower levels, the gyrations are partly cyclonic, and partly anti-cyclonic, the former increasing and the latter decreasing in area, and the barometric depression in the centre becoming deeper as the altitude is diminished, until we reach the earth's surface.

" Since the horizontal component of the motion of the air, in its vertical circulation, is gradually retarded as the air approaches the centre of the cyclone, and becomes o at the centre, and small 'even at a considerable distance from the centre, while the gyratory component of motion here is usually large, the inclination of the resultant or cyclonic motion near the centre is small in comparison with what it is in the outer part of the area of violence and of the cyclone proper, and so much more nearly circular.

" At and near the earth's surface, just beyond the ring of highest pressure, there is an exception to the inclination toward the centre. In consequence of the anti-cyclonic gyratory motion being retarded here by the greater amount of friction, the deflecting force toward the centre here is so much weakened that it is not equal to that of the pressure gradient, and the air, instead of flowing in toward the centre, is forced out in the contrary direction from beneath the high pressure. Hence here the resultant direction is anti-cyclonic and outward and the inclination negative, as in the upper part of the atmosphere.

" On the other side of the ring of highest pressure the tendency, for the same reason, is for the air to be forced out from beneath toward the centre, and consequently this force combines with that of the general gradient near the surface arising from differences of temperature, upon which the vertical circulation depends, and so increases this component of motion very much and causes the inclination here to be much greater than it otherwise would be, and also greater than that in the higher strata immediately above.

" On both sides of the ring of highest pressure the gyrations are strengthened by the outflow of air beneath on each side, for the deflecting force depending upon the earth's rotation arising from this outflow is in both cases in the right direction for this. In fact, on the outside of this ring there could be no anti-cyclonic gyration if it were not for this force."

Calms in Cyclones.—There are two regions of calms in cyclones. Directly at the centre of the cyclone there is little or no horizontal motion, as the force which keeps up the gyratory motion becomes small there. Also, Ferrel says :—

" Under the ring of highest pressure there is no gyratory motion, for we have seen that the gyratory velocities must necessarily vanish and change sign at some distance from the centre, and it has been shown that here is the place where there is the highest pressure. And as the air also flows out from beneath this highest pressure, on the one hand toward the interior and on the other toward the exterior, it is evident that there is here no radial motion. There being, therefore, neither gyratory nor radial motion, there must be here a ring of calms."

Gradual Enlargement of Cyclones.—

" In the preceding consideration of cyclones it has been supposed that the whole system of circulation has a definite limit, and comprises at all times the same air, which, by means of the vertical circulation, is being continually interchanged between the interior and exterior parts within this limit. This, however, is far from being the case in nature. As long as the vertical circulation is maintained with increasing, or at least sufficient energy, the tendency of the cyclone is to extend farther and farther from the centre, and so to continually extend the gyrations and the whole system of circulation over a greater area. But unless the energy which sustains it increases likewise, it must reach a limit, for otherwise the temperature gradient and the forces upon which the vertical circulation depends become barely sufficient to overcome the frictional resistances, when further enlargement must cease ; and then, as the energy begins to fail, either through a diminution of the supply of aqueous vapour or a change in the relative temperatures of the lower and upper strata by means of the interchange of air in the vertical circulation, the whole system of circulation must gradually become weaker and finally entirely subside.

" It may be that cyclones mostly commence over a small area and gradually enlarge, but this is not necessarily so. If the atmosphere is in an unstable state, and the temperature conditions are such that the initial upward motion takes place over a large area at once, then the vertical and gyratory circulations begin at once over a large area and gradually grow in strength until the frictional resistance from increase of velocity becomes equal to the force, when further increase in violence, as well as extent of limits, must cease. There is, therefore, a certain limit to the extent and violence of a cyclone somewhat proportional to the amount of energy; and if the initial temperature conditions are such as to start it over a small area only, it gradually increases in both extent and violence until this limit is reached."

Progressive Motions of Cyclones.—Concerning the progressive movements of cyclones, Ferrel says :—

" Ordinary cyclones are never stationary, and the directions in which their centres move and their velocities vary not only in different latitudes and regions of the globe, but in the same place at different times. In general their tendency in lower latitudes is westerly, and in the middle and higher latitudes easterly. There are several circumstances which control, to a greater or less extent, the progressive motions of cyclones. The principal one of these is undoubtedly the general motion of the atmosphere in which they exist, not at and near the earth's surface merely, but at high altitudes where the centre of energy is. This carries them along as a stream of water carries along the small whirling eddies which are formed in it. This idea was first suggested by the writer [Ferrel] more than a quarter of a century ago. [Fig. 111 shows Ferrel's first diagram of paths of cyclones, and also the direction of the air movement within the cyclone.] Tropical cyclones move westward, or at least have a large west component, because the general motion of the atmosphere there, up to a certain altitude, is westerly, while in the higher latitudes cyclones move in an easterly direction with much greater velocities, because there the general motion at all altitudes is easterly, and the velocity, especially at high altitudes, is comparatively great.

" If the general motions of the atmosphere have a controlling influence upon the progressive motions of cyclones, then we would expect that these motions would not be only westerly in lower latitudes and easterly in the higher latitudes, but also, if there is an annual inequality in the one, there must also be such in the other. According to the computation the easterly velocity of the air in the middle latitudes is more than twice as great in January as in July ; but the averages for the winter half and the summer half of the year would be nearly as 29 to 23 respectively, the same as the ratio between the winter and summer velocities of the progressive motions of cyclones given above. This velocity, however, is much greater than that of the easterly motion of the atmosphere in the lower strata, even up to a considerable altitude.

But the energy of the cyclone is mostly above, where the condensation of the aqueous vapour occurs, and where the air temperature in the cyclone differs most from that of the surrounding atmosphere ; and so the progressive motion is controlled mostly by that of the general motion of the atmosphere up at that altitude.   Theoretical computation shows that at the altitude of 2·5 miles in the United States the east component of the general motion of the air is about 26 miles, the same as the easterly velocity of the cyclones.

" As the energy of the cyclone is mostly in the aqueous vapour con-densed, and without this we rarely have the conditions of more than an

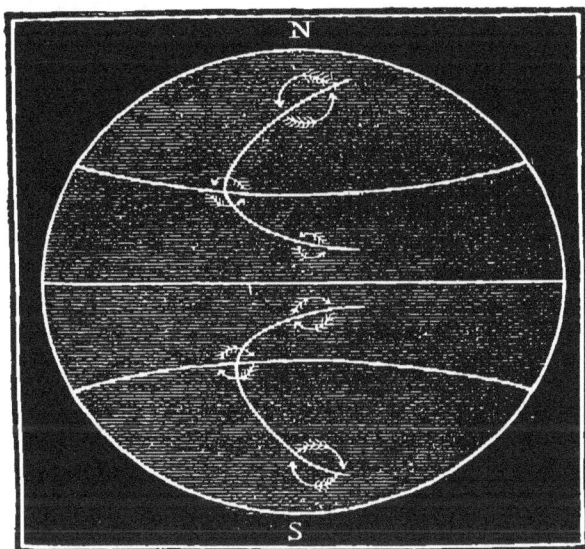

FIG. 111.

initial cyclonic action, the velocity and direction of the progressive motion of a cyclone depends, to some extent at least, upon the distri-bution of this vapour in the region in which the cyclone exists, for the cyclone is likely to be drawn somewhat in the direction in which there is the most vapour, and to pass around those regions in which there is little vapour.   It is for this reason, perhaps, that the chain of lakes between Canada and the United States seems to be a great highway for cyclones.

" In actual cases of cyclones there is a great difference in tempera-ture and a steep temperature gradient in an east and west direction, where, before cyclonic disturbance, there was none. Not only does the cold and dry air on the west side tend to press the warmer and moister air on the east side still farther toward the east, but the latter is continually forming a new centre of temperature disturbance a little in advance, which becomes a new cyclone centre, and so there is an apparent progressive motion from the tendency to continually form a new cyclone a little in advance of the old one. Such an effect would evidently take place in an atmosphere without any drifting progressive motion; but this effect must be subordinate to, and much less than, that of the progressive motion of the atmosphere, for if it were not, since its action is always in a direction from west to east, cyclones in the tropical latitudes would never move westwardly."

§ 6.—*Hann's Recent Ideas concerning Cyclones and Anti-Cyclones.*

The synchronous weather charts show us the distribution of the meteorological "elements" at the ground in cyclones and anti-cyclones, but they do not point out with such certainty the nature of this distribution aloft. Much discussion has been caused among meteorologists by their various interpretations of the results of observations made at some few isolated mountain stations during the passage of cyclones and anti-cyclones. This has resulted in giving a very confusing idea of the conditions which are supposed to prevail in the free air at high levels. Hann has recently done a great service to meteorology by his investigations of this problem, which are published in the *Sitzungsbericht d. k. Akad. d. Wissenschaften* at Vienna, and the *Meteor. Zeitschrift*. Using the observations made at the Alpine observatory on the Sonnblick and surrounding stations, and judiciously applying the theorems which he considers to have been established concerning the dynamic changes in the circulating air, he has been

26

able to advance our conceptions of the conditions and changes going on above the earth's surface. Among other results given in a late[1] communication to the Vienna Academy, he finds that cyclones possess a cold air mass, and anti-cyclones a relatively warm air mass, which is occasioned by dynamic causes which arise from the general circulation of the atmosphere, and not by purely thermic causes, as has generally been supposed.

Hann finds that at high levels in the Alps a barometric maximum is accompanied by a relatively high temperature, low relative humidity, and small amount of cloud. At low levels there exist at the same time a relatively lower temperature, a moister air, and more cloud ; and a downward current of air is acting. The barometric maxima in winter are accompanied by maxima at the earth's surface, the amplitude of oscillations being about the same ; however, in summer the amplitude aloft is double that below, but in the same direction. The temperature anomaly is likewise positive both above and below (and of equal amounts at the summit of Sonnblick, and at Ischl, a base station). The air is relatively warm, therefore, up to an altitude of 3 kilometres, and in the cases investigated by Hann no trace of the usually accepted cold centre was found in the anti-cyclone.

The monthly minima of air pressure on the Sonnblick are accompanied by relatively low temperatures, in winter as well as in summer, both aloft and below, but with the greatest departures aloft ; the relative humidity is large, and is at about saturation aloft, and there is a great intensity of cloud. There is, for the

[1] April, 1891.

great mass of air, a cold centre in the cyclone, instead of the usually assumed warm centre ; and Hann thinks the cyclonal action is far higher reaching in altitude than is generally supposed.

It is seen that the relations observed at the earth's surface are no criterion by which to judge those for the great mass of air, for at the surface in anti-cyclones, in clear weather in winter, there is a central temperature deficiency, and in summer an excess, while for cyclones there is in winter an excess, and in summer a deficiency of heat.

At the altitude of the Sonnblick observatory the temperature rises and falls *with* the air pressure all the year around ; but the times of the maximum and minimum temperatures lag a day behind those of the maximum and minimum air pressures. This is opposite to the relation observed at the earth's surface in the passage of an anti-cyclone in the winter season. The influence of cloudiness on the temperature aloft is also exactly the reverse of that below, at this season ; the highest temperature below accompanies the greatest amount of cloud, but the highest temperature aloft accompanies the least amount of cloud. These phenomena, Hann ascribes to the downward motion of the air mass in the anti-cyclone. It is found to be the case that cyclones are usually accompanied by a relatively low temperature aloft, both in summer and in winter. The following table shows quite clearly the variation of the temperature with the altitude, under various conditions of air pressure distribution in the Eastern Alps :—

VERTICAL DISTRIBUTION OF TEMPERATURE IN THE EASTERN ALPS IN WINTER
OBSERVED FOR VARIOUS PREVAILING CONDITIONS OF AIR PRESSURE.

PLACE.	ALTI-TUDE.	TEMPERATURES IN THE QUADRANTS OF ANTI-CYCLONES, OR BAROMETRIC MAXIMA.				TEMPERATURES IN THE BAROMETRIC		TEMPERATURES IN THE CENTRE OF THE BAROMETRIC	
	Metres.	E.	S.	W.	N.	Max.	Min.	Max.	Min.
Sonnblick	3100	−18·9° C.	−18·1° C.	−11·0° C.	−11·8° C.	−9·8° C.	−17·5° C.	−7·8° C.	−13·4° C.
Säntis	2500	−12·2	−11·7	−6·5	−7·3	−4·3	−13·4	−2·3	−10·4
Obir............	2046	−11·0	−12·5	−5·0	−5·9	−4·8	−10·2	−3·1	−6·8
Schmittenhöhe	1974	−11·9	−12·0	−3·9	−5·0	−3·6	−10·6	−2·0	−6·6
Kolm	1600	−8·5	−10·0	−0·9	−3·2	−3·1	−7·7	−1·5	−3·3
Haller Salzberg	1490	−7·1	−9·0	−1·0	−1·9	−1·5	−7·1	−0·2	−3·5
Stelzing	1410	−6·5	−8·1	−2·4	−2·6	−1·7	−6·3	−0·3	−2·4
Rauris............	940	−4·7	−7·5	−3·0	−3·0	−5·7	−4·5	−5·6	−0·2
Zell am See.....	766	−3·9	−7·1	−4·1	−3·8	−6·9	−4·9	−7·1	+0·3
Lienz	680	−3·1	−6·0	−3·3	−3·4	−5·9	−3·6	−5·3	+0·2
Ober-Drauburg	610	−3·6	−5·7	−3·1	−3·3	−6·5	−4·0	−6·3	+0·3

Column 1 gives the name of the observing station; col. 2, the altitude in metres; cols. 3, 4, 5, 6, the temperature in degrees Centigrade in the E., S., W., N., parts of areas of high air pressure, or anti-cyclone; cols. 7, 8, the temperatures when areas of maximum and minimum air pressure occupy Central Europe; cols. 9, 10, the temperatures when the areas of maximum and minimum air pressure are central at the Sonnblick. This table of Hann's is worthy of very careful study.

Hann finds that, according to the results obtained by him in the continental Alpine region, that is, warm anti-cyclones and cold cyclones, we cannot assign thermic causes as the origin of cyclones and anti-cyclones. These atmospheric whirls are, then, due to a dynamic cause, and are brought about by the primary atmospheric currents which flow from the equator polewards, and from the poles equatorwards. This view is substantiated by the frequently-observed cases in which the wind is directed against the air pressure gradient, both below and aloft.

With a purely thermic origin this would not be possible.

Hann also finds it impossible to believe that the convection theory (that ascribed to Ferrel and others) of cyclones can be applied to such vast areas of air as we find occupied by the cyclones and anti-cyclones of the middle latitudes, and in which the relation of the height to the diameter is so small. The usually cited principle of the draught in a chimney may be applied to the case of small atmospheric whirls, but it can have little connection with the great cyclonic disturbances. Another strong argument brought forward is the fact that the annual periodicity of the frequency and intensity of cyclones shows that in the summer they are really least, while the convective action would naturally be greatest at this time. Also, the fact that one cyclone succeeds another in about the same path would seem to show that the equilibrium above and below, which they would tend to restore, is continually broken by some cause extraneous to them. The theoretical deductions of von Helmholtz strengthen Hann's position in this matter.

§ 7.—*Von Bezold's Attempt at Reconstruction of Existing Theories of Cyclones.*

Main Differences in Older Theories of Cyclones.— While the older form of ideas of the Trade-wind theory, as to the prime cause of nearly all atmospheric motions, may still be of value in attempting to explain the phenomena of the lower latitudes, yet some of the later views for explaining the great cyclonal and anti-cyclonal systems of the middle latitudes seem to point out the more local causes of heating up and cooling and the existing conditions of humidity as the main points to be studied, and that

the whirling motion of the air which accompany these
areas of high and low pressure are natural conse-
quences of the differences in pressure caused by these
local conditions. Von Bezold thinks that in the rush
from one extreme to the other in seeking to explain
these phenomena, the advocates of the last-mentioned
theories have gone too far in isolating the local from
the general atmospheric disturbances. He calls
attention to the almost forgotten, and, in fact, apparently almost unnoticed, paper by Hann,[1] which he
thinks presents views which are likely to be recognised
as of great value when we shall have reached the final
outcome of the very recent study which has been
given to the problem by able investigators. In the
present connection due prominence must be given to
the theoretical deduction of von Helmholtz,[2] that
"by the continuous action of forces, surfaces of dis-
continuity can be formed in the air, and the anti-
cyclonal motion in the lower layers, and the great
and gradual increasing cyclone of the upper layers,
which was to be expected at the pole, will be broken
up into a great number of irregularly moving cyclones
and anti-cyclones, but with a preponderance of the
former." The main differences between this idea
and the one advanced many years earlier by Ferrel
seems to be that Ferrel is credited with laying too
great stress on the local conditions necessary to form
these secondary whirls within the larger whirl ; but
it must be remembered that when he first formulated
his theory, what is now known concerning vortex

[1] *Einige Bemerkungen zur Lehre von den allgemeinen atmosphä-
risches Strömungen.* Von J. Hann. *Oster. Zeit. f. Meteorologie,* 1879.
[2] *Ueber atmosphärische Bewegungen.* Von Helmholtz, *Sitzungsber.
d. Preuss. Akad.,* 1888.

and discontinuous motion was not then available as a help in explaining this puzzling matter. It was found from Loomis's investigations that the latent heat given out in rain formation was not necessary for the formation of weak cyclones ; but the theorem of von Helmholtz goes much farther than this, and will probably save much discussion in trying to reconcile this and other phenomena with the old condensation theory of cyclone formation.

Hann's recent investigations of the relations of air pressure and temperature on the Alpine peaks during the presence of cyclones and anti-cyclones have convinced him that, in some cases at least, the conditions are such that they could not be due to any relations of specific gravity of the mass of air composing them ; and he sees in them the effects of the action of the general atmospheric circulation.

On the poleward side of the ring of high pressure, which has been mentioned in speaking of the general air circulation, there exist at *considerable elevations* the conditions which, according to Helmholtz's view, are sufficient cause for the origination of the cyclones which follow each other in their easterly march ; the series being broken by the ridges of high pressure which lie between them. The anti-cyclones, or regions of high pressure, are merely portions of the great ring of high pressure which project out from the main body in irregular forms, the central or culminating regions of which depend essentially on the temperature distribution ; the region of the greatest cold being most favourable to their existence. Their much greater frequency and magnitude in the northern hemisphere is due to the greater temperature differences introduced by the excess of land ;

and, correspondingly, a uniformity of the great ring of high pressure is found in the southern hemisphere. In the northern hemisphere the culminating centre of highest pressure is found in the extreme case of Central Asia to be 25° north of the average latitude of the ring of high pressure (35° N.), or at a distance of 60° from the equator.

Von Bezold's Discussion of Centred Cyclones.—While Hann has recently studied the cyclone mainly from the data given by direct observations and with special regard to the vertical temperature distribution, von Bezold has endeavoured to throw further light on the subject from the theoretical side, in which the effects of the air pressure and the wind shall be given the greater prominence. At the very outset the query comes up as to which is the cause and which the effect. Are the cyclonal motions of the air due to the lower pressure at the centre, or is the latter the result of the former?

The simplest case of the cyclone is that in which the isobars are circular and the wind blows along them, that is, makes an angle of 90° with the radius of curvature at any point. Such cyclones von Bezold designates as "*centrirte cyclonen,*" or "centred" cyclones. Supposing the " centred " cyclone to be in existence, then there are three forces at work ; the centrifugal force Pc due to the rotary motion of the cyclone, the deviatory force Pi due to the earth's axial rotation, and the gradient force T due to the differences of air pressure. In order that the assumed condition be maintained, the sum of these forces must be equal to zero ; or, if we express this in a simple general algebraic form,

$$Pc + Pi + T = O,$$

This permits, however, the occurrence of four combinations or special cases. It is necessary to state these separately, and they are as follows :—

1. When there is a true cyclonal gradient, that is, inwards towards the centre, such as exists near the surface of the earth in a barometric depression, T is opposed to the other two forces, and the equation is written,

$$Pc + Pi - T = O.$$

2. When the gradient is outward, such as exists for an area of high barometric pressure, and such a condition as is to be found in the air above a cyclone with warm centre, in which the cyclonal motion is due to the moment of rotation which the air masses have brought with them from the lower regions, and which cannot at once disappear. This applies, however, to only the lower part of the anti-cyclonic area superimposed on the surface cyclonic system. In this region the assumed condition of the whirl being " centred," cannot be fulfilled.

3. When anti-cyclonal rotation exists for an outward gradient such as is found in the case of an area of high pressure at the earth's surface. In this case

$$Pc - Pi + T = O ;$$

but the only place where this could occur would be in the upper part of a cyclone (with warm centre) at the place where the moment of rotation brought to the region by the air masses coming from below, is completely used up in overcoming resistances. This is probably well within the region of high pressure which must exist superimposed on the area of low pressure existing at and near the earth's surface.

4. There still exists, at least in theory, one more form ; that of anti-cyclonal rotation with an inward

directed gradient, that is, with an area of low pressure
at the centre.   The equation in this case takes the
form :

$$Pc—Pi—\varGamma=O.$$

This case certainly does not occur at the earth's surface, but is *supposed* to be the condition which prevails
in the air above a surface area of high pressure.   In
order to account for the downward current necessary
to feed the outflow of air from the surface anti-cyclone,
theory superimposes on it a cyclonic or low pressure
area for the upper air ; but von Bezold remarks that
observations are still needed to make its existence a
certainty, and he thinks it may not be the true
condition.

In the cyclones in which we have observations
showing a cold centre, it is highly improbable that
this relative cold extends very far up above the surface
of the earth.   The inflow of air above the anti-cyclone
is most probably due to a dynamic cause, and is more
likely the result of a damming or heaping up of the air,
rather than a result of an inwardly directed gradient.
The small moment of rotation which arises from the
transfer of the rotation in an anti-cyclone is so weak
in comparison with that of a cyclonal rotation, that it
would not cause the abnormal conditions which have
just been discussed for the cyclonal rotation, and in
which the original motion is carried to a considerable
distance within the region where we should expect
the opposite conditions to prevail.

The most Important Form of Centred Cyclone.—Now
while the four cases which have been noticed exist
theoretically, the first one is the most important one
that we have to consider, for it is the one which our
accurate observations at the earth's surface show us

when they are brought together on the synchronous weather charts. We have still too little data con- cerning the motions of the air at even a short distance above the earth's surface to allow of any- thing like a true presentation of the facts concerning them. We will take, then, the ordinary case in which we have an atmospheric whirl with a centre of low pressure, and consequently a gradient directed in- wardly, but with the simplifying proviso that the isobars are circular and the wind circulation cyclonal, and with the wind direction always tangent to the isobar. The acceleration of the air due to the gradient force is the same as that which would be imparted to a particle sliding frictionless down an inclined plane which has the same inclination to the horizontal as the surface of equal atmospheric pres- sure, only when this inclination is so small that the sine and tangent of the angle are practically equal. Möller[1] seems to have been the first to put in its proper light this important application to the atmo- spheric motions. The accelerating force in the descent is gravity × sine of the angle, while the resisting force to the descent is (reduced to the horizontal direction) gravity × tangent of the angle. An example from actual observations will show what an exceedingly small inclination these surfaces of equal pressure in reality have ; an extreme case taken from observations made in England on Oct. 14, 1881, showed the angle to be only 0° 1′ 36″ for the isobaric surface of 730 mm. This gradient is so slight in absolute measures that in the distance of 18 kilo- metres the isobaric surface will change its absolute

[1] Max Möller, *Der Kreislauf d. atmosp. Luft. Archiv der Deutsche Seewarte*, Band X., 1887.

level only 12 metres, and yet this was sufficient to produce violent winds.

Theoretically, there is a direct relation between the pressure distribution and the velocities of the wind ; and we find that in the one case the wind moving circularly around a centre causes a "centred" cyclonal whirl to be maintained, or in the other reverse case symmetrical circular pressure distribution causes the maintenance of these wind velocities.   In practice the layers of air have different velocities, and the cause of these must be looked for elsewhere than in the system we have considered ; and the force which overcomes friction (which we have not considered) must also arise outside of the system.   Concerning this von Bezold says : " In no case can these resistances within the 'centred' whirl be overcome by forces which arise from the pressure distribution."

The question arises as to whether this ideal condition is ever fulfilled for the atmosphere at the earth's surface ; and in case it should be so, if these same relations extend upward through a layer of air of considerable thickness.   If it is found that the centred cyclone is not to be observed in its perfect form, then we must see what conclusions may be drawn from those in which the deviations have certain marked characteristics.

With the approach towards the centre of the cyclone the inclination of the isobaric surfaces (or the gradient) must increase, unless the velocities of the wind decrease ; and if the wind does not decrease, but, on the contrary, increases, then this inclination would increase more rapidly.   Even in the case of an unvarying wind velocity the gradient would become infinite unless some compensatory action occurred.   But we find that the increase of the centrifugal force as the centre is approached, can be so entirely overbalanced by a corresponding decrease of velocity that the gradient may even decrease in the immediate region of the centre. It will be best to take from nature an example illustrative of the occurrence of this class of phenomena. Sprung has given (*Lehrbuch d. Meteorologie*, p. 150) a sort of average or composite picture of the pressure distribution in several cyclones, which will serve to show the computed conditions, supposing it to be a centred cyclone.

Distances from the centre of cyclones in kilometres ... ... } 100, 200, 300, 400, 600, 800, 1,000.

Corresponding wind velocities, in metres per second, at these distances } 10·8, 20·7, 21·4, 23·0, 18·0, 13·3, 10·4.

And these velocities are not too great to make them contradicted by the observed velocities.

Von Bezold has computed, that in a centred cyclone, in the latitude of 45°, having a barometric pressure of 730 mm. and temperature of 10° C, there would exist for wind velocities of 10 m.p.s. at 100, 10, 1 kilometres distance from the centre corresponding differences of 2 mm., 11 mm., 101 mm. pressure per degree of arc.

For wind velocities of 20 metres per second the gradients would be nearly four times as great, but these enormous gradients are prevented by the centrifugal motions becoming centripetal. It is usual, however, to find that in observed cases the wind velocities as well as the gradients decrease in the immediate vicinity of the centre of the cyclone, as was computed for the "centred" cyclone. It is further

more not at all certain that the change from centripetal to centrifugal motion does not extend to the
currents at a considerable altitude, and that the
ascending central current below may be a descending
one above, as is indicated by the frequently observed
clear space (cloudless) at the centre of cyclones, and
which is known as the " eye " of the storm.

We have found that the wind velocities near the
earth's surface are sometimes such as would occur
there for a centred cyclone, and now the same relations in the higher atmosphere must be examined.
Von Bezold gives a mathematical proof that in the
case of *any* symmetrical circular distribution of air
pressure there is a possible system of velocities which
will make the existing whirl a centred whirl, and that
the converse is also true : that in the case of any
system of uniform circular motions (motions around
a common axis), in which one motion is gradually
merged into the next, a definite distribution of air
pressure (that of a "centred" whirl) is produced, which
causes a continuation of these motions in the case
when friction is omitted.

If we suppose the distribution of air pressure to be
that of a "centred " whirl, and conceive the isobaric
surfaces to be rigid, and the air to be able to glide
frictionless down them, this air will move in horizontal
circles and will continue so to move, because, as has
been seen, the inwardly directed force, gravity × sine
of the angle of inclination of the surface ($g \times$ sine a'),
is just counterbalanced by the horizontal component
gravity × tangent of the angle of inclination (or
$g \times$ tangent a), of the outwardly directed force
which tends to drive the surface upward, and which
is gravity × tangent of the angle of inclination ×

cosine angle of inclination (or g × tangent a × cosine a'), but the angle being practically so small the cosine is equal to unity. When, however, the wind velocity at any circle is either too great or too small to preserve the original conditions of equilibrium, then there will arise an ascending or descending motion along the supposed rigid sloping isobaric surface.

Von Bezold proposes to call the velocities which will exist theoretically for a given maintained pressure distribution, the "critical velocities"; and the pressure surfaces which would result theoretically from the maintenance of given wind velocities, the "critical surfaces."

The actual gradient is conveniently termed the "effective gradient," while that for the "critical surface" is called the "critical gradient." Von Bezold puts this idea into the following theorem :—

"In centred whirls the surfaces of pressure must coincide with the critical surfaces, and the effective gradients be equal to the critical gradients."

It is hardly probable that when this condition exists at the earth's surface, the same relations extend to any great distance upward.

Since, in general, the variations in the distance between two isobaric surfaces is dependent on the differences of absolute temperature, then these variations must be very slight, and any two surfaces will be nearly parallel. On account of the decrease of temperature with the elevation, the isobaric surfaces in a cyclone will crowd somewhat closer together with increasing distance from the central axis : this is independent of the decrease of temperature at the earth's surface with increase of distance from the warm centre of the cyclone. The critical surfaces, however, are raised higher with the increase of distance from the centre, and the velocities increase with the elevation.

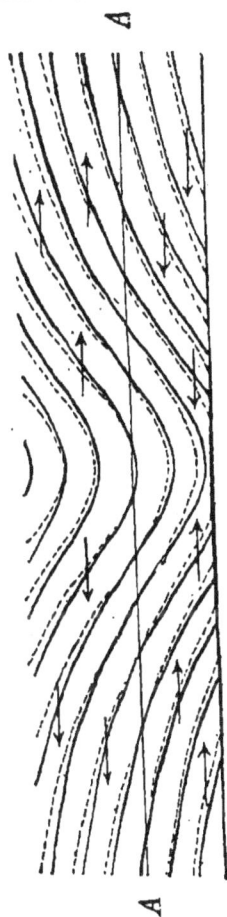

Theory shows that the inclination of the isobaric surfaces (the gradient) increases nearly as the square root of the velocity. In the accompanying Fig. 112 we have a vertical section of a cyclone; the lowest horizontal line represents the earth's surface, the curved full lines represent the isobaric surfaces, and the curved dotted lines the critical surfaces. We will suppose that at a certain distance above the earth's surface, at the line AA, the cyclone is a "centred cyclone"; but above and below this line the "centred" condition does not obtain, but the velocities increase with the elevation. It is seen that the critical surfaces and the pressure surfaces coincide at points along AA; above this surface the former are the steepest, and below not so steep; above, the centrifugal force is greater than the gradient force directed towards the axis, and consequently motions must exist against the gradient, that is, centrifugal motions. The lower centripetal air motions are shown by the inwardly directed arrows, and the upper centrifugal motions by the outwardly directed arrows;

FIG. 112.

the surface, AA, which separates these two regions, need not necessarily be a plane.

In a case like that shown here, we have a change in direction of the winds without a corresponding change in the sign of the gradient, the motions at the earth's surface being nearly circular. The velocities increase very rapidly with the altitude, near the earth's surface, while the inclination of the pressure surfaces (gradients) decrease ; and this would give the slight centripetal motion to the lowest layers, which becomes circular a little above and centrifugal still higher up, as is shown. Accordingly, it is quite evident that the centred condition is to be found existing for only a limited vertical distance.

Since the inclination of the surfaces of equal pressure measures the amount of the gradient force, while the inclination of the critical surfaces measures the combined amount of the centrifugal force and that arising from the earth's rotation, acting outwardly from the centre, then, when the inclination of the critical surface is less than that of the pressure surface, the resultant motion is centripetal, that is, inward towards the axis ; and when the inclination of the critical surface is the greater, the resultant motion is outward, or centrifugal. It must be remembered, however, that the critical surfaces for a symmetrical circular pressure distribution become surfaces of rotation only when the air moves in circles perpendicular to the axis and when the centres of these circles coincide with this axis ; and under these conditions the condition is one of stable equilibrium when the critical and the pressure surfaces do not coincide throughout.

Actual Cases of Centred Cyclones.—Von Bezold

thinks it is quite probable that tornadoes and other spout phenomena have a nearly circular motion of the air, at any rate near the axis ; and if the results found for centred cyclones are applied to these, then with the approach to the axis of the tornado the critical surfaces become very much inclined and the gradient must become very great in order to keep the circular motions from becoming centrifugal in direction. He computes, that at a distance of 10 metres from the centre and for a wind velocity of 30 metres per second, there would be a decrease of 0·81 mm. in a horizontal distance of 1 metre ; the inclination of the critical surface amounting to about 84°. If, now, this motion is kept up by some energy extraneous to the tornado, then we have a marked decrease in the atmospheric pressure in the immediate region near and around the axis, which can be computed when we know (or assume) the decrease in velocity with the distance away from the axis. Ferrel has carried out this computation on assumed conditions of velocity for the tornado as a centred whirl in which the form of the pressure surfaces are the same as the critical surfaces of von Bezold ; but the latter believes that the enormous gradients which must exist close to the axis, if no centrifugal motion ensues, make it very improbable that the outer air is drawn into this inner region as a feeding current and has a motion towards the axis, that is, centripetal ; because in order to have this possible, the effective gradient, resulting from the inclination of the existing pressure surfaces, must be greater than the critical gradient, caused by the inclination of the critical surfaces, which is so great. But if there is no continued inflow of air, then of course there can be no ascending current at the axis.

Von Bezold thinks that in the axial region there is
no considerable vertical motion, but that this region
is a progressively moving space in which the pressure
of the air is considerably lowered, but in which new
air is constantly to be found, due to the change in
location of the whirl. Ferrel's computations have
shown that the decrease in the air pressure due to the
centrifugal force is amply sufficient to produce condensation without the conduction of heat. And the
apparent lowering of the spout from the clouds, or
growth of the spout downwards, is sufficiently
explained by the fact that this thinning of the air
takes place first up above where the velocities are
greatest and where the air is more nearly saturated.
Still it must not be considered that the cause of the
whole phenomenon is to be referred to this action.
When the condition of unstable equilibrium is brought
about as a result of the heating up of the surface of
the earth, and consequently of the lower air layers,
even then the winds will reach great velocities up
above earlier than at and near the earth's surface.
Since on the resumption of stable equilibrium, the
accelerating forces increase with the elevation, then,
not only the generally distributed vertical currents,
but also the horizontal currents must increase in
velocity, since with the increasing altitude the resistances decrease. That the cloud is first seen above
and then descends, cannot be taken in itself as signifying where the originating cause is to be found,
whether above or below, nor the direction of the flow
of any vertical currents.

The formation of the cloud by condensation shows,
in itself, that no powerful descending current can
exist, as then the adiabatic compression would prevent the condensation of the vapour.

In the case of the larger cyclones, von Bezold thinks that when they originate in the middle layers of the air, caused by the general circulation of the atmosphere, as well as by local causes, there can be a descending current above (which explains the "eye of the storm" phenomena), and that in special cases this may even descend to the earth's surface. He explains the great dryness at the centre of the Manilla Hurricane (October 22, 1882, and mentioned in Sprung's *Meteorologie*) as being the result of such a motion.

This reasoning seems to apply to Hann's cases of cyclones with cold centre and anti-cyclones with warm centre, which have been described and which have lately created much discussion as to whether they are or are not abnormal phenomena. Von Bezold also considers that he has, in a measure, explained the phenomena which have led Faye to conclude that a descending air current is to be found at the centre of all cyclones.

Here we must leave the subject of air motions, although only a casual mention has been made of a great class of phenomena, mainly of local occurrence, and in which are included tornadoes and all spout phenomena, thunder-squalls, derechos (straight blows), and other intense but limited atmospheric disturbances. In these classes of storms there is such a variety of phases that any adequate presentation of them could not be given in a single short chapter. A word must be added, however, concerning their connection with the cyclonic phenomena treated in the present chapter. In the south-east quadrant of the cyclone, there is usually a region, at some distance from the centre, where there are abnormally intense local atmospheric commotions of greater or less relative magnitude.   It is in this section that the torna-does and thunder-squalls most frequently occur ; and their investigation, both as local storms and in connection with the greater cyclones, has been very assiduously prosecuted during the past few years by European and American meteorologists.

CHAPTER VI.

APPLIED METEOROLOGY.

§ 1.—Application of Meteorological Observations made during Long Periods to the determination of Oscillations in Climate.

THE main results of direct meteorological observations, and the laws which are deduced from them, have furnished most of the material gathered together in the text-books and elementary works on meteorology, and the later editions of these publications are in some cases brought well up to their dates. Van Bebber's *Meteorologie*, in German, and Scott's *Meteorology*, in English, both present excellent brief statements of these results ; although the former undoubtedly gives a rather better idea of the actual methods of meteorologists, as it is not written in such a popular manner as the latter. There is, however, one very important matter which has not been at all freely dealt with in such works, because it had not already received a comprehensive treatment by the original investigator, who must precede the bookmaker. I refer to the climatic oscillations during long periods.

Professor Brückner,[1] of Berne, has given a very complete discussion of a vast amount of unused

---

[1] *Klimaschwankungen*, Vienna, 1891.

material, and has brought together important partial investigations made by others. I will here summarise very briefly some of the results of his study of the selections from the vast mass of data accumulated in various portions of the globe since the institution of meteorological observations. These results are of special importance, as it is mainly to obtain them that the long, costly, and laborious series of observations are so carefully and uninterruptedly continued. Brückner's monograph is certainly a remarkable book, and should be read by every one who is really interested in the subject of variations and changes of climate. While the results are of great scientific value, yet Brückner's method of presenting them is such that any one who is at all accustomed to reading meteorology will have no difficulty in comprehending, not only the results, but also the methods employed in obtaining them. Besides his original investigations, Brückner has made mention of the most important literature treating of oscillations of climate, and, taking the book as a whole, it must be placed in the front rank of those devoted to the presentation of accumulated data.

Fluctuations of Rainfall.—Brückner has made use of observations at 321 points on the earth's surface, distributed as follows : Europe, 198 ; Asia, 39 ; North America, 50; Central and South America, 16; Australia, 12; Africa, 6. It is seen that most of these are in the northern hemisphere, but only those stations could be used where the observations extended over a sufficiently long period. While for most of the stations the data are for the period 1830–85, yet for many they extend farther back, and at Paris commence in 1691–95. In dealing with such immense masses of data, it would be too burdensome to use single years, and so the averages for five-year periods are used ; thus 1691–95 refers to the mean for the years 1691–92–93–94–95. The following table contains the average result of these data since 1830 in terms of excess and deficiency of rainfall :—

PERIODS OF RAINFALL.

	Deficiency.	Excess.	Deficiency.	Excess.	Deficiency.
Europe..........	1831–40	1841–55	1856–70	1871–85	
Asia	1831–40	1841–55	1856–70	1871–85	
Australia	–45	1846–55	1856–65	1866–75	1876–85
N. America...	1831–40	1841–55	1856–65 (71–75)	1866–70 (76–85)	
Cent. and S. America ..	1831–45	1846–60	1861–75	1876–85	
Taken all to-gether	1831–40	1846–55	1861–65	1876–85	

The average amplitude of oscillation, expressed in per cent. of the total average amount of rainfall, is as follows : Europe, 16 per cent. ; Asia, 30 per cent. ; Australia, 22 per cent. ; North America, 26 per cent. ; Central and South America, 28 per cent. ; the average for all being 24 per cent. That is, in the driest period the rainfall is only three-fourths of that in the rainiest. It appears also that this oscillation is true for the whole of the land surface, and that a deficiency in one section is not counterbalanced by an excess in another section. What takes place on the water surface of the globe we do not know with certainty, as the rainfall observations at sea are not suitable for such investigations, but the sea-coast stations probably indicate fairly well the results for the open ocean.

It has been suggested that with a shifting towards

the eastward a progressive retardation of the time of maximum or minimum exists; but Brückner shows that no such relation exists either with change of longitude or latitude. As regards the amplitude of the oscillation, it can be said that while it does not vary very much for the same place, yet for different regions of the earth's surface it is by no means the same. In fact, there is found to be the general law that the intensity of the oscillations of rainfall increases with the continentality of the region. The average ratios of maximum : minimum, and the average rainfall at the maximum and minimum for the period 1830–80, according to the continental distribution, and also arranged for progressive longitudes, are as follows :—

	Ratio. Max. : Min.	Average Min.	Average Max.
		mm.	mm.
Eastern part of England	1·18	599	744
North Germany	1·23	573	705
South-West Russia	1·26	447	570
South-East Russia...	1·40	273	384
Ural Mountain Region	1·36	350	480
West Siberia	2·31	149	347
East Siberia	1·59	355	564
Deccan	1·24	...	...
U.S. America, West Coast ...	1·38	379	517
,, far Western interior	1·42	483	684
,, Southern part ...	1·36	...	...
,, Eastern interior (Ohio, &c. ?)	1·20	890	1,059

Thus we see that in West Siberia 2·3 times as much rain falls in the rainy period as in the dry period, while in England the difference is relatively slight.

The observations along the east coast of the United States, some of the islands of the Atlantic Ocean, and the Irish coast, show minima at about the time of the maxima at the interior; and if these be taken to represent the Atlantic Ocean rainfall, then there may exist in the case of this ocean a compensatory oscillation of rainfall, the reverse of the oscillations on the land. We have in reality during long periods a shifting back and forth of the isohyetal lines (lines of equal rainfall). The amount of this for various isohyetals is shown in the following table; in the first columns for the period from the minimum of 1861–65 to the maximum of 1881–85, and in the second column for the average time of maximum and minimum :—

DISPLACEMENT IN KILOMETERS.
Counting + towards the inland.

Isohyetals.	OLD WORLD. West side.	West side.	NORTH AMERICA.	
1,000 mm.	—1000	—600	200	700
900	—500	—200	300	500
800	+300	100	400	600
700	+1,500	1,000	500	500
600	+400	1,600	1,200	1,100
500	+600	1,000	2,000	2,000
400		800	Vanishes.	Vanishes.
300	Vanishes.	Vanishes.		

It can be readily seen what an important factor this displacement becomes in investigating questions concerning any actual increase or decrease of rainfall.

There have been found periods of fluctuations in rainfall with the following times of excess and deficiency, the former being given in large type and the later in small type :—

1691–1715.	1806–25.
(1716–35.)	(1826–40.)
1736–55.	1841–55.
(1756–70.)	(1856–70.)
1771–80.	1871–85.
(1781–1805.)	

For the present century more exact dates can be assigned to the times of actual maximum and minimum of rainfall, and they are as follows :—

1815.
(1831–35.)
1846–50.
(1861–65.)
1876–80.

From this it would appear probable that we are now entering upon another period of very low rainfall during 1890–95.

Fluctuations in the Elevation of Water Surfaces.— An indirect measure of the long period variations in rainfall, is the fluctuation of the level of water surfaces of inland seas, lakes, and rivers; and Brückner has given this question a careful study. Of the inland seas, the Caspian Sea has been the most observed, with a broken record going back to the tenth century, and having a quite complete record since the early part of the eighteenth century.

There are, for the Caspian Sea, maxima about 1743, between 1780–1809, 1847 and 1879 ; and minima about 1715, 1766, 1845, and 1856–60. Brückner places the average duration of an oscillation at about from thirty-four to thirty-six years. That these oscillations are due to large sectional climatic influences is very evident ; and a comparison with the oscillations of rainfall and temperature shows that for the period since 1840 these are accompanied by oscillations in this sea level.

For the region west and north of the Caspian Sea there occurred wet, cold periods, about 1745, 1775, 1810, 1845, 1880 ; and dry, warmer periods, about 1715, 1760, 1795, 1825, 1860. These changes affect the amount of water in the inflowing rivers, which, in turn, change the level of the sea. A comparison with the table of sea levels shows the increase with cold, wet periods, and decrease with the dry, warm periods. There are also found to exist long periods of oscillations, in addition to the short periodic fluctuations.

As the result of an investigation of a number of lakes Brückner gives the following results :—

1. The oscillations of the true river-lakes are small, and follow without retardation the oscillations of the rivers.

2. The oscillations of lakes having no outlet are large, and the epochs show a retardation in relation to the corresponding epochs of the inflowing streams ; and the maximum height of the lake may occur at the time when the receding inflowing streams have reached an average height.

3. Lakes having no outlet have a slighter retardation in the case where the inflowing streams have large oscillations than for those where the oscillations are small ; and the same is true of lakes with low banks in contrast with those having high banks.

4. Small secondary oscillations of the influx in lakes having no outlet, are without marked effect when the difference of the inflow and out-go still retains the same sign. The only effect is to accelerate or retard the rise or fall of the water.

5. In these particulars the partial river-lakes stand somewhere between the true river-lakes and the lakes having no outlet.

Brückner gives a list of the times of maxima and minima before 1800 for the Alpine glaciers and 7 lakes ; and for the present century for the Alpine glaciers and 10 lakes in Europe ; Caucasian glaciers and 12 lakes in Asia ; 10 lakes in North America ; 2 lakes in South America ; 6 lakes in Africa ; and 3 lakes in Australia. As a result of a careful study of this data, Brückner finds that there is no law of retardation of epochs (phases) as regards longitude or latitude ; but in general the periods of high and low water occur simultaneously over the whole earth. The period from maximum to maximum, or minimum to minimum, varies between 30 and 40 years, with an average of 35·6 years. If inferences as to periodic changes in climate are to be made from the oscillations of lakes without outlet, the following little table is obtained :—

CLIMATE.		CLIMATE.	
Dry or warm, or dry and warm.		Wet or cold, or wet and cold.	
Before and about 1720		Before and about 1740	
,,	1760	,,	1780
,,	1800	,,	1820
,,	1835	,,	1850
,,	1865	,,	1880

(Min. water.) (Max. water.)

A corresponding table of the times of maxima and minima of water in thirteen rivers and thirteen river-lakes is as follows :—

Minima.		Maxima.	
About 1760.		About 1740.	
,,	1795.	,,	1775.
,,	1831–35.	,,	1820.
,,	1861–65.	,,	1850.
		,,	1876–80.

It is seen that these periods are about the same for all three classes of bodies of water, and the natural inference is that a common climatic cause must exist on all continents.

Oscillations of Atmospheric Pressure.—The study of synoptic weather charts, and the theoretical studies of Ferrel, Guldberg and Mohn, and others already mentioned, have shown us the controlling influence of the distribution of air-pressure on climate for short periods, and the desirability of a comparative investigation for long periods was unquestioned ; but the data for such a work was in such a condition that the chance was exceedingly slight for obtaining any results deserving of notice. Brückner has boldly attacked the problem, however, and has deduced some conclusions which the agreement of so many different series of observations forbids our criticising as being based on deviations within the possibility of errors for single or a few places of observation. These investigations are based mainly on the matter presented· by Hann in his recent great work *Luftdruck in Europa ;* and in fact without this work as a basis, Brückner would have been obliged either to forego this investigation, or else to spend a year or more in the critical discussion of his preliminary data. As it is, his results refer mainly to Europe, and cannot be made to include other continents, until some one has done for the observations of the air pressure on them what Hann spent two years (and many more in preparation) in doing for Europe.

The oscillations of the air-pressure are discussed by Brückner under two heads—those of the annual averages, and those of the seasons. The table given below shows the lustra deviations of the air-pressure in millimeters from the average for 1851–1880, and also the rainfall in per cents.

		1826–1830.	1831–1835.	1836–1840.	1841–1845.	1846–1850.	1851–1855.	1856–1860.	1861–1865.	1866–1870.	1871–1875.	1876–1880.	1881–1885.
N. Atlantic. Cent. Europe.	Pressure...	+·08	·92	·02	—·29	—·19	—·35	·17	·42	—02	·10	—·33	·34
	Rainfall ...	—1	—·10	—1	4	1	4	—4	—10	0	0	10	6
N. Atlantic.	Pressure...				—·08	—·18	·54	·51	—·78	—·12	—·13	·32	—73
	Rainfall ...				1	2	0	0	3	5	2	—6	—10
W. Siberia.	Pressure...				—·23	∞	—·09	—·04	—·46	—·49	—·08	·09	·24
	Rainfall ...				24	31	1	13	—34	24	14	54	74
E. Siberia.	Pressure...					·45	—·06	—·04	—·10	—.45	—·29	·36	·21
	Rainfall ...				26	15	0	—20	—10	—5	9	23	28

We see here that for the North Atlantic Ocean a relation exists the reverse of that in Central Europe.

The following arrangement shows the pressure variations in winter and summer for the dry and wet periods :—

RELATIVE BAROMETRIC PRESSURE.

	North Atlantic.		W. & Central Europe.		E. Europe & N. Asia.	
	Winter.	*Summer.*	*Winter.*	*Summer.*	*Winter.*	*Summer.*
Dry period.	Lower.	Higher.	Higher.	Higher.	Higher.	Lower.
Wet period.	Higher.	Lower.	Lower.	Lower.	Lower.	Higher.

The discussion of these oscillations shows that for the dry period as compared with the rainy period there exist : (1) a deepening of the constant cyclone which the annual averages show for the North Atlantic Ocean ; (2) an increase of the high pressure which extends from the Azores to the interior of Russia ; (3) a deepening of the low pressure in the northern part of the Indian Ocean and China Sea ; (4) a decrease of the high pressure which exists, for the yearly means, over Siberia ; (5) a general increase of the amplitude of the annual oscillation, which causes in the dry period in winter a relatively high pressure in Europe and Siberia, and a relatively low pressure over the North Atlantic Ocean ; and in summer a relatively low pressure in Central and Western Europe, and on the North Atlantic Ocean.

Each rainy period is, then, accompanied by a smoothing out of the differences of air-pressure, and each dry period by a sharpening of them, not only for annual averages from place to place, but also for seasonal averages at the same place. But it is to the amounts and directions of the gradients of air-pressure that we owe the general rainfall conditions, and it is to these long period oscillations of air-pressure that we owe the simultaneous long oscillations of rainfall. We need, however, a similar comparison for other portions of the globe before these relations can be said to be completely proven to have a general applicability.

Oscillations of Temperature.—The much more intimate temporary relations which exist between air-pressure and temperature than between pressure and rainfall, lead us to confidently expect that the long period relations of the former will be more readily recognised than those just found for the latter. For the temperature oscillations Brückner has computed lustra-means for many places in twenty-two extended regions, and he has also made use of Köppen's collected data for twenty-nine regions. The observations which have been used do not extend much back of the beginning of the present century, except for a very limited number of places, and in nearly all cases they are for the northern hemisphere.

These investigations give the following periods of relative heat and cold :—

KÖPPEN.			BRÜCKNER.		
Warm	...	... 1791–1805	Warm	...	... 1791–1805
Cold	...	... 1806–1820	Cold	...	... 1806–1820
Warm	...	... 1821–1835	Warm	...	... 1821–1835
Cold	...	... 1836–1850	Cold	...	... 1836–1850
Warm	...	... 1851–1870	Warm	...	... 1851–1870
			Cold	...	... 1871–1885

Only 5 per cent. of the material used by Köppen, and 8 per cent. of that used by Brückner, showed the opposite relation to the ones given ; but perhaps 15 per cent. gave no definite results the one way or the other. The following table shows the amounts of excess (+) and the deficiency (−) of temperature referred to the average temperature as determined by many years' observation.

	1821–1825.	1836–1840.	1866–1870.	1821–1825.	1836–1840.	1866–1870.
Tropics ...	+·34°C.	−·37°	(−·10°)	+·20″	−·05″	(−·09°)
Sub-tropics	+·65	−·40	+·10	+·59	−·26	(+·03)
Warm temperate	+·49	−·56	+·21	+·37	−·35	+·16
Cold temperate ...	+·47	−·33	+·20	+·40	−·17	+·14
Cold zone or sub-Arctic	+·81	−·20	+·37	+·69	−·23	19

In the first half of the table are given the actual
28

differences of the lustra averages from the mean ; and
when these maxima and minima are determined by
taking into account the values for the lustra on
either side, the amounts are somewhat diminished,
as shown by the numbers given in the last half of the
table. For the average of the whole earth (so far as
observed at the various dates) the times and relative
amounts of the maxima and minima are :—

1736–40	...	... —·43° C.	1821–25	...	... +·56° C.
1746–50	...	... +·45	1836–40	...	... —·39
1766–70	...	... —·42	1851–55	...	... +·11
1791–95	...	... +·46	1866–70	...	... +·11
1811–15	...	... —·46	1881–85	...	... —·08

We see, then, that the amount of change in temperature during the thirty-six years' period of oscillation
is about 1° C., or nearly 2° Fahrenheit. For Central
Europe this would mean a displacement of isotherms
through a distance of 3° latitude, or about 200 miles,
during the time of oscillation ; and this is no inconsiderable climatic change, for it places the temperature of the warmest portion of the period at Riga
equal to the coldest at Königsberg.

Oscillations of Ice Periods.—We have next to follow
Brückner in considering the climatic oscillations as
shown for Northern Euro-Asia by the duration of
time when the waters (rivers) are free from ice, and
the dates of their becoming free from ice in the
spring ; the secular oscillations of the dates of the
grape harvest for Central Europe ; and the remarkably
cold winters of which we have records. The following table gives the deviations in days, at the times
of maxima and minima, from the average for 1816–
80, of the lustra periods of the days free from ice ;

the averages being smoothed out, and the negative
sign denoting the colder periods :—

			Days.					Days.
1736–40	...	...	—9·6	1811–15	...	...	—8·6
1766–70	...	...	+4·0	1821–25	...	...	+4·5
1781–85	...	...	—3·6	1836–40	...	...	—5·0
1791–95	...	...	+0·6	1876–80	...	...	+3·0

This data refers to the average of the observations in
Euro-Asia, and for a short period for the Hudson
River.

Brückner gives the corrections, in days, to reduce
the dates of breaking up of the ice to the average
for 1816–1880 for the separate regions, but does
not seem to have been able to unite them in a
general average, suitable for reproduction here, so
that the reader must be referred to the original source
for this distributed data. On the average the differences in the number of days free from ice between
the cold periods and the warm periods amounted
to about 16 days for Siberia, 18 days for Central
Russia, 25 to 32 days for West and South-west
Russia respectively, and 24 days for the Hudson
River. The dates for breaking up of winter ice
varied about half these amounts for these Euro-
Asiatic regions. Rykatschew's great memoir on the
times of opening and closing of the rivers of Russia
and Siberia serves as the chief basis for this research
on the ice-conditions.

Oscillations of Times of Grape Harvest.—The record of the dates giving the times of the grape harvest
in France, South Germany, and Switzerland extends
back for several hundred years, and the material has
been used to determine the forwardness and back-
wardness of the seasons which are, to a certain

extent, representative of the character of the year. The data used by Brückner extend back to the year 1496. There are regular oscillations from early to late harvests, and these periods obtained from smoothed-out curves of the tabulated data are given in the following table:—

Early harvest.	Late harvest.	Early harvest.	Late harvest.
1501-05	1511-15	1681-85	1696-1700
1521-25	1546-50	1725-30	1741-45
1555-50	1566-70	1756-60	1766-70
1586-90	1591-95	1781-85	1816-20
1601-05	1626-30	1826-30	1851-55
1636-40	1646-50	1866-70	1886-88
1656-60	1671-75		

While for single years the amount of variation in the time of grape harvest may vary by many days, and for the five years' lustra reaches even three weeks, yet for the smoothed-out lustra averages the extremes are about a week earlier than the average for the minima, and a week later for the maxima times of harvest. A comparison of these periods with similar periods of rainfall and temperature shows that in general the earlier period of harvest is identical with that of high temperature and small rainfall, and conversely the times of late harvest and low temperature and excessive rainfall occur together. By means of the recorded dates of the grape harvest it is possible, therefore, to trace back to about the year 1400, for Central Europe, the periods of unusual warmth and dryness, and excessive cold and wetness.

Times of Severe Winters.—The occurrences of severe winters have been catalogued still further back, and Brückner commences his list (using that of Pilgram) with the year 800, but thinks that the records before the year 1000 are of little value. He

finally finds that for the period 1020 to 1190 there
exists a 34 years' period of oscillation; from 1190
to 1370, a 36 years' period; from 1370 to 1545, a
35 years' period; from 1545 to 1715, a 34 years'
period; and from 1715 to 1890, a 35 years' period.

The relation to the glacial changes has also been
investigated, and a close connection established
between them and the periods of abnormal heat and
cold.

An average time of about 35 years is, then, found
to intervene between one period of excess or deficiency of warmth and the next, accompanied by the
opposite relative condition of moisture; and this
shows itself in all of the various data and methods
which Brückner has used in considering the question.

§ 2.—*Meteorology applied to Agriculture.*

Introductory Remarks on Agricultural Meteorology.—
There are in Agricultural Meteorology an older and a
newer school of workers. The results obtained by
the former are well represented in England by the
prize essay of Nicholas Whitley, entitled *On the
Climate of the British Islands in its effects on Cultivation*, and published in the Journal of the Royal Agricultural Society of England, 1850; in America, by
the chapters devoted to the relation of vegetation
and climate in the great work on *Climatology of the
United States* by Lorin Blodgett, 1857; and by the
more recent valuable work of Lieut. Dunwoody in
his Professional Paper, No. 10, of the United States
Signal Service, bearing the title *Signal Service
Tables of Rainfall and Temperature compared with
Crop Production* (for the years 1875–80), 1882.

The general method of procedure in these papers is to compare the averages of climate over a large territory, and using averages for a considerable length of time, with the results of crops from the same territory and for the same time.

The more modern way is to compare the continuous growth with the current meteorological conditions for the same times, for a number of scattered but limited fields of observation, and generalise from a combination of these individual results. Instead, as in the former method, of making thermic energy and rainfall, and to a certain degree, humidity also, the only important factors entering into the growth of economic plants, the newer method lays great stress on the actinic energy, and the hygrometry of circulation and transpiration as developed in the more recent investigations in vegetable physiology. These latter methods, in conjunction with the former, allow of a much more accurate estimate to be made of the conditions necessary for the maturity (or any phases of growth) of plants than could possibly be obtained by the former methods alone ; and it is especially these more recent studies that need to be brought to the attention of the practical agriculturalist.

By the progressive school of agriculture the application of meteorology by the newly suggested methods is brought under the general division of Agricultural Physics, and the principal results of work done so far in this line are found, mentioned at least, in the successive volumes of Wollny's *Forschungen auf dem Gebiete der Agricultur-Physik* and Deherain's *Annales Agronomiques*. In the papers given in these and other journals we find much that

will be of use in practical farming, but in many cases
the results are not yet past the laboratory stage, and
will still require the attention and study of the practical meteorologist and agriculturalist before they
can be said to have economic value. In the present
brief sketch only a few results can be enumerated.
In this connection I must mention that it is unfortunate that in the preparation of this chapter I have
not had access to reports of the great work done
by Lawes and Gilbert in England.

In the ordinary uses of meteorological observations
the meteorologist has a clear idea of what object is in
view, and he can accordingly regulate the form of
observations and their reductions to suit the desired
ends. In the application of meteorology to agriculture and to the study of plant life in general,
investigators in these sciences seem for the most part
to have waited for the meteorologists to decide for
themselves what observations should be undertaken
in order to show accurately the relations existing
between meteorology and plant growth. To this
cause more than anything else is due the backward
condition of this extremely important application
of meteorology. It is seldom that the meteorologist
is a botanist, a mycologist, an ornithologist, a pomo-
logist, an entomologist, an arboriculturist, a biologist
or a vegetable physiologist, and yet all of these
branches of agriculture (broadly considered) have a
direct dependence on meteorological conditions and
have individual uses for meteorological data.

It is evident that, as the meteorologist cannot hope
to know the needs of all of these separate professions,
each of these departments should make known to
him its wants ; and being familiar with what can be

done with the instruments employed in his science, he can carry out such suggestions with accuracy and certainty, and furnish them to the sub-departments of agricultural science desiring them. The meteorologist, then, awaits the demands of the agriculturalist, while the latter seems to be waiting to see what the meteorologist can do for him. The only plan for securing the co-operation of the two is for the various agricultural specialists to formulate their needs, and for meteorologists to co-operate with them and secure for them the desirable data.

After the observations are made, the next step to be taken is their reduction and grouping, and in this again the meteorologist needs the advice and assistance of the specialists in agriculture, for these last alone can determine what are the times most important in the development of plants. But in order that the most may be gained from the comparative observations, they should be made so that they will be available for current use ; and if each step in plant growth can be compared *at the time of observation* with the action of the chief agents governing their growth (after the methods of the plant physiologists) a much more exact knowledge will be obtained than if the working up of the results is put off until the data has accumulated for a season or a year. Let us now consider some of the special cases in which can be recognised the importance of meteorological conditions to agriculture.

PHENOLOGICAL OBSERVATIONS.

Animal Phenology.—It is often remarked among

farmers that the time of the reappearance of certain forms of animal life indicates the nature of the coming season of vegetation ; or that the time of departure of these gives us an idea of the severity or mildness of the coming winter. These and other ideas have gained such a foothold among the agricultural people that many proverbs have come into use in relation to the habits of birds, animals, and insects. It is probable, however, that few of these proverbs are founded on the facts as they have been found to exist by actual observation ; they are, in the main, merely the impressions of some people of reputed wisdom in such matters. As instances of these proverbial relations such sayings as the following are given : Cranes follow the last frost ; When cranes fly southward early, expect a cold winter ; A heavy plumage of geese indicates an approaching cold winter ; When martins appear, winter has broken, and there will be no more frost ; An early appearance of woodcock indicates the approach of a severe winter, &c. A very extensive list of these sayings has been published by Marriott, and by the United States Signal Service, under the title *Weather Proverbs*. One of the practical objects of making observations of the times of arrival and departure of birds and insects is that we may compare the dates with recorded meteorological observations, and by doing this for a succession of years arrive at some of the true relations between the periodic climatic changes and the phenology of animal life, and also incidentally to cause many of the so-called weather proverbs to be cast aside, and show how worthless they are as a class. Some observations of this nature have been made in older lands ; but in the newly-settled and still uncultivated countries we have scarcely any scientific data concerning the average dates of changes in the conditions of animal life, so that observations there are particularly desirable.

Plant Phenology.—The practical benefits to be derived from a careful study of plant phenology as compared with meteorology is more evident, to every one, than corresponding relations for animal · life. The whole of plant life is affected by the immediate past and the current meteorological conditions. There is evidently a condition of the ground caused by atmospheric conditions when it is best to plant seeds, followed by a succession of conditions which will cause earliest and fullest budding, blossom, bloom, and maturity. Now it is only by carefully comparing the occurrence of all of these phenomena with the observed meteorological conditions that we can find out this law, which will evidently vary somewhat for different plants. A careful study of these combined observations will undoubtedly allow a very close estimate to be made of the percentage of normal crops to be calculated from the current conditions of all crops; and such calculations are now made in a rough way. Moreover, it is very probable that a series of years of comparison of the times of the different phases in the growth of crops with the meteorological conditions will allow local agricultural stations to notify the local farmers when the best time for planting crops in the various exposures and soils has arrived, as shown by the recorded accurate observations for previous years. At present such questions are left entirely to the judgment of the farmers based upon what can be remembered of previous years, and there can be no doubt but that a reliance upon the results of carefully recorded series of comparative observations will give on the average better crops than can be obtained from the remembered experience of individuals.

It is also highly important for us to have a knowledge of the effect of differences in exposure in the same localities, particularly in mountainous districts, and also in different parts of the country, on the phenology of the same plants. It is absolutely necessary to have these data carefully recorded for a number of years in order to find the widest possible distribution of the profitable cultivation of valuable vegetable products. In the *Petermann's Geographischen Mittheilungen* of January, 1881, Professor Hoffmann has given a comparative plant phenological chart of the April blooming for Central Europe, the epochs being referred to Giessen (lat. $=50\frac{1}{2}°$ N., long. $=6\frac{1}{2}°$ E.). The regions in which the time of April bloom is before and after that of Giessen is shown by the differently coloured portions of the map, successive five-day intervals being represented by distinct colours. From the latitude of the Adriatic to that of the southern coast of the Baltic the time of occurrence extends from about twenty days earlier to twenty days later than Giessen for the low lands ; but between these points the contour of the country causes such extremes as forty days earlier and fifty days later than Giessen. Such charts, with the accompanying tables, as Hoffmann has given for Europe, are greatly to be desired for other countries, and the comparison of this data with the corresponding meteorological data would be of the greatest benefit to the agriculturalist in helping him to decide what crops will thrive best in his locality. In a fairly recent little book by Egon Ihne, on the history of phenological observations, is found the best statement of the condition of this subject in Europe. That author has given not only a complete, but condensed, history of the rise of this important department of natural science in Sweden, and its subsequent introduction into other lands, but has also given a very valuable list of reference books on the subject, with some account of the most important contributions ; moreover a complete list of all the European stations where plant-phenological observations are made is to be found in the volume.

Comparison of Animal and Plant Phenology.—The comparison of animal phenology, as shown by the reappearance of migratory birds, with plant phenology, as shown by the times of attainment of certain phases of growth, is a work which is likely to prove of great interest to the student of phenological phenomena. A comparison of this nature made by von Reichenau for the middle Rhine region is indicative of what may be expected from a wider study of the question.

The spring-time is divided into five periods:

1. The early spring, when the following plants bloom—snowbells (*leucoium*), hepatica, daphne, pulsatilla, &c.

2. The blooming of the stone fruits—apricots, peaches, cherries, plums, &c.

3. The blooming of seed fruits—pears, apples.

4. Full spring, as shown by the bloom of *Sambucus nigra, Robinia pseudo-acacia.*

5. The end of spring, as shown by the blooming of roses, the summer linden, and the vine.

Observations during the years 1878–88 showed extreme differences for

1st period, as shown by the elm, 39 days.
2nd   ,,   ,,   apricot, 39 days.
3rd   ,,   ,,   pear, 34 days.
4th   ,,   ,,   acacia, 24 days.
5th   ,,   ,,   vine, 19 days.

For seven species of birds the extreme differences were, wagtail or seed bird, 24 days ; black redstart, 24 days ; chimney swallow, 11 days ; sailor bird, 16 days ; nightingale, 17 days ; turtle dove, 16 days ; oriole, 10 days.

As the extreme limits for the birds are much less than for the plants, we may say that the former are influenced more by the astronomical spring than the latter, which follow more closely the actual weather conditions.

Temperature.—While the action of heat is of great importance to all organic life, yet plants and cold-blooded animals are the most directly affected by it ; and the indirect effects of heat are the more important to the warm-blooded animals. It is this direct action of heat on plant life that we will consider. Sachs first showed that the germination, and probably also the other growth of plants, proceeds most rapidly at a certain temperature, and that this growth is retarded in proportion to the excess or deficiency of the actual as compared with the required heat ; and the growth may be brought to a standstill, or the plant even lose all life, if there occurs a considerable deficiency or excess of heat as compared with this standard.

The French botanist, Adanson, early expressed this

theorem:—the development of buds is determined by the sum of the daily mean temperatures counted from the beginning of the year. Bouissingault, however, did not take into consideration the time when growth was not in progress, and counted the sums for the period of vegetation only, and he concluded that the length of the period of vegetation stands in inverse ratio to the mean temperature. C. Linsser found, however, that, the sums of the temperatures above zero which are necessary for a certain development of similar plants in two places, are in a direct proportion to the sum total of the temperatures above zero at the two stations. Sachs, the greatest of the plant physiologists, who has studied this question by direct experiments on individual plants, finds that for each form of plant life there is a minimum, an optimum, and a maximum temperature.

A number of European meteorologists have attempted to derive a quantitative formula which shall show the law of the influence of heat (or temperature) on the rapidity with which plants develop. Sonee stated the "phase of development" to be proportional to the square or square root of the excess of the existing temperature above the freezing point; while others, again, think that the product given by the elapsed time from a chosen epoch multiplied by the excess of temperature above a certain assumed minimum, gives more accurately the phases of growth. Still another, and perhaps more satisfactory assumption, is to take into consideration the length of time during which the temperature exceeds a certain chosen minimum, or during which the temperature oscillates between certain limits; the growth of particular plants being found to cease where this length of time does not at least equal the time found necessary for the full development of the plants.

Now, whatever way of counting is adopted, it is very evident that the heat necessary for maximum development (Sach's optimum), and also the excess or deficiency which may exist and still allow profitable cultivation, can only be determined by experiment. While the adaptability of plants is so great as to allow of their wide distribution either by means of seeds or transplanting, yet there is probably a fixed law of relation between heat and growth which must exist in all cases where maturity is reached, and it is evident that nothing should be planted in localities which do not receive the minimum amount of heat necessary for maturity. It is well known that in the extreme northern limits of cultivation (of any one plant) the rate of growth is more rapid than at the extreme southern limits, but these relations have only been meagrely studied with reference to the relative amounts of heat received. So, too, in a mountainous country, the heat received in one valley may be sufficient to mature certain plants, while another adjacent valley with a greater elevation will not receive enough heat, and will need more northerly growing plants to make their cultivation pay.

Minimum Temperatures.—But the length of the season and the amount of heat received are not the only elements of heat entering into the successful growth of plants. The question of the minimum temperatures which the plants can be subjected to and not be retarded or stopped in their growth must also be taken into consideration. This is especially necessary in those portions of a country which are subject to a great range of temperature during the months of plant growth.

De Blanchis has carefully investigated the action of low temperature on special forms of vegetation. He shows, as others have also done, that the vegetation temperature is different from the air temperature, and it can best be determined by means of a vegetation thermograph, which, as used by him, consists of a mercurial minimum thermometer having its bulb enveloped in a green muslin, which latter dips into a vessel of pure water. He assumes the normal exposure to be $1\frac{1}{2}$ metres above the sod. This thermometer will read several degrees lower than the ordinary exposed dry bulb minimum thermometer. This is a most important idea as to the method of obtaining the minimum temperatures to which plants may be exposed without harm. De Blanchis finds also that the capability of resisting cold increases with the age of a plant, but the more water it contains the less will it resist cold. Such questions, however, require the co-operation of the vegetable physiologist and the meteorologist in order to observe accurately cause and effect.

Heat Constants of Plants.—Probably no one has given more attention to the determination of vegetation thermal constants than Professor Hoffmann, of Giessen, Germany. He has published nearly every year extensive comparative observations, and his results agree as well as can be expected. Take, for instance, his comparison of the Giessen and Upsala observations of *Syringa vulgaris.*

PLACE.	FIRST BLOOM.	INSOLATION SUMS.
Upsala	June 17	1433°
Giessen	April 29	1482°

The insolation sums are determined by adding up the daily excess above freezing of the maximum temperature observed when the thermometer is exposed to the direct sunlight. His results show a very gratifying success of the two measures of the same species, at such widely different places.

In this case the relative amounts are—

Giessen : Upsala : : 100 : 97.

A comparison of twelve different plants gave an average relation of—Giessen : Upsala : : 100 : 88. The insolation sums for the first ripening of fruits gave for eleven species the relation—Giessen : Upsala : : 100 : 82. For the interval between the first bloom and the first ripening of the fruit the insolation sums had the relation—Giessen : Upsala : : 100 : 93, eleven kinds of plants entering into the computation.

Professor Hoffmann makes a very interesting comparison between his method and that pursued by Professor Fritsch. The latter summed up all of the positive average shade temperatures for each day, and used these sums as Hoffmann used the sums of insolations. This comparison is mentioned to show the agreement of the two methods.

The following table gives the relative amounts of temperature sums necessary for first bloom at Upsala as compared with those at Giessen, the latter being represented by 100.

PLANTS.	RELATIVE AVERAGE SHADE TEMPERATURE SUMS.	RELATIVE INSOLATION SUMS.
Betula alba	63	96
Crataegus oxyacantha	91	100
Lonicera alpigena	59	89
Lonicera tartarica	81	94
Prunus avium	71	103
Prunus padus	76	102
Ribes aureum	69	102
Rosa alpina	106	104
Syringa vulgaris	94	102

29

Hoffmann judges from such results that since his own method, as represented by the third column, shows a better agreement than that shown by the second, his is the proper one to use.

In a two years' comparison of the bright bulb sun thermometer with the black bulb (*in vacuo*) thermometer, the former gave the best results. Hoffmann found that the black bulb was so sensitive that it gave changes of temperature which could not possibly affect the plants on account of the short duration of action. These comparative observations are by no means to be considered as a ground for rejecting the black bulb, however, as recent investigations have given us better ways of using the black-bulb thermometer than were known before his results were published : this instrument will undoubtedly play an important part in the future of our knowledge of the effects of solar heat on plant life.

We may notice another point in connection with the effects of the transportation of plants from a place having a low mean temperature to one having a higher, or *vice versa*. Linsser, and later, in a more complete manner, Hoffmann, showed that the seeds of plants produced in the north if planted in the south were earlier than local plants, while the southern-grown seeds planted in the north were later than the local plants. The same has also been found true of the interchange of seeds grown in the mountains in the neighbouring valleys. This shows that the difference of temperature is the main cause of the observed differences. The same species require different sum totals of heat for different latitudes and corresponding phases require proportional amounts of heat. Suppose a certain plant to require 4,000° of total heat at Venice, it may require only 2,000° at St. Petersburg to bring it to maturity. Then it would require perhaps only 1,000° at the former place and 500° at the latter to produce bloom. But it cannot be supposed that the fullest growth can be attained by such greatly differing temperature sums, and there will be a temperature producing a maximum growth and fruit, which occurs for some particular relation of the time to the heat received.

Average Temperatures and Plant Growth.—It has been found that the average daily temperature in the shade must be counted from 4° C. upwards for sugar beets, and from 8° C. for potatoes, in order to observe the effects of heat on the seed germination ; that is, those average temperatures below these are to be considered as nothing in the total of growth. Also, the higher the temperature the sooner the germination. For instance, Huberlandt found that at a temperature of 9·4° C. it took twenty-two days for the beet seeds to germinate, while only 3¾ days were necessary at an average temperature of 15·8° C.

The table on the next page, from experiments made in Austro-Hungary, shows for two staple crops the advantages of knowing the relation between the observed temperature and the plant growth.

As a result of these and some other experiments Briem finds that :—

1. A too early planting of beets and potatoes is not to be advised, for the appearance above ground, the blossoming and the ripening, coincided with that of the planting which took place when the ground temperature had exceeded the minimum temperature of germination of the plant under consideration.

| SOWING TIME. | AVERAGE WEIGHT IN GRAMMES OF | | NO. OF DAYS FROM PLANTING TO OCT. 20. | TEMPE-RATURE TOTALS °C. | AVER-AGE TEM-PERA-TURES. | RAINY DAYS. | RAIN-FALL TOTAL. |
	a Beet.	a Potato stock.					
							mm.
March 1	298	196	234	3271°	14°·0	108	519
16	231	222	219	3209	14°·7	108	506
April 1	207	272	203	3151	15°·5	102	496
16	304	257	188	3020	16°·0	94	453
May 1	306	302	173	2881	16°·6	87	417
16	266	228	158	2726	17°·3	80	373
June 1	211	217	142	2469	17°·3	68	294
16	82	173	127	2197	17°·3	55	169
July 1	75	158	112	1890	16°·8	48	154
16	52	86	97	1627	16°·7	37	122
August 1	14	47	81	1331	16°·4	31	99
16	13	22	66	1026	15°·5	35	76

2. If the necessary earth temperature has been reached, then the seeds should be planted, and the maximum production may be expected.

3. Too late a planting diminishes the productiveness.

It is seen from these conclusions how important it is for each farmer to know just when the ground is warm enough, as shown by observations, for him to plant his seed.

Rainfall.—The rainfall, both as to amount and frequency, is, next to temperature, the most important climatic factor influencing agriculture. For most of the civilised portions of the earth data have been accumulated and published concerning the monthly and annual amounts of rain, and these can easily be compared in a general way with the various crops for corresponding seasons, as has been done by a number of writers. But these are not the only rain data which we should know and make use of. It is also of great importance to know the frequency with which the rain falls, and the average amounts for

single rainfalls. Four inches of rain may fall during
the month at two different places, and it is obvious
that the place in which the amount is distributed
evenly at proper intervals through the month receives
more benefit from it than another place in which all
the rain may have fallen during a couple of days.
There are also other points which we should know, and
which will find application either in informing us of
the desirability of any particular place for raising any
crops in question, or else will allow of watching the
growth through all its phases, and predicting at any
stage the probable ultimate results to be obtained.
Among these points may be mentioned, the absolute
probability of rain ; the probable total duration of
rain for all the days with rain ; the average amount of
rain for all the days with rain ; the average rainfall
during an hour of rain. Information on all of these
points is necessary in order to follow out all the
effects of rain in producing vegetation, and to know
the condition of the ground as regards the moisture
in it. In the United States the monthly and annual
amounts of rainfall have been investigated, but until
recently very little had been done in the way of
obtaining results which should give information concerning the questions of frequency just mentioned.
These relations have been determined for some of the
regions of Europe ; but even in those in which the
most thorough investigations have been made there
still remains much to be done in this line of work.

The amounts and frequency of rainfall are of great
variability in the temperate zone, and we can never
arrive at such accurate results in studying them as
have been obtained for temperatures. Moreover, the
amount of rainfall being accumulative, it can never be

known whether a rain will be beneficial (except in regions of small rainfall) unless the condition of the ground is taken into consideration. In sections of the country having a moderately wet and dry season, as, for instance, portions of California, and in climates having less rainfall than is necessary for raising good crops and requiring irrigation, as is the case in Colorado, an increase of rainfall nearly always means fuller crops.

In Sweden it has been found that the time of ripening of fruits occurs with the greatest rainfall ; and it is possible that in a country where the rainfall is neither so small as to prevent plant growth, nor so large as to destroy it, the time of ripening at different places is influenced somewhat by the rain amount at each individual place. The nature and condition of the ground has such a controlling influence on the results accomplished by the fallen rain, that a somewhat lengthy mention must be made of it : this is given a little farther along under the heading of Meteorology of the Ground.

Evaporation of Water.—The evaporation of fallen rain is also of great importance, but unfortunately there are very few places where in the past this has been made a subject of observation.

In order to show a little the relation of rainfall to evaporation from soil and from a vessel of water exposed to the air, the result of some experiments made in Paris in 1877 is given. For the period Aug. 14–Sept. 14 the following relations were observed :—

RAINFALL.	EVAPORATION.	
	From ground	From vessel
61·9	41·66	81·71

and for the period Sept. 15–Oct. 15 :—

	10·1	31·52	80·18
Giving a total ......72·0		73·18	161·89

That is, just about as much was evaporated from the
ground as it received, and the earth was in the same
condition at the end of two months as it was at the
beginning. The loss by evaporation from the vessel
was nearly the same for the two periods, but this was
not the case with the earth. Directly after each rain
the evaporation from the earth was very rapid, and in
some cases exceeded that from the vessel of water,
but diminished rapidly in proportion as the upper
layers became dry. The time of the maximum of
evaporation from the clouded ground is between
9 A.M. and noon ; but that for plants is noon or a
little after ; while that for the evaporimeter (Piche's)
is between noon and 3 P.M., and that from a sheet of
water is at about 3 P.M.

In these various surfaces and exposures the evaporation takes place by different channels, as, for instance, the capillary forces take the water from one
layer of soil to another ; the influence of light is felt
in the first two, and the temperature in all cases.

It is as yet scarcely possible to make any very
accurate estimate of the amount of water consumed
by plants, on account of the difficulties of the method
of procedure in making the determination, and the
rough experiments usually made can give only very
approximate results. The amount of water contained
in the plant at any one time, or the weight of hydrogen which is found to have been added in the formation of tissue, are neither of them any indication of

the amount of water used.   The method of isolating a
plant and measuring the water fed to it is also inexact
as usually carried on.

METEOROLOGY OF THE GROUND.

The meteorology of the ground may be divided
into the two chief divisions, those relating to temperature and moisture, and in both of these there are so
many points to be considered that the number of
special problems to be worked out is very great.

Temperature of the Ground.—In Europe especially,
there have been made numerous long series of under-
ground temperature observations, and at present it is
an important feature of the larger meteorological
observatories to carry on such observations.   In
America very few attempts have been made to obtain
the ground temperatures, although they exert such a
strong influence on plant development.   The difficulty of making these observations greatly increases
with the depth of exposure of the thermometer.   In
order to supply the information desired by the agri-
culturist, it is not necessary to go deeper than 50 cm.,
or at most 1 metre, but by far the most important
temperatures are those at the depth of about 5 cm.,
and when the earth just covers the thermometer bulb.
The earth temperatures near the surface depend to a
large extent upon the nature of the ground, both as
to its slope and the composition and colour of the
particles of earth, because these conditions influence
the amount of heat absorbed from the solar rays.
Much also depends on the short period fluctuations
of the atmospheric temperatures.   For the greater
depths the most important question to be determined
is the distance to which the frosts descend.   The

average daily variation of the ground temperatures is greatest at the surface, about half as much at a slight depth of 0·02 metres, about one fourth that of the surface at 0·10 metres, very slight at 0·30 metres, and scarcely perceptible at a depth of 1·00 metre.

The average monthly temperatures vary considerably with the seasonal changes, and this variation even at a depth of 1 metre may amount to perhaps 10° C. With increasing depth the times of neither the daily nor monthly maxima and minima agree with the times for the surface or the outer air. These relative values apply to temperate regions.

The growth of plants for an oscillating moderate temperature, and at various fixed temperatures of the ground, is shown by the following table for barley :—

Temperature of the ground.	Oscillating moderate temperature.	10° C.	20° C.	30° C.	40° C.
Weight of plant in milligrams.	8142	7638	8221	3854	928

Those plants growing at an oscillating moderate temperature showed an even strong growth of good colour. Those at 20° C. were likewise well developed with nearly the same rapidity, but those at 10° C. had a much slower development, although they had a healthy growth. Those at 30° C. had at first a very rapid growth, but this was not as substantial as for the other temperatures, and the plants were not sturdy. At the temperature of 40° C., although the plants lived, yet they grew slowly and were dwarfed, the roots suffering as well as the stalk.

It has also been found that the sheltered barley plant could be exposed to a temperature of 40° C. for a short time (as in the middle of May) without harm to the plant; and even when subjected to a heat of

57° C. the plant was not killed, although greatly injured.

Wollny, who has investigated the question of the influence of the structure of the ground on its temperature, finds that :

1. The temperature of the dry ground during the warmer part of the year increases with the size of the earth particles and the crummy condition, to a certain fixed limit ; and then decreases again.

2. The temperature oscillations increase and decrease in general in the same degree.

3. The temperature of a mixture of different-sized particles of earth is midway between those of the extremes.

4. The differences of 1 and 2 are in general very slight.

If the ground is *moist*, then in the *warm* season :

1. The temperature of the ground increases, according to the size of the earth particles, up to a certain diameter, but beyond this it decreases again.

2. The ground in a crummy condition is warmer than in a pulverized.

3. The temperature oscillations increase and decrease in the same directions as the average temperatures.

4. The mixture of different-sized particles has a middle temperature between the extremes.

5. The characteristic differences shown in 1 and 2 are considerably greater for the moist conditions of the earth than for the dry.

In the *cold* season of the year the following relations hold good : the coarser the particles of earth the colder the ground, on the average ; the more rapidly does the cold enter the ground when the frosts commence ; and the more does it become warmed up again after it thaws.

Moisture of the Ground.—The question of the moisture of the ground is one of the most unsatisfactory problems with which the meteorologist has to deal, because the different conditions which may exist are so numerous. We have fewer observations on this subject than on almost any other that we are considering, and this is no doubt due to the difficulty of obtaining results that are comparable. The experiments which have furnished valuable information belong in the main to the branch of physics of the ground which is outside the province of the meteorologist. Such questions as the following must be studied : the depth to which rain penetrates ; the moisture at various depths for different soils and varying slopes ; the relative evaporation of soils as compared with water surface (this latter must always be used as a standard of reference) for different parts of the world. Some few experiments on these relations have been made in America, as well as in Europe, by agricultural investigators, but as the same conditions do not always hold good for different climates, it is necessary to have the same experiments performed at a number of points, each of which shall represent some special climatic conditions.

The moisture of the ground depends quite as much on the relative frequency of rainy days as on the amount of rain, except where the rainfall is very deficient in amount; and the humidity of the air above ground is also a very important factor entering into any discussion of underground humidity.

The amount of moisture in the ground being somewhat under control, it is of great practical import-

ance to know the humidity which gives the best
results of plant growth. The following experiment
was made with barley in a sandy soil ; the table
showing the relative humidity and the resulting
harvest.

HUMIDITY IN PER CENT. OF SATURATION.	STRAW AND CHAFF MILLIGRAMS.	GRAIN MILLIGRAMS.
80–60	6941	6143
60–40	6053	6130
40–20	4671	5245
20–10	3156	696

Under 5 per cent. of moisture, there was no plant
growth ; and for 10–20 per cent., after a period of six
weeks the plant had not made its appearance above
the ground. Also throughout the whole growth the
same relative development seems to be maintained
that is shown at the harvest.

The following results of experiments on the relation of ground moisture to the harvest for several
plants are also of interest.

PLANT NAME.	GROUND HUMIDITY GIVING HIGHEST RETURNS AT HARVEST.
Summer rye.	Number of ears and grains, amount by weight of straw, chaff, and grains 60–80 per cent. ; average weight of grains 40–60 per cent.
Pease.	For all of the same parts mentioned above, 40–60 per cent.
Winter turnips.	For the same parts, 40–60 per cent., except for the straw, which was 60–80 per·cent., thus making the maximum total weight of the plant come be-tween 60–80 per cent.
Horse beans.	For the same parts, 80 per cent., except that the highest average weight of grains occurred at 60 per cent.
Mixed grasses.	60–80 per cent.

Wollny finds the following relations between the structure of the ground and its humidity :

1. The power of the ground for containing water increases in general with the fineness of the particles of earth, and in the pulverised condition is considerably greater than in the crummy ; because with the decrease of the size of the granulations, that is through pulverising, the surface moistened by the water and the number of the capillary acting interstices increase, and also the downward motion of the surface water into the ground is considerably retarded.

2. The ground evaporates the greater amount of water the less the size of the particles of earth, since the capacity for water and the capillary conduction of the water to the surface increase in the same degree.

3. The amount of water that can pass through the ground is in direct proportion to the size of the particles of earth, because the resistances to the downward motion of the water (adhesion, friction, and capillarity) become the smaller, the larger the particles of earth, and the more non-capillary spaces the earth contains.

4. The action of evaporation explained in number 2 is especially marked in the case of continued dryness, high temperatures and strong winds, in which case (according to 1) the difference in the amount of water in the various conditions of the earth becomes less, and under some circumstances vanish, or even the usual order may be reversed.

5. The penetrability and the evaporation are frequently (and principally in the case of different-sized particles of earth) in inversely proportional relations,

so that an equalising of the amount of moisture exists
in spite of the fact that the inner structure of the
earth is different.

Sun-light.—The amount of sunshine has usually
been determined indirectly by noting the cloudiness at
intervals during the day, but in comparatively recent
years other more direct methods already described
are coming into use. The records derived by these
will surely play an important part in the future study
of the action of light and heat on vegetation, but no
memoirs have been noticed in which comparative
observations have been studied, although in England
a number of sunshine-observing stations have been in
operation for some years : but there have not been
enough of such observations made to allow charts
to be drawn, showing the geographical distribution of
the amount of sunlight for any extended region,
although we know pretty well the comparative cloudiness of many countries.

Not only does the sunlight, as a whole, influence plant life very strongly, but the various
qualities must be taken into consideration. The
actinic action is of special importance. If heat is
necessary to assist the action of light in the development of plants, yet this last produces in itself a large
part of their useful development.

In order to arrive at some definite knowledge of
this action, some limited experiments were made in
Paris a few years ago. Four tufts of the kidney bean
plant were transplanted into pots. These were placed
on scales, the plants being exposed to the sun but
not to the rain, and the experiments were made in
July. The watering was accomplished by artificial
sprinkling. The amount of water transpired was

carefully compared each day with the average temperature, the maximum temperature, the degree of intensity of sunlight, and the evaporation from the Piche evaporimeter. By far the closest relation was found to exist between the transpiration and the degree of sunlight. The amount of transpiration in a single day was very great, and seemed to depend directly on the amount of sunlight, the maximum being at noon. The quantity transpired in twenty-four hours was 1·27 kilogr. while that evaporated from a water surface, and earth, was only ·26 kilogr. and ·10 kilogr. respectively.

This shows the enormous amount of water necessary to convey to the plants those minute particles of matter that are taken from the soil, and which make up the true substance of the plant ; so that the greater part of the water used is only a means towards an end. The following little table for the region of Paris, contains the *résumé* of the five years' observations for the months March – July, these being the months in which grains are in growth. The years are arranged in the order of successful growth, both quality and quantity being considered.

	1874	1876	1875	1877	1873
Rainfall ...	166mm.	197mm.	207mm.	320mm.	307mm.
Evaporation ...	582mm.	558mm.	508mm.	448mm.	537mm.
Amount of heat	2096°	1995°	2053°	2007°	2029°
Amount of sunlight	6621°	6450°	6249°	6008°	6201°

Here we find the increased amount of sunlight was accompanied by an increase in the yield of grain, except in the year 1873, and this year was bad on account of unfavourable weather during the latter part of the time, June especially being a bad month,

as more rain fell in that month than in any other in
.the five years.

Winds.—Various .effects of winds have been
observed by the agriculturalist, but always in a qualita-
tive manner ; but now that meteorologists have accumulated a vast amount of data, concerning both
wind-direction and amount of wind, it is time
for the data to be applied for the benefit of the
practical farmer. The direction from which winds
are most likely to blow at certain seasons of the year
has already been pretty well ascertained, but the
question of wind force is of equal, if not more,
importance than wind direction, and yet it has
received much less attention from the meteorologist,
and has scarcely been considered at all, at least in a
quantitative manner, by persons practically interested
in agriculture. There is a very great difference in the
force and amount of wind in the various regions and
exposures, and it is obvious that plants which cannot
bear much wind will not be raised most successfully
in the most exposed regions ; and especially is cau-
tion needed in the planting of fruit trees where the
violent winds are sure to blow off most of the fruit
just as it is ripening and when it will fall easily.
The wind charts of Kiersnowsky and Waldo show
that there can be pointed out in a general way the
regions most and least adapted to products affected
by the force of the winds. The various exposures
to winds in the same locality are also being
studied, and the probability is that eventually it will
be possible to foretell about the average amount of
winds to be expected for any exposure (hill-tops or
valleys) in any locality of Europe or the United
States. This will not only be of advantage in regard

to crops, but it will also greatly aid the farmers in the selection of places for obtaining wind power for driving wind wheels used in running mills, operating wells, and also for irrigating purposes. It will be a great convenience to know just how high the wind-wheel should be placed in any locality in order to be sure of a certain amount of power for any month and for any region. This can be worked out with considerable certainty for some countries from the data at present collected. The probable amount of wind cannot be stated for any particular day or week, but for so long a period as a month the minimum amount to be expected can be pretty accurately fore-told where observations of wind force extend over a few years. By this means the minimum actual amount of work that can be accomplished each month of the year can be counted on with a great degree of certainty ; and if the loss of power in the running of each form of wind-mill is determined by a few experiments, then the amount of monthly or annual foot tons of power of the wind can be given. For instance, at Dodge City, Kansas, U.S., at a height of perhaps fifty feet above the ground the wind exerts on the average, on a plate a foot square and placed perpendicular to the wind, a pressure equal to the number of pounds shown in the following table :

JAN.	FEB.	MAR.	APR.	MAY.	JUNE.	JULY.	AUG.	SEPT.	OCT.	NOV.	DEC.
0·6	0·7	1·0	1·1	1·0	0·8	0·8	0·6	0·7	0·7	0·5	0·5

Of course these amounts are to be multiplied by the number of square feet of surface exposed by the wind-wheel, and also by a factor showing the relation of actual power per square foot of the wheel compared with the power of a plate a foot square placed perpendicular to the wind, and pressed upon by it.   This factor should be furnished by the maker of the wind-wheel.

Concerning the irregularly occurring high winds, it may be said that there is an enormous loss each year by storms, and insurance in Storm Insurance Companies will probably become as usual as insurance with Fire Companies.   At the present time, in the United States, there is a system of insurance against loss by tornadoes.   But there is still much to be done in the way of securing data of storms, and losses by them before anything like a fair percentage can be charged for such insurance.

Entomology.—The immense losses of crops due to the attack of insects, whether they come only in their usual numbers, or whether they come in such multitudes as to devastate the portion of the country in which they appear, has made it necessary to study their habits and to provide means for their destruction.   No such study can be complete unless the meteorological conditions of the air and ground are taken into consideration ; and this, for the most part, has been done only in the general way that we have noticed for some of the other applications of climatological knowledge.   There can be no doubt but that insect life and development is strongly influenced by climatic causes and conditions, and the only way that the entomologist can make the necessary comparative observations, is to have the meteorological observations for the period and region of investigation of insect life so computed and arranged as to allow him to use them with the minimum labour to himself.   If he is left to do this work himself, it will most probably not be done, as it takes much time to put original meteorological records in good shape for use.

Particularly in the study of the migrations of insects should the meteorological records be carefully considered in order to find out the conditions causing or accelerating their movements. For instance, one theory of the flight of the Rocky Mountain locust or grasshopper is based on meteorological causes.

Mycology.—In the majority of mycological studies in which the climatic influences are considered, there is very little attention paid to the meteorological data as represented by actual numbers, such expressions as " a wet season " or " a dry period " being considered sufficiently accurate to convey the idea of the weather conditions.

There can be no doubt but that many of the diseases to which plants are liable can be located as occurring between certain definite climatological limits, which latter can be represented numerically on the scales chosen to represent various degrees of moisture, heat, sunlight, or the other elements which may be found to have some connection with the diseases. By a careful study of these conditions it may be possible to warn the farmer of the beginning of periods in which the meteorological conditions are favourable for the development of various diseases. In the case of those diseases which are spread by the transfer of germs from one plant to another, the winds, and the flow of surface water after hard rains must be the important atmospheric elements causing the spread of the disease germs. A careful comparative study of such points as these would doubtless result in getting at the causes of diseases now imperfectly understood.

I will add that there are two very important additional applications of meteorological data which I regret not being able to include in the present chapter.   These are—the aid given to commerce by weather predictions, storm warnings, and wind charts for the oceans ; and the use of meteorological data as an aid in the practice of medicine.   The former of these has been fully described in various books treating of storms, storm warnings, and weather ; but the latter is an almost unworked field for the investigator, although great masses of statistics have been accumulated concerning the relation of disease to climate, and are awaiting the industry and patience of the original workers who may enter this field.

THE END.

INDEX.

———◆◇◆———

Air, maximum, pressure in high altitudes at the equator, 295
Air pressure at the earth's surface in general, 295
Air, interchange of, between the two hemispheres, 296
Air, Sprung's table of meridianal distribution of the, pressure and temperature, 298
Air, graphical representation of the circulation of the, as outlined by Ferrel, 302
Air, computation of east-westerly velocities of the, from the pressure gradient, 304
Air, general circulation of the, according to von Siemens, 309
Air, cause of meridianal circulation of the, 309
Air, cause of vertical currents of, 310
Air, cause of maxima and minima, pressures, 311
Air, general circulation of the, according to Möller, 311
Air, tropical circulation of the, as outlined by Möller, 313
Air, energy developed by the, at various altitudes, 314
Air motions without axial rotation of the earth, 317
Air motions with axial rotation of the earth, 317
Air, theory of the origin of the east-westerly motions of the, 323
Air, vertical components of motions of the, 323
Air, meridianal components of motions of the, 323
Air, general remarks on friction of horizontal currents of, 324
Air, conduction of heat from one layer of, to another, 326
Air, stable equilibrium of a layer of, 327
Air, motions of horizontal currents of, 328
Air, mingling of layers of, 331
Air, Polar cold as a source of, motion, 333
Air, hindrance to the formation of very rapid currents of, 334

Anemometers, 106
Anemometers, pendulum, 111
Anemometers, cup, 113
Anemometer, the Robinson cup, 114
Anemometer, standard, 118
Anemometers, Combe's whirling apparatus for testing, 118
Anemometers, recording apparatus for, 124
Anemometers and anemoscope register, 126
Anti-cyclones, application of thermodynamics to interchange of air between cyclones and, 229
Anti-cyclones, temperature gradient in cyclones and, 251
Anti-cyclones, formation of, 332
Anti-cyclones, Hann's recent ideas concerning, 385
Atmosphere (see also Air)
Atmosphere, moisture of the, 135
Atmosphere, thermodynamics of the, 204
Atmosphere, general motions of the, 268
Atmosphere, development of modern theories of motions of the, 268
Atmosphere, historical review of theories of circulation of the, 268
Atmosphere, Ferrel's first paper on circulation of the, 271
Atmosphere, Tracy's paper on circulation of the, 272
Atmosphere, Ferrel's second paper on circulation of the, 273
Atmosphere, comparison of primary and secondary motions of the, 274
Atmosphere, Guldberg and Mohn's studies on motions of the, 277
Atmosphere, general treatise on motions of the, 279
Atmosphere, outline of Ferrel's theory of the general circulation of the, 280
Atmosphere, primary cause of motions of the, 280
Atmosphere, influences modifying an ideal circulation of the, 281
Atmosphere, modifying effects of the

THE SCOTT LIBRARY.

Cloth, Uncut Edges, Gilt Top.　Price 1s. 6d. per Volume.

London : WALTER SCOTT, LIMITED, Paternoster Square.

THE SCOTT LIBRARY—continued.

14 GREAT ENGLISH PAINTERS. SELECTED FROM Cunningham's *Lives*. Edited by William Sharp.

15 BYRON'S LETTERS AND JOURNALS. SELECTED, with Introduction, by Mathilde Blind.

16 LEIGH HUNT'S ESSAYS. WITH INTRODUCTION AND Notes by Arthur Symons.

17 LONGFELLOW'S "HYPERION," "KAVANAH," AND "The Trouveres." With Introduction by W. Tirebuck.

18 GREAT MUSICAL COMPOSERS. BY G. F. FERRIS. Edited, with Introduction, by Mrs. William Sharp.

19 THE MEDITATIONS OF MARCUS AURELIUS. EDITED by Alice Zimmern.

20 THE TEACHING OF EPICTETUS. TRANSLATED FROM the Greek, with Introduction and Notes, by T. W. Rolleston.

21 SELECTIONS FROM SENECA. WITH INTRODUCTION by Walter Clode.

22 SPECIMEN DAYS IN AMERICA. BY WALT WHITMAN. Revised by the Author, with fresh Preface.

23 DEMOCRATIC VISTAS, AND OTHER PAPERS. BY Walt Whitman. (Published by arrangement with the Author.)

24 WHITE'S NATURAL HISTORY OF SELBORNE. WITH a Preface by Richard Jefferies.

25 DEFOE'S CAPTAIN SINGLETON. EDITED, WITH Introduction, by H. Halliday Sparling.

26 MAZZINI'S ESSAYS: LITERARY, POLITICAL, AND Religious. With Introduction by William Clarke.

27 PROSE WRITINGS OF HEINE. WITH INTRODUCTION by Havelock Ellis.

28 REYNOLDS'S DISCOURSES. WITH INTRODUCTION by Helen Zimmern.

29 PAPERS OF STEELE AND ADDISON. EDITED BY Walter Lewin.

30 BURNS'S LETTERS. SELECTED AND ARRANGED, with Introduction, by J. Logie Robertson, M.A.

London: WALTER SCOTT, LIMITED, Paternoster Square.

London: WALTER SCOTT, LIMITED, Paternoster Square.

THE SCOTT LIBRARY—continued.

48 STORIES FROM CARLETON. SELECTED, WITH INTRO-
duction, by W. Yeats.

49 JANE EYRE. BY CHARLOTTE BRONTË. EDITED BY
Clement K. Shorter.

50 ELIZABETHAN ENGLAND. EDITED BY LOTHROP
Withington, with a Preface by Dr. Furnivall.

51 THE PROSE WRITINGS OF THOMAS DAVIS. EDITED
by T. W. Rolleston.

52 SPENCE'S ANECDOTES. A SELECTION. EDITED,
with an Introduction and Notes, by John Underhill.

53 MORE'S UTOPIA, AND LIFE OF EDWARD V. EDITED,
with an Introduction, by Maurice Adams.

54 SADI'S GULISTAN, OR FLOWER GARDEN. TRANS-
lated, with an Essay, by James Ross.

55 ENGLISH FAIRY AND FOLK TALES. EDITED BY
E. Sidney Hartland.

56 NORTHERN STUDIES. BY EDMUND GOSSE. WITH
a Note by Ernest Rhys.

57 EARLY REVIEWS OF GREAT WRITERS. EDITED BY
E. Stevenson.

58 ARISTOTLE'S ETHICS. WITH GEORGE HENRY
Lewes's Essay on Aristotle prefixed.

59 LANDOR'S PERICLES AND ASPASIA. EDITED, WITH
an Introduction, by Havelock Ellis.

60 ANNALS OF TACITUS. THOMAS GORDON'S TRANS-
lation. Edited, with an Introduction, by Arthur Galton.

61 ESSAYS OF ELIA. BY CHARLES LAMB. EDITED,
with an Introduction, by Ernest Rhys.

62 BALZAC'S SHORTER STORIES. TRANSLATED BY
William Wilson and the Count Stenbock.

63 COMEDIES OF DE MUSSET. EDITED, WITH AN
Introductory Note, by S. L. Gwynn.

64 CORAL REEFS. BY CHARLES DARWIN. EDITED,
with an Introduction, by Dr. J. W. Williams.

London: WALTER SCOTT, LIMITED, Paternoster Square.

London : WALTER SCOTT, LIMITED, Paternoster Square.

THE SCOTT LIBRARY—continued.

London : WALTER SCOTT, LIMITED, Paternoster Square.

GREAT WRITERS.

A NEW SERIES OF CRITICAL BIOGRAPHIES.

Edited by ERIC ROBERTSON and FRANK T. MARZIALS.

A Complete Bibliography to each Volume, by J. P. ANDERSON, British Museum, London.

Cloth, Uncut Edges, Gilt Top. Price 1/6.

VOLUMES ALREADY ISSUED—

LIFE OF LONGFELLOW. By PROF. ERIC S. ROBERTSON.
"A most readable little work."—*Liverpool Mercury*.

LIFE OF COLERIDGE. By HALL CAINE.
"Brief and vigorous, written throughout with spirit and great literary skill."—*Scotsman*.

LIFE OF DICKENS. By FRANK T. MARZIALS.
"Notwithstanding the mass of matter that has been printed relating to Dickens and his works . . . we should, until we came across this volume, have been at a loss to recommend any popular life of England's most popular novelist as being really satisfactory. The difficulty is removed by Mr. Marzials's little book."—*Athenæum*.

LIFE OF DANTE GABRIEL ROSSETTI. By J. KNIGHT.
"Mr. Knight's picture of the great poet and painter is the fullest and best yet presented to the public."—*The Graphic*.

LIFE OF SAMUEL JOHNSON. By COLONEL F. GRANT.
"Colonel Grant has performed his task with diligence, sound judgment, good taste, and accuracy."—*Illustrated London News*.

LIFE OF DARWIN. By G. T. BETTANY.
"Mr. G. T. Bettany's *Life of Darwin* is a sound and conscientious work."
—*Saturday Review*.

LIFE OF CHARLOTTE BRONTË. By A. BIRRELL.
"Those who know much of Charlotte Brontë will learn more, and those who know nothing about her will find all that is best worth learning in Mr. Birrell's pleasant book."—*St. James' Gazette*.

LIFE OF THOMAS CARLYLE. By R. GARNETT, LL.D.
"This is an admirable book. Nothing could be more felicitous and fairer than the way in which he takes us through Carlyle's life and works."—*Pall Mall Gazette*.

London : WALTER SCOTT, LIMITED, Paternoster Square.

GREAT WRITERS—continued.

LIFE OF ADAM SMITH. By R. B. HALDANE, M.P.

"Written with a perspicuity seldom exemplified when dealing with economic science."—*Scotsman.*

LIFE OF KEATS. By W. M. ROSSETTI.

"Valuable for the ample information which it contains."—*Cambridge Independent.*

LIFE OF SHELLEY. By WILLIAM SHARP.

"The criticisms . . . entitle this capital monograph to be ranked with the best biographies of Shelley."—*Westminster Review.*

LIFE OF SMOLLETT. By DAVID HANNAY.

"A capable record of a writer who still remains one of the great masters of the English novel."—*Saturday Review.*

LIFE OF GOLDSMITH. By AUSTIN DOBSON.

"The story of his literary and social life in London, with all its humorous and pathetic vicissitudes, is here retold, as none could tell it better."—*Daily News.*

LIFE OF SCOTT. By PROFESSOR YONGE.

"This is a most enjoyable book."—*Aberdeen Free Press.*

LIFE OF BURNS. By PROFESSOR BLACKIE.

"The editor certainly made a hit when he persuaded Blackie to write about Burns."—*Pall Mall Gazette.*

LIFE OF VICTOR HUGO. By FRANK T. MARZIALS.

"Mr. Marzials's volume presents to us, in a more handy form than any English or even French handbook gives, the summary of what is known about the life of the great poet."—*Saturday Review.*

LIFE OF EMERSON. By RICHARD GARNETT, LL.D.

"No record of Emerson's life could be more desirable."—*Saturday Review.*

LIFE OF GOETHE. By JAMES SIME.

"Mr. James Sime's competence as a biographer of Goethe is beyond question."—*Manchester Guardian.*

LIFE OF CONGREVE. By EDMUND GOSSE.

"Mr. Gosse has written an admirable biography."—*Academy.*

LIFE OF BUNYAN. By CANON VENABLES.

"A most intelligent, appreciative, and valuable memoir."—*Scotsman.*

LIFE OF CRABBE. By T. E. KEBBEL.

"No English poet since Shakespeare has observed certain aspects of nature and of human life more closely."—*Athenæum.*

LIFE OF HEINE. By WILLIAM SHARP.

"An admirable monograph . . . more fully written up to the level of recent knowledge and criticism than any other English work."—*Scotsman.*

London: WALTER SCOTT, LIMITED, Paternoster Square.

GREAT WRITERS—continued.

LIFE OF MILL. By W. L. COURTNEY.

"A most sympathetic and discriminating memoir."—*Glasgow Herald.*

LIFE OF SCHILLER. By HENRY W. NEVINSON.

"Presents the poet's life in a neatly rounded picture."—*Scotsman.*

LIFE OF CAPTAIN MARRYAT. By DAVID HANNAY.

"We have nothing but praise for the manner in which Mr. Hannay has done justice to him."—*Saturday Review.*

LIFE OF LESSING. By T. W. ROLLESTON.

"One of the best books of the series."—*Manchester Guardian.*

LIFE OF MILTON. By RICHARD GARNETT, LL.D.

"Has never been more charmingly or adequately told."—*Scottish Leader.*

LIFE OF BALZAC. By FREDERICK WEDMORE.

"Mr. Wedmore's monograph on the greatest of French writers of fiction, whose greatness is to be measured by comparison with his successors, is a piece of careful and critical composition, neat and nice in style."—*Daily News.*

LIFE OF GEORGE ELIOT. By OSCAR BROWNING.

"A book of the character of Mr. Browning's, to stand midway between the bulky work of Mr. Cross and the very slight sketch of Miss Blind, was much to be desired, and Mr. Browning has done his work with vivacity, and not without skill."—*Manchester Guardian.*

LIFE OF JANE AUSTEN. By GOLDWIN SMITH.

"Mr. Goldwin Smith has added another to the not inconsiderable roll of eminent men who have found their delight in Miss Austen. . . . His little book upon her, just published by Walter Scott, is certainly a fascinating book to those who already know her and love her well; and we have little doubt that it will prove also a fascinating book to those who have still to make her acquaintance."—*Spectator.*

LIFE OF BROWNING. By WILLIAM SHARP.

"This little volume is a model of excellent English, and in every respect it seems to us what a biography should be."—*Public Opinion.*

LIFE OF BYRON. By HON. RODEN NOEL.

"The Hon. Roden Noel's volume on Byron is decidedly one of the most readable in the excellent 'Great Writers' series."—*Scottish Leader.*

LIFE OF HAWTHORNE. By MONCURE CONWAY.

"It is a delightful *causerie*—pleasant, genial talk about a most interesting man. Easy and conversational as the tone is throughout, no important fact is omitted, no valueless fact is recalled; and it is entirely exempt from platitude and conventionality."—*The Speaker.*

LIFE OF SCHOPENHAUER. By PROFESSOR WALLACE.

"We can speak very highly of this little book of Mr. Wallace's. It is, perhaps, excessively lenient in dealing with the man, and it cannot be said to be at all ferociously critical in dealing with the philosophy."—*Saturday Review.*

London: WALTER SCOTT, LIMITED, Paternoster Square.

SELECTED THREE-VOL. SETS

IN NEW BROCADE BINDING.

6s. PER SET, IN SHELL CASE TO MATCH.

Also Bound in Roan, in Shell Case, Price 9s. per Set.

O. W. Holmes Set—
Autocrat of the Breakfast-Table.
Professor at the Breakfast-Table.
Poet at the Breakfast-Table.

Landor Set—
Landor's Imaginary Conversations.
Pentameron.
Pericles and Aspasia.

Three English Essayists—
Essays of Elia.
Essays of Leigh Hunt.
Essays of William Hazlitt.

Three Classical Moralists—
Meditations of Marcus Aurelius.
Teaching of Epictetus.
Morals of Seneca.

Walden Set—
Thoreau's Walden.
Thoreau's Week.
Thoreau's Selections.

Famous Letters Set—
Letters of Byron.
Letters of Chesterfield.
Letters of Burns.

Lowell Set—
My Study Windows.
The English Poets.
The Biglow Papers.

Heine Set—
Life of Heine.
Heine's Prose.
Heine's Travel-Sketches.

Three Essayists—
Essays of Mazzini.
Essays of Sainte-Beuve.
Essays of Montaigne.

Schiller Set—
Life of Schiller.
Maid of Orleans.
William Tell.

Carlyle Set—
Life of Carlyle.
Sartor Resartus.
Carlyle's German Essays.

London: WALTER SCOTT, LIMITED, Paternoster Square.

IBSEN'S PROSE DRAMAS.

EDITED BY WILLIAM ARCHER.

Complete in Five Vols. Crown 8vo, Cloth, Price 3/6 each.

Set of Five Vols., in Case, 17/6; in Half Morocco, in Case, 32/6.

"We seem at last to be shown men and women as they are; and at first it is more than we can endure. . . . All Ibsen's characters speak and act as if they were hypnotised, and under their creator's imperious demand to reveal themselves. There never was such a mirror held up to nature before: it is too terrible. . . . Yet we must return to Ibsen, with his remorseless surgery, his remorseless electric-light, until we, too, have grown strong and learned to face the naked—if necessary, the flayed and bleeding—reality."—SPEAKER (London).

VOL. I. "A DOLL'S HOUSE," "THE LEAGUE OF YOUTH," and "THE PILLARS OF SOCIETY." With Portrait of the Author, and Biographical Introduction by WILLIAM ARCHER.

VOL. II. "GHOSTS," "AN ENEMY OF THE PEOPLE," and "THE WILD DUCK." With an Introductory Note.

VOL. III. "LADY INGER OF ÖSTRÅT," "THE VIKINGS AT HELGELAND," "THE PRETENDERS." With an Introductory Note and Portrait of Ibsen.

VOL. IV. "EMPEROR AND GALILEAN." With an Introductory Note by WILLIAM ARCHER.

VOL. V. "ROSMERSHOLM," "THE LADY FROM THE SEA," "HEDDA GABLER." Translated by WILLIAM ARCHER. With an Introductory Note.

The sequence of the plays *in each volume* is chronological; the complete set of volumes comprising the dramas thus presents them in chronological order.

"The art of prose translation does not perhaps enjoy a very high literary status in England, but we have no hesitation in numbering the present version of Ibsen, so far as it has gone (Vols. I. and II.), among the very best achievements, in that kind, of our generation."—*Academy.*

"We have seldom, if ever, met with a translation so absolutely idiomatic."—*Glasgow Herald.*

London : WALTER SCOTT, LIMITED, Paternoster Square.

THE CANTERBURY POETS.

EDITED BY WILLIAM SHARP. IN 1/- MONTHLY VOLUMES.

Cloth, Red Edges - **1s.**	Red Roan, Gilt Edges, 2s. 6d.
Cloth, Uncut Edges - **1s.**	Pad. Morocco, Gilt Edges, 5s.

London : WALTER SCOTT, LIMITED, Paternoster Square.

THE CANTERBURY POETS—continued.

Quarto, cloth elegant, gilt edges, emblematic design on cover, 6s. May also be had in a variety of Fancy Bindings.

THE

MUSIC OF THE POETS:

A MUSICIANS' BIRTHDAY BOOK.

EDITED BY ELEONORE D'ESTERRE KEELING.

THIS is a unique Birthday Book. Against each date are given the names of musicians whose birthday it is, together with a verse-quotation appropriate to the character of their different compositions or performances. A special feature of the book consists in the reproduction in fac-simile of autographs, and autographic music, of living composers. Three sonnets by Mr. Theodore Watts, on the "Fausts" of Berlioz, Schumann, and Gounod, have been written specially for this volume. It is illustrated with designs of various musical instruments, etc.; autographs of Rubenstein, Dvorák, Greig, Mackenzie, Villiers Stanford, etc., etc.

London : WALTER SCOTT, LIMITED, Paternoster Square.

www.ingramcontent.com/pod-product-compliance
Lightning Source LLC
Chambersburg PA
CBHW020857210326
41598CB00018B/1697